# An Introduction to Nonlinear Analysis

## Martin Schechter
*University of California, Irvine*

CAMBRIDGE
UNIVERSITY PRESS

CAMBRIDGE UNIVERSITY PRESS
Cambridge, New York, Melbourne, Madrid, Cape Town, Singapore, São Paulo, Delhi

Cambridge University Press
The Edinburgh Building, Cambridge CB2 8RU, UK

Published in the United States of America by Cambridge University Press, New York

www.cambridge.org
Information on this title: www.cambridge.org/9780521843973

First published 2004

*A catalogue record for this publication is available from the British Library*

ISBN 978-0-521-84397-3 hardback

Transferred to digital printing 2009

# An Introduction to Nonlinear Analysis

The techniques that can be used to solve nonlinear problems are very different from those used to solve linear problems. Many courses in analysis and applied mathematics attack linear cases simply because they are easier to solve and do not require a large theoretical background in order to approach them. Professor Schechter's book is devoted to nonlinear methods using the least background material possible and the simplest linear techniques.

An understanding of the tools for solving nonlinear problems is developed while demonstrating their application to problems first in one dimension, and then in higher dimensions. The reader is guided using simple exposition and proof, assuming a minimal set of prerequisites. To complete, a set of appendices covering essential basics in functional analysis and metric spaces is included, making this ideal as a text for an upper-undergraduate or graduate course, or even for self-study.

MARTIN SCHECHTER is Professor of Mathematics at the University of California, Irvine.

# CAMBRIDGE STUDIES IN ADVANCED MATHEMATICS

All the titles listed below can be obtained from good booksellers or from Cambridge University Press. For a complete series listing visit:
http://publishing.cambridge.org/stm/mathematics/csam

BS" D

To my wife, Deborah, our children,
our grandchildren (all twenty one of them so far)
and our extended family.
May they enjoy many happy years.

# Contents

# Preface

The techniques that can be used to solve nonlinear problems are very different from those that are used to solve linear problems. Most courses in analysis and applied mathematics attack linear problems simply because they are easier to solve. The information that is needed to solve them is not as involved or technical in nature as that which is usually required to solve a corresponding nonlinear problem. This applies not only to the practical material but also to the theoretical background.

As an example, it is usually sufficient in dealing with linear problems in analysis to apply Riemann integration to functions that are piecewise continuous. Rarely is more needed. In considering the convergence of series, uniform convergence usually suffices. In general, concepts from functional analysis are not needed; linear algebra is usually sufficient. A student can go quite far in the study of linear problems without being exposed to Lebesgue integration or functional analysis.

However, there are many nonlinear problems that arise in applied mathematics and sciences that require much more theoretical background in order to attack them. If we couple this with the difficult technical details concerning the corresponding linear problems that are usually needed before one can apply the nonlinear techniques, we find that the student does not come in contact with substantive nonlinear theory until a very advanced stage in his or her studies. This is unfortunate because students having no more background in mathematics beyond that of second year calculus are often required by their disciplines to study such problems. Moreover, such students can readily understand most of the methods used in solving nonlinear problems.

During the last few years, the author has been giving a class devoted to nonlinear methods using the least background material possible and the simplest linear techniques. This is not an easy tightrope to walk. There are times when theorems from Lebesgue integration are required together with theorems from functional analysis. There are times when exact estimates for the linear problem are needed. What should one do?

My approach has been to explain the methods using the simplest terms. After I apply the methods to the solving of problems, I then prove them. True, I will need theorems from functional analysis and Lebesgue integration. At such times I explain the background theorems used. Then, the students have two options: either to believe me or to consult the references that I provide.

This brings me to the purpose of the text. I was unable to find a book that contained the material that I wanted to cover at the level that I wanted it presented. Moreover, I wanted to include a concise presentation (without proofs) of all of the background information that was needed to understand the techniques used in the body of the text. The writing of this book gave me the opportunity to accomplish both. If I think the students can handle it, I do prove background material in the body of the text. Otherwise, I explain it in four appendices in the back of the book. This also applies to topics that require a whole course to develop. This approach is intended to accommodate students at all levels. If they do not wish to see proofs of background materials, they can skip these sections. If they are familiar with functional analysis and Lebesgue integration, they can ignore the appendices.

The purpose of the course is to teach the methods that can be used in solving nonlinear problems. These include the contraction mapping theorem, Picard's theorem, the implicit function theorem, the Brouwer degree and fixed point theorem, the Schauder fixed point theorem, the Leray–Schauder degree, Peano's theorem, etc. However, the student will not appreciate any of them unless he or she observes them in action. On the other hand, if the applications are too complicated, the student will be bogged down in technical details that may prove to be extremely discouraging. This is another tightrope.

What surprised me was the amount of advanced background information that was needed to understand the methods used to attack even the easiest of nonlinear problems. I quote in the appendices only that material that is needed in the text. And yet, an examination of these appendices will reveal the substantial extent of this background knowledge. If we waited until the student had learned all of this, we would not be able to cover the material in the book until the student was well advanced. On the other hand, students with more modest backgrounds can understand the statements of the background theorems even though they have not yet learned the proofs. In fact, this approach can motivate such students to learn more advanced topics once they see the need for the material. In essence, I am advocating the cart before the horse. I want the student to appreciate the horse because it can be used to transport all of the items in the cart.

Equipping the student with the tools mentioned above is the main purpose of the book. Of course, one can first present a theorem and then give some applications. Most books function this way. However, I prefer to pose a problem and then introduce a tool that will help solve the problem. My choice of problems could be vast, but I tried to select those that require the least background. This outlook affected my choice: differential equations. I found them to require the least preparation. Moreover, most students have a familiarity with them. I picked them as the medium in which to work. The tools are the main objects, not the medium. True, the students do not know where they are headed, but neither does a research scientist searching for answers. It is also true that no matter what medium I pick, the nonlinear problems that the students will encounter in the future will be different. But as long as they have the basic tools, they have a decent chance of success.

I begin by posing a fairly modest nonlinear problem that would be easy to solve in the linear case. (In fact, we do just that; we solve the linear problem.) I then develop the tools that we use in solving it. I do this with two things in mind. The first is to develop methods that will be useful in solving many other nonlinear problems. The second is to show the student why such methods are useful. At the same time I try to keep the background knowledge needed to a minimum. In most cases the nonlinear tools require much more demanding information concerning the corresponding linear problem than the techniques used in solving the linear problem itself. I have made a concerted effort to choose problems that keep such required information to a minimum. I then vary the problem slightly to demonstrate how the techniques work in different situations and to introduce new tools that work when the original ones fail.

I then introduce new problems and new techniques that are used to solve them. The problems and techniques become progressively more difficult, but again I attempt to minimize the background material without ignoring important major nonlinear methods. My goal is to introduce as many nonlinear tools as time permits. I know that the students will probably not be confronted with the problems I have introduced, but they will have a collection of nonlinear methods and the knowledge of how they can be used.

In the first chapter we confront a seemingly simple problem for periodic functions in one dimension and go about solving it. The approach appears not to be related to the problem. We then fit the technique to the problem. (It is not at all obvious that the technique will work.) The student then sees how the techniques solve the problem. The chapter deals with the differentiation of functionals, Fourier series, finding minima of functionals and Hilbert space methods. I try to explain why each technique is used.

In the second chapter we consider the same problem for the cases when the functionals used in the first chapter have no minima. We begin with a simple algebraic problem in two dimensions. I introduce methods that can be used to solve it by producing saddle points, and then generalize the methods to arbitrary Hilbert spaces. We then apply them to the original problem. The tools used include the contraction mapping principle, Picard's theorem in a Banach space, extensions of solutions of differential equations in a Banach space and the sandwich theorem.

The third chapter leaves periodicity and deals with boundary value problems. I introduce mollifiers and test functions. As expected, different and stronger techniques are required.

The fourth chapter studies saddle points of functionals using such properties as convexity and lower semi-continuity. Conditions are given which produce saddle points, and these are applied to various problems. Partial differentiation is introduced, and the implicit function theorem is proved.

The fifth chapter discusses the calculus of variations, the Euler equations and the methods of obtaining minima. Necessary and sufficient conditions are given for the existence of minima. Many examples are presented.

In the sixth chapter I cover degree theory and its applications. Topics include the Brouwer and Schauder fixed point theorems, Sard's theorem, Peano's theorem and the Leray–Schauder degree. Applications are given.

The seventh chapter is devoted to constrained minima, of both the integral (iso-perimetric) and finite (point-wise) types. The Lagrange multiplier rule is proved and a more comprehensive type of differentiation is introduced.

The eighth chapter discusses mini-max techniques and gives examples.

In the ninth chapter I present the method of solving semi-linear equations which are sub-linear at infinity. We solve them by relating them to the Dancer–Fučík spectrum.

The tenth chapter is by far the largest; it is the first to tackle problems in higher dimensions. Even so, I limit the discussions to periodic functions. We consider spaces of distributions, Sobolev inequalities and Sobolev spaces. As expected, a lot of preparation is needed, and the proofs are more difficult. We generalize the one-dimensional results to higher dimensions.

There are four appendices. The first assembles the definitions and theorems (without proofs) from functional analysis that are needed in the text. I was surprised that so much was required. The second appendix deals with the theorems concerning Lebesgue integration required by the text. Again, proofs are omitted with one exception. We prove that Carathéodory functions are measurable. This theorem is not well known and hard to find in the literature. The third appendix describes what is needed concerning metric spaces. The fourth shows how pseudo-gradients can be used to strengthen some of the theorems in the text.

It is hoped that this volume will fill a need and will allow students with modest backgrounds to tackle important nonlinear problems.

TVSLB″O

# 1

# Extrema

## 1.1 Introduction

One of the most powerful tools in dealing with nonlinear problems is **critical point theory**. It originates from the fact in calculus that the derivative of a smooth function vanishes at extreme points (maxima and minima). In order to apply this basic reasoning, the given problem must be converted to one in which we look for points where the derivative of a function vanishes (i.e., critical points). This cannot always be arranged. But when it can, one has a very useful method. The easiest situation is when the function has extrema. We discuss this case in the present chapter. We present a problem and convert it into the desired form. We then give criteria that imply that extrema exist. The case when extrema do not exist will be discussed in the next chapter.

## 1.2 A one dimensional problem

We now consider the problem of finding a solution of

$$-u''(x) + u(x) = f(x, u(x)), \quad x \in I = [0, 2\pi], \quad (1.1)$$

under the conditions

$$u(0) = u(2\pi), \ u'(0) = u'(2\pi). \quad (1.2)$$

We assume that the function $f(x, t)$ is continuous in $I \times \mathbb{R}$ and is periodic in $x$ with period $2\pi$. If $u(x)$ is a solution, then we have

$$(u', v') + (u, v) = (-u'' + u, v) = (f(\cdot, u), v)$$

for all $v \in C^1(I)$ satisfying (1.2). Here,

$$(u, v) = \int_0^{2\pi} u(x)v(x)\, dx,$$

1

$C^1(I)$ is the set of continuously differentiable periodic functions on $I$, and we used the fact that no boundary terms arise in the integration by parts. The expression

$$(u, v)_H = (u', v') + (u, v) \tag{1.3}$$

is a scalar product corresponding to the norm

$$\|u\|_H = (\|u'\|^2 + \|u\|^2)^{1/2}. \tag{1.4}$$

(See Appendix A for the definition of a scalar product and related terms.) Thus, a solution of (1.1),(1.2) satisfies

$$(u, v)_H = (f(\cdot, u), v) \tag{1.5}$$

for all $v \in C^1(I)$ satisfying (1.2).

As we mentioned before, our approach will be that of critical point theory. It begins by asking the question, "Does there exist a differentiable function $G$ from a Hilbert space $H$ to $\mathbb{R}$ such that (1.1),(1.2) is equivalent to

$$G'(u) = 0?" \tag{1.6}$$

(See Appendix A for the definition of a Hilbert space.) The reason for the question is that in elementary calculus it is known that a critical point of a differentiable function does satisfy such an equation. It is therefore hoped that one can mimic the methods of calculus to find critical points and thus solve

$$G'(u) = 0. \tag{1.7}$$

Anyone who thinks that this should be easy is due for a rude awakening. Actually, we are asking the following: Does there exist a mapping $G$ from a Hilbert space $H$ to $\mathbb{R}$ such that

(a) $G$ has an extremum at a "point" $u$
(b) $G$ has a derivative at $u$
(c) $G'(u) = -u'' + u - f(x, u(x))$?

(The reason we want $H$ to be a Hilbert space will become clear as we proceed.)

As we shall see, none of these is easily answered. Moreover, a lot of explaining has to be done. The motivation is clear. If $g(x)$ is a differentiable function on $\mathbb{R}$ and

$$g(a) = \min_{\mathbb{R}} g,$$

then it follows that

$$g'(a) = 0.$$

Thus, if we can find a "function" $G(v)$ from $H$ to $\mathbb{R}$ (such a "function" is called a "functional") which is differentiable and satisfies

$$G(u) = \min_H G, \tag{1.8}$$

we should be able to conclude that

$$G'(u) = 0. \tag{1.9}$$

However, we do not know what it means for a functional to be differentiable. We know that

$$g'(a) = \lim_{h \to 0} [g(a+h) - g(a)]/h. \tag{1.10}$$

But if we try to use this definition for a functional, we get

$$G'(u) = \lim_{v \to 0} [G(u+v) - G(u)]/v, \tag{1.11}$$

which is ridiculous since we cannot divide by elements of a vector space. As an alternative definition, we have

$$b = g'(a) \iff [g(a+h) - g(a) - hb]/h \to 0 \ \text{ as } \ h \to 0. \tag{1.12}$$

The corresponding statement for a functional is

$$w = G'(u) \iff [G(u+v) - G(u) - vw]/v \to 0 \ \text{ as } \ v \to 0. \tag{1.13}$$

This does not appear to be any better. In fact, it is worse because one cannot multiply two elements of vector spaces. Or can one? If $w$ is an element of a vector space such that a "product" $\langle v, w \rangle$ can be defined (and have whatever properties we need), we might be in business. So we try

$$w = G'(u) \iff [G(u+v) - G(u) - \langle v, w \rangle]/v \to 0 \ \text{ as } \ v \to 0. \tag{1.14}$$

But we still have the same objection as before, namely our inability to divide by an element of a vector space. However, there is a difference between (1.10) and (1.12) in that we can replace (1.12) by

$$b = g'(a) \iff [g(a+h) - g(a) - hb]/|h| \to 0 \text{ as } h \to 0, \qquad (1.15)$$

while we cannot replace the denominator in (1.10) by $|h|$. "But what does that gain?" you object. "We do not have absolute values of elements." True, but we do have a substitute which sometimes serves the same purpose, namely the norm. Thus we can replace (1.14) by

$$w = G'(u) \iff [G(u+v) - G(u) - \langle v, w \rangle]/\|v\|_H \to 0 \text{ as } \|v\|_H \to 0. \qquad (1.16)$$

We want the "product" $\langle v, w \rangle$ to be linear and symmetric in $v$ and $w$, and we want the derivative of $G$ at $u$ to be unique. So suppose $w_1$ also satisfied (1.16). Then we would have

$$\langle v, w - w_1 \rangle / \|v\|_H \to 0 \text{ as } \|v\|_H \to 0.$$

If we take $v = \varepsilon h$ and let $\varepsilon \to 0$, we find that

$$\langle h, w - w_1 \rangle = 0 \quad h \in H.$$

Thus, we would like

$$\langle h, w \rangle = 0 \ \forall h \in H \implies w = 0. \qquad (1.17)$$

However, this is not essential. Suppose (1.17) does not hold. If $\langle g, h \rangle$ satisfies

$$\langle g, h \rangle \leq C\|g\|_H \cdot \|h\|_H, \quad g, h \in H, \qquad (1.18)$$

then for each $w \in H$ there is a unique $g \in H$ such that

$$\langle h, w \rangle = (h, g)_H, \quad h \in H. \qquad (1.19)$$

To see this, note that for each fixed $w$, $\langle h, w \rangle$ is a bounded linear functional on $H$ (for definitions see Appendix A). We apply the Riesz representation theorem (Theorem A.12) to obtain (1.19). This is one advantage of having $H$ be a Hilbert space; we shall see others later. Instead of defining $G'(u)$ to be $w$, we can define it to be $g$. This gives

$$(G'(u), v)_H = \langle w, v \rangle, \quad v \in H. \qquad (1.20)$$

Thus, even though $w$ need not be unique, $G'(u)$ is unique, as we now show.

To summarize, we define the derivative as follows. If $G(v)$ is a functional defined on a Hilbert space $H$ and there is a symmetric bilinear form $\langle g, h \rangle$ on $H$ satisfying (1.18) such that

$$[G(u + v) - G(u) - \langle v, w \rangle]/\|v\|_H \to 0 \quad \text{as} \quad \|v\|_H \to 0, \quad v \in H, \quad (1.21)$$

holds for some $u, w \in H$, then $G'(u) \in H$ exists and is given by

$$(G'(u), v)_H = \langle w, v \rangle, \quad v \in H. \quad (1.22)$$

This definition is independent of the bilinear form $\langle \cdot, \cdot \rangle$ and the element $w$. By this we mean that if $\langle \cdot, \cdot \rangle_1$ is another bilinear form on $H$ satisfying (1.18) and $w_1$ is another element in $H$ such that (1.21) holds with $\langle v, w \rangle$ replaced by $\langle v, w_1 \rangle_1$, then

$$\langle v, w \rangle = \langle v, w_1 \rangle_1, \quad v \in H.$$

To see this, we note that there are elements $g, g_1 \in H$ such that

$$(v, g)_H = \langle v, w \rangle, \quad (v, g_1)_H = \langle v, w_1 \rangle_1, \quad v \in H$$

by the Riesz representation theorem (Theorem A.12). Then we must have

$$(v, g)_H = (v, g_1)_H, \quad v \in H,$$

which implies $g = g_1$ by taking $v = g - g_1$. Thus, we may use any convenient bilinear form $\langle \cdot, \cdot \rangle$ as long as it satisfies (1.18).

The derivative as defined above is called the **Fréchet derivative**. There are other definitions. We shall discuss some others later on.

We also note that as in the case of functions of a single variable, differentiability implies continuity. For if $G'(u_0)$ exists, then (1.21) implies that there is a $\delta_1 > 0$ such that

$$|[G(u_0 + v) - G(u) - (v, G'(u_0))_H]/\|v\|_H| < 1, \quad \|v\|_H < \delta_1.$$

Thus,

$$|G(u_0 + v) - G(u_0)| \leq \|v\|_H[\|G'(u_0)\|_H + 1], \quad \|v\|_H < \delta_1.$$

Let $\varepsilon > 0$ be given. Take $\delta > 0$ such that $\delta < \delta_1$ and $\delta < \varepsilon/[\|G'(u_0)\|_H + 1]$. Then $\|v\|_H < \delta$ implies

$$|G(u_0 + v) - G(u_0)| \leq \|v\|_H[\|G'(u_0)\|_H + 1] < \varepsilon.$$

Continuity of the derivative is no problem to define. It is the same as that for the functional itself. That is, $G'$ is continuous at $u$ if

$$G'(v) \to G'(u) \ \text{as} \ v \to u.$$

As expected, we have

**Lemma 1.1.** *If $G'(u)$ exists and (1.8) holds, then $G'(u) = 0$.*

*Proof.* By (1.8),

$$G(u) \le G(u+v), \quad v \in H.$$

Therefore,

$$-(v, G'(u))_H / \|v\|_H \le [G(u+v) - G(u) - (v, G'(u))_H]/\|v\|_H$$
$$\to 0 \ \text{as} \ \|v\|_H \to 0.$$

Take $v = -\varepsilon h$. Then

$$(\varepsilon h, G'(u))_H / \|\varepsilon h\|_H \le [G(u - \varepsilon h) - G(u) + (\varepsilon h, G'(u))_H]/\|\varepsilon h\|_H$$
$$\to 0 \ \text{as} \ \varepsilon \to 0, \quad h \in H.$$

Consequently,

$$(h, G'(u))_H \le 0, \quad h \in H.$$

Since $-h \in H$ whenever $h \in H$, this implies

$$(h, G'(u))_H = 0, \quad h \in H.$$

Hence $G'(u) = 0$ (just take $h = G'(u)$).                                        □

Now that we know what it means for a functional to be differentiable, we want to find a functional which will satisfy (c) above. This is not a trivial matter. In fact, we have no logical reason to believe that such a functional exists. But this does not deter us. We first look for a functional such that $G'(u) = -u''$. At this point we will have to think seriously about the Hilbert space $H$ in which we will work. Since we are looking for a solution of (1.1),(1.2), it would be natural to take $H$ to have the norm given by (1.4). Since we want

$$(v, u'') = -(v', u'), \quad u, v \in H \cap C^2(I), \tag{1.23}$$

this suggests

$$\langle v, h \rangle = (v', h'). \tag{1.24}$$

Thus, we want

$$[G(u+v) - G(u) - (v', u')]/\|v\|_H \to 0 \quad \text{as} \quad \|v\|_H \to 0.$$

We can make use of the identity (1.23). This suggests trying

$$G(u) = \|u'\|^2, \quad u \in H. \tag{1.25}$$

It looks good because

$$[G(u+v) - G(u) - 2(v', u')]/\|v\|_H = \|v'\|^2/\|v\|_H \to 0 \quad \text{as} \quad \|v\|_H \to 0.$$

Hence we have

**Theorem 1.2.** *If $G$ is given by (1.25), then*

$$(G'(u), v)_H = 2\,(u', v'), \quad u, v \in H. \tag{1.26}$$

*If $u \in H \cap C^2(I)$, then*

$$(G'(u), v)_H = 2\,(-u'', v), \quad v \in H \cap C^1(I). \tag{1.27}$$

Note that we are off by a factor of 2, but this should not cause any indigestion. The same procedure will give us a functional satisfying

$$(G'(u), v)_H = 2\,(u, v), \quad u, v \in H. \tag{1.28}$$

We merely take $G(u) = \|u\|^2$ in this case.

Next we turn to finding a functional $G$ which will satisfy

$$(G'(u), v)_H = (f(\cdot, u), v), \quad u, v \in H. \tag{1.29}$$

Here it is not obvious how to proceed. Going back to the definition, we want to find a functional $G$ such that

$$[G(u+v) - G(u) - (v, f(\cdot, u))]/\|v\|_H$$

converges to 0 as $\|v\|_H \to 0$. It is not easy to guess what $G$ should be, but if we refer back to the comparison with a real valued function, we want to find a function $F(t)$ such that

$$F'(t) = f(x, t).$$

This would suggest that we try something of the form

$$G(u) = \int_I F(x, u(x))\, dx = (F(\cdot, u), 1). \tag{1.30}$$

If we now apply our definition, we want

$$[G(u+v) - G(u) - (v, f(\cdot, u))]/\|v\|_H$$

$$= \int_I [F(x, u+v) - F(x, u) - vf(x, u)] \, dx/\|v\|_H. \qquad (1.31)$$

But

$$F(x, u+v) - F(x, u) = \int_0^1 [dF(x, u+\theta v)/d\theta] \, d\theta$$

$$= \int_0^1 F_t(x, u+\theta v)v \, d\theta. \qquad (1.32)$$

Thus, the left-hand side of (1.31) equals

$$\int_I \int_0^1 [F_t(x, u+\theta v) - f(x, u)]v \, d\theta \, dx/\|v\|_H.$$

We want this expression to converge to 0 as $\|v\|_H \to 0$. This would suggest that we take $F_t(x, t) = f(x, t)$. Let us try

$$F(x, t) = \int_0^t f(x, s) \, ds. \qquad (1.33)$$

Then the expression above is bounded in absolute value by the square root of

$$\int_I \int_0^1 |f(x, u+\theta v) - f(x, u)|^2 \, d\theta \, dx. \qquad (1.34)$$

In order to proceed, we must make an assumption on $f(x, t)$. The easiest at the moment is that it is continuous in $I \times \mathbb{R}$ and satisfies

$$|f(x, t)| \leq C(|t| + 1), \quad x \in I, \ t \in \mathbb{R}, \qquad (1.35)$$

for some constant $C$. We have

**Theorem 1.3.** *Under the above hypothesis, if $G$ is given by (1.30), then it satisfies (1.29).*

*Proof.* The theorem will be proved if we can show that the expression (1.34) converges to 0 together with $\|v\|_H$. If this were not true, there would be a sequence $\{v_k\} \subset H$ such that $\|v_k\|_H \to 0$ and

$$\int_I \int_0^1 |f(x, u+\theta v_k) - f(x, u)|^2 d\theta \, dx \geq \varepsilon > 0. \qquad (1.36)$$

Since $\|v\| \leq \|v\|_H$, we have $\|v_k\| \to 0$. Moreover by Theorem B.25, there is a renamed subsequence (i.e., a subsequence for which we use the same notation) such that $v_k(x) \to 0$ a.e. The integrand of (1.36) is majorized by

$$C(|u|^2 + |v_k|^2 + 1),$$

which converges in $L^1(I)$ to

$$C(|u|^2 + 1).$$

Moreover, the integrand converges to 0 a.e. Hence, the left-hand side of (1.36) converges to 0 by Theorem B.24, contradicting (1.36). (See Appendix B for the elements of Lebesgue integration.) This proves the theorem. □

In order to solve the problem we will have to

(a) find $G(u)$ such that

$$(G'(u), v)_H = (u, v)_H - (f(\cdot, u), v), \quad u, v \in H. \qquad (1.37)$$

holds,

(b) show that there is a function $u(x)$ such that $G'(u) = 0$,
(c) show that $u''$ exists in $I$,
(d) show that $G'(u) = 0$ implies (1.1).

This is a tall order, and some of these steps are not easily carried out.

But first things first. We must decide on the space $H$ where we will be working. Our initial impulse is to look for a space contained in $C^2(I)$, since (1.1) involves second order derivatives. However, the use of such a space is not suitable for our approach since the scalar product (1.3) looks very tempting in light of (1.37), and second derivatives do not appear in it. One might then think that $C^1(I)$ would be suitable. It turns out that even this is not the case because $C^1(I)$ is not complete with respect to the norm (1.4).

"So what?" you exclaim, "Why do we need completeness?" The reason is simple. We want to find a function $u(x)$ which satisfies $G'(u) = 0$. This means that $u$ is a critical point of $G$. How do we find critical points? In most cases in infinite dimensional spaces, this is carried out by a limiting process, and one would like to know that this process has a limit satisfying the desired equation. In particular, if we want to use the scalar product (1.3), it would be very prudent on our part to use a space which is complete with respect to the norm (1.4).

How do we find such a space? Clearly, the norm (1.4) applies to all $u \in C^1(I)$. We also want to restrict it to functions which satisfy (1.2) (i.e., functions that are periodic with period $2\pi$). However, this space is not complete with respect to this norm. The easiest thing to do is to "complete" the space with respect to the norm (1.4). This is not a trivial process, but it is rather straightforward. Descriptions can be found in many textbooks. In the next section we shall accomplish this in a very easy way. Moreover, we shall know precisely what functions are added. We call the resulting space $H$.

What do we get? We allow in functions that are not in $C^1(I)$, but possess "derivatives" in some sense (to be described later). However, when we are dealing with such functions we can make use of the fact that they are the limits in $H$ of functions in $C^1(I)$. This usually suffices for our purposes. Here we make use of the fact that $C^1(I)$ is dense in $H$, but $H$ is a Hilbert space while $C^1(I)$ is not. This advantage will make itself clear as we proceed.

## 1.3    The Hilbert space $H$

Let $C^1(I)$ denote the set of continuously differentiable functions on $I$ satisfying (1.2). For $u \in C^1(I)$ the norm of $H$ is given by

$$\|u\|_H^2 = \|u\|^2 + \|u'\|^2. \tag{1.38}$$

However, if $\{u_k\}$ is a Cauchy sequence in $H$ of functions in $C^1(I)$, then

$$\|u_j - u_k\| \to 0, \ \|u_j' - u_k'\| \to 0.$$

This means that there are functions $u, h \in L^2(I)$ such that

$$u_k \to u, \ u_k' \to h \quad \text{in } L^2(I). \tag{1.39}$$

If $H$ is to be complete, we must allow $u$ to be a member of $H$. But it is unclear what role the function $h$ plays. In particular, is it unique? Does it have any relationship to $u$? To help us understand this process, note that

$$(u_k, v') = -(u_k', v), \quad v \in C^1(I), \tag{1.40}$$

by integration by parts. Thus in the limit,

$$(u, v') = -(h, v), \quad v \in C^1(I). \tag{1.41}$$

Now, if $u \in C^1(I)$, $h \in L^2(I)$, and (1.41) holds, then $h = u'$ a.e. This follows from

**Lemma 1.4.** $C^1(I)$ *is dense in* $L^2(I)$.

We shall prove Lemma 1.4 a bit later. To show that $h = u'$ a.e., we note that by applying integration by parts to (1.41), we have

$$(u' - h, v) = 0, \quad v \in C^1(I).$$

Since $C^1(I)$ is dense in $L^2(I)$, there is a sequence $\{v_n\} \subset C^1(I)$ such that

$$\|v_n - (u' - h)\| \to 0.$$

Consequently,

$$0 = (u' - h, v_n) \to \|u' - h\|^2.$$

This shows that $h = u'$ a.e.

Now, suppose that $u \in L^2(I)$ and there is an $h \in L^2(I)$ such that (1.41) holds. Then I claim that $h$ is unique. For, if (1.41) also holds with $h$ replaced by $g$, then

$$(h - g, v) = 0, \quad v \in C^1(I).$$

Again, by Lemma 1.4 we see that $g = h$ a.e. Even though $u$ is not in $C^1(I)$ and we do not know whether or not it has a derivative at any point, we define the "weak" derivative of $u$ to be $h$ and denote it by $u'$. We define it this way because it behaves like a derivative with respect to integration by parts. We have

**Lemma 1.5.** *If $H$ is the completion of $C^1(I)$ with respect to the norm given by (1.38), then every function in $H$ has a weak derivative in $L^2(I)$.*

*Proof.* If $\{u_k\}$ is a Cauchy sequence in $H$ of functions in $C^1(I)$, then there are functions $u, h$ satisfying (1.39). By (1.40), we see that (1.41) holds. Thus, $h$ is the weak derivative of $u$. If $\{u_k\}$ is a Cauchy sequence in $H$ of functions in $H$, then there are functions $u, h$ satisfying (1.39). By (1.40), we see that (1.41) holds. Thus, $h$ is the weak derivative of $u$. $\square$

Actually, we have

**Theorem 1.6.** *If $H$ has the norm given by (1.38) and consists of those functions $u$ in $L^2(I)$ which have a weak derivative in $L^2(I)$, then $H$ is complete.*

*Proof.* Let $\{u_k\}$ be a Cauchy sequence in $H$. Then there are functions $u, h \in L^2(I)$ such that (1.39) holds. Since $u'_k$ is the weak derivative of $u_k$, (1.40) holds. By (1.39), this implies (1.41). Consequently, $u$ has a weak derivative $u'$ in $L^2(I)$ and $h = u'$ a.e. This means that $u \in H$, and

$$\|u_k - u\|_H \to 0. \tag{1.42}$$

This "completes" the proof.                                                  □

Since we want $H$ to be complete, we are well advised to define it to be the set described in Theorem 1.6. We state this as

**Definition 1.7.** *We define $H$ to be the set of those periodic functions $u \in L^2(I)$ which have weak derivatives in $L^2(I)$. By Theorem 1.6, $H$ is a Hilbert space with norm given by (1.38).*

We started off by defining the norm of $H$ on functions in $C^1(I)$. However, the space $C^1(I)$ is not complete with respect to this norm. To make $H$ complete we added periodic functions in $L^2(I)$ which have weak derivatives in $L^2(I)$. The norm of $H$ makes sense for such functions. Then we showed that $H$ now becomes complete. We might ask if we made $H$ too large, that is, did we have to add all functions having weak derivatives in order to make $H$ complete. Perhaps we could have started with a set smaller than $C^1(I)$ such as $C^\infty(I)$, the set of infinitely differentiable periodic functions on $I$, and completed it with the norm given by (1.38). Perhaps we could have got by without adding so much. The answer is negative as we see from

**Theorem 1.8.** *The space $C^\infty(I)$ is dense in $H$.*

Thus, if $u$ is any element of $H$, then there is a sequence $\{\varphi_k\} \subset C^\infty(I)$ converging to $u$ in $H$. If we remove $u$ from $H$, the Cauchy sequence $\{\varphi_k\}$ will not have a limit in $H$, and $H$ will not be complete.

In proving Theorem 1.8 we shall make use of

**Lemma 1.9.** *For $u \in L^2(I)$, define*

$$\alpha_k = (u, \bar\varphi_k), \quad k = 0, \pm 1, \pm 2, \ldots, \tag{1.43}$$

*where*

$$\varphi_k(x) = \frac{1}{\sqrt{2\pi}} e^{ikx}, \quad k = 0, \pm 1, \pm 2, \ldots \tag{1.44}$$

*Then*

$$\left\| u - \sum_{|k| \leq n} \alpha_k \varphi_k \right\| \to 0 \text{ as } n \to \infty. \tag{1.45}$$

**Remark 1.10.** *The constants $\alpha_k$ and the functions $\varphi_k$ are complex valued. However, they satisfy*

$$\alpha_{-k} = \bar{\alpha}_k, \quad \varphi_{-k} = \bar{\varphi}_k.$$

*Consequently, any sum of the form*

$$\sum_{|k| \leq n} \alpha_k \varphi_k$$

*is real valued. Recall that*

$$(u, v) = \int_0^{2\pi} u(x) v(x) \, dx.$$

Note that Lemma 1.4 is a consequence of Lemma 1.9. We shall prove Lemma 1.9 in the next section. Now we show how it implies Theorem 1.8.

*Proof.* Let $u$ be any function in $H$. It has a weak derivative $u' \in L^2(I)$. Define $\alpha_k, \varphi_k$ by (1.43) and (1.44). Then (1.45) holds by Lemma 1.9. Let

$$\beta_k = (u', \bar{\varphi}_k), \quad k = 0, \pm 1, \pm 2, \ldots$$

By (1.41),

$$\beta_k = -(u, \bar{\varphi}_k') = -(u, -ik\bar{\varphi}_k) = ik(u, \bar{\varphi}_k) = ik\alpha_k.$$

Consequently,

$$\beta_k \varphi_k = ik\alpha_k \varphi_k = \alpha_k \varphi_k', \quad k = 0, \pm 1, \pm 2, \ldots$$

Thus,

$$\left\| u' - \sum_{|k| \leq n} \alpha_k \varphi_k' \right\| = \left\| u' - \sum_{|k| \leq n} \beta_k \varphi_k \right\| \to 0 \text{ as } n \to \infty$$

by Lemma 1.9. If we define

$$u_n(x) = \sum_{|k| \leq n} \alpha_k \varphi_k, \quad n = 0, 1, 2, \ldots,$$

we see that $u_n(x)$ is a real valued periodic function in $C^\infty(I)$ for each $n$, and it satisfies

$$\|u - u_n\| \to 0, \quad \|u' - u_n'\| \to 0.$$

Thus,

$$\|u - u_n\|_H \to 0 \quad \text{as} \quad n \to \infty.$$

Hence, $C^\infty(I)$ is dense in $H$.                                      $\square$

Although functions in $H$ need not be in $C^1(I)$, we shall show that they are in $C(I)$, the set of continuous functions on $I$. In fact we have

**Lemma 1.11.** *All functions in $H$ are in $C(I)$, and there is a constant $K$ such that*

$$|u(x)| \le K\|u\|_H, \quad x \in I, \ u \in H. \tag{1.46}$$

*Moreover,*

$$u(0) = u(2\pi). \tag{1.47}$$

*More precisely, every function $u \in H$ is almost everywhere equal to a function in $C(I)$. Inequality (1.46) holds for the continuous function equal to $u$ a.e. It holds for $u$ if we adjust it on a set of measure zero to make it continuous. The same is true of (1.47). Moreover, if the sequence $\{u_k\}$ converges in $H$, then it converges uniformly on $I$.*

*Proof.* First we prove inequality (1.46) for $u \in C^1(I)$. We note that

$$\left|u(x)^2 - u(x')^2\right| = \left|\int_{x'}^{x} 2u'(y)u(y)\,dy\right| \le \left|\int_{x'}^{x} |2u'(y)u(y)|\,dy\right|$$

$$\le \int_I (|u'(y)|^2 + |u(y)|^2)\,dy = \|u\|_H^2, \quad x, x' \in I.$$

We pick $x' \in I$ so that

$$2\pi u(x')^2 = \int_I u(y)^2\,dy.$$

This can be done by the mean value theorem for integrals. Hence,

$$u(x)^2 \le \frac{1}{2\pi}\|u\|^2 + \|u\|_H^2 \le \left(1 + \frac{1}{2\pi}\right)\|u\|_H^2.$$

This gives inequality (1.46) for $u \in C^1(I)$. Now, I claim that it holds for any $u \in H$. To see this, let $u$ be any function in $H$. Then there is a sequence $\{u_k(x)\}$ of functions in $C^1(I)$ such that

$$\|u_k - u\|_H \to 0.$$

By the inequality (1.46) for functions in $C^1(I)$,

$$|u_j(x) - u_k(x)| \leq K\|u_j - u_k\|_H \to 0, \quad j, k \to \infty.$$

This shows that $\{u_k\}$ is a Cauchy sequence of functions in $C^1(I)$ converging uniformly in $I$ to a function $\tilde{u}$. Since they converge to $u$ in $H$ (and consequently in $L^2(I)$), they converge to both $u$ and $\tilde{u}$ in $L^2(I)$. Thus, we must have $u = \tilde{u}$ a.e. Since

$$|u_k(x)| \leq K\|u_k\|_H,$$

we see in the limit that (1.46) holds. To prove (1.47), note that $u_k(0) = u_k(2\pi)$. Since the sequence converges uniformly, we obtain (1.47) in the limit. $\qquad\square$

**Remark 1.12.** *Note that functions in $L^2(I)$ need not be defined on a set of measure zero. Thus two functions in $L^2(I)$ are considered equal if they differ only on such a set. We shall consider a function $u \in L^2(I)$ to be in $C(I)$ if it is equal a.e. to a function in $C(I)$. In particular, it can be made continuous by changing its definition on a set of measure zero. Any inequality it will be reputed to satisfy will be valid after this change has been made.*

What if $u$ has a weak derivative which is continuous in $I$? That $u$ is continuously differentiable and $u'$ is its derivative in the usual sense follow from

**Lemma 1.13.** *If $u$ has a weak derivative $h$ which is continuous on $I$, then $u$ is differentiable at each point of $I$ and $h$ is the derivative of $u$ in the usual sense.*

*Proof.* Since $u \in H$ and $C^\infty(I)$ is dense in $H$ (Theorem 1.8), there is a sequence $\{u_n\} \subset C^\infty(I)$ such that

$$\|u_n - u\| \to 0, \quad \|u_n' - h\| \to 0.$$

Thus,

$$u_n(x) - u_n(0) = \int_0^x u_n'(t)dt.$$

By Lemma 1.11, $u_n$ converges to $u$ uniformly in $I$. Taking the limit, we have

$$u(x) - u(0) = \int_0^x h(t)dt.$$

This shows that $u$ is differentiable at each point and its derivative equals $h$. $\qquad\square$

We also have

**Theorem 1.14.** *If* $f \in L^2(I)$, $u \in H$ *and*

$$(u, v)_H = (f, v), \quad v \in C^1(I), \tag{1.48}$$

*then* $u' \in H$ *and* $u'' = (u')' = u - f$. *In particular,* $u'$ *is continuous in* $I$ *and is the derivative of* $u$ *in the usual sense.*

*Proof.* By (1.3), $u$ satisfies

$$(u', v') = -(u - f, v), \quad v \in C^1(I). \tag{1.49}$$

This means that $u'$ has a weak derivative equal to $u - f$. Hence, $u' \in H$. Thus $u'$ is continuous in $I$ (Lemma 1.11). Apply Lemma 1.13. $\qquad\square$

We also have

**Theorem 1.15.** *If, in addition,* $f$ *is in* $C(I)$, *then* $u''$ *is continuous in* $I$, *and* $u'' = u - f$ *in the usual sense.*

*Proof.* By Theorem 1.14, $u'$ has a weak derivative equal to $u - f$. Since $u$ is in $H \subset C(I)$, and it is now assumed that $f \in C(I)$, we see by Lemma 1.13 that $u''$ is the derivative of $u'$ in the usual sense. $\qquad\square$

**Theorem 1.16.** *If, in addition,* $f$ *is in* $H$, *then* $u'' \in H$, $u''$ *is continuous in* $I$, $u'' = u - f$ *in the usual sense and* $(u'')' = u' - f'$.

*Proof.* Suppose $v \in C^2(I)$. By (1.49),

$$(u', v'') = -(u - f, v').$$

Since $u' \in H$ and $u - f \in H$, this equals

$$(u'', v') = -(u' - f', v). \tag{1.50}$$

Thus, (1.50) holds for all $v \in C^2(I)$. Now suppose $v$ is only in $C^1(I)$. Then there is a sequence $\{v_k\} \subset C^\infty(I)$ such that

$$\|v_k - v\|_H \to 0$$

(Theorem 1.8). Thus,

$$\|v_k - v\| \to 0, \ \|v_k' - v'\| \to 0.$$

Since

$$(u'', v_k') = -(u' - f', v_k)$$

for each $k$, we see that (1.50) holds in the limit. Thus (1.50) holds for all $v \in C^1(I)$. This means that $u'' \in H$, that $u''$ is continuous, that $u'' = u - f$ in the usual sense, and that $(u'')' = u' - f'$. □

## 1.4    Fourier series

In this section we give the proof of Lemma 1.9. In order to prove it, we shall need

**Lemma 1.17.** *If $f(x)$ is continuous in I and*

$$(f, \bar{\varphi}_k) = 0, \quad k = 0, \pm 1, \pm 2, \ldots, \tag{1.51}$$

*then $f(x) = 0$ in I.*

Before we prove Lemma 1.17, we note that a consequence is

**Corollary 1.18.** *If $f(x)$ is continuous in I and*

$$(f, \bar{\varphi}_k) = 0, \quad k \neq 0, \tag{1.52}$$

*then $f(x)$ is constant in I.*

*Proof.* Let

$$\alpha_0 = \frac{1}{\sqrt{2\pi}} \int_0^{2\pi} f(x) \, dx,$$

and take $g(x) = f(x) - (\alpha_0/\sqrt{2\pi})$. Then

$$\beta_k = (g, \bar{\varphi}_k) = (f, \bar{\varphi}_k) - (\alpha_0\sqrt{2\pi}) \int_0^{2\pi} \bar{\varphi}_k \, dx = 0$$

for $k \neq 0$. Moreover,

$$\beta_0 = (g, 1)/\sqrt{2\pi} = \alpha_0 - \alpha_0 = 0.$$

Thus, $g(x) \equiv 0$ by Lemma 1.17. Hence, $f(x) \equiv \alpha_0/\sqrt{2\pi}$. □

Another consequence is

**Theorem 1.19.** *If $f \in L^2(I)$ satisfies (1.51), then $f(x) = 0$ a.e. in I.*

*Proof.* Define

$$F(x) = \int_0^x f(t) \, dt. \tag{1.53}$$

Then $F(x)$ is continuous in $I$ (Theorem B.36), and $F(2\pi) = \sqrt{2\pi}\alpha_0 = 0 = F(0)$. Hence $F$ is periodic in $I$. Let

$$\gamma_k = (F, \bar{\varphi}_k), \quad k = 0, \pm 1, \pm 2, \ldots$$

Then

$$\gamma_k = (F, e^{-ikx}/\sqrt{2\pi}) = (F, (e^{-ikx}/\sqrt{2\pi})'/(-ik))$$
$$= -(F', \bar{\varphi}_k)/(-ik) = (f, \bar{\varphi}_k)/ik = 0$$

for $k \neq 0$. Hence, $F(x) \equiv$ constant by Corollary 1.18. Consequently, $f(x) = F'(x) \equiv 0$. □

We can now give the proof of Lemma 1.9.

*Proof.* Let

$$u_n(x) = \sum_{|k| \leq n} \alpha_k \varphi_k.$$

Since

$$(\varphi_j, \bar{\varphi}_k) = \frac{1}{2\pi} \int_0^{2\pi} e^{i(j-k)x} dx = \delta_{jk} = \begin{cases} 0, & j \neq k, \\ 1, & j = k, \end{cases}$$

we have

$$\|u - u_n\|^2 = \|u\|^2 - 2(u, u_n) + \|u_n\|^2 = \|u\|^2 - 2\sum_{|k| \leq n} |\alpha_k|^2 + \sum_{|k| \leq n} |\alpha_k|^2$$
$$= \|u\|^2 - \sum_{|k| \leq n} |\alpha_k|^2.$$

Consequently,

$$\sum_{|k| \leq n} |\alpha_k|^2 \leq \|u\|^2.$$

Since this is true for every $n$, we have

$$\sum_{k=-\infty}^{\infty} |\alpha_k|^2 \leq \|u\|^2. \tag{1.54}$$

Moreover, if $m < n$, then

$$\|u_n - u_m\|^2 = \sum_{m < |k| \leq n} |\alpha_k|^2 \to 0 \quad \text{as } n \to \infty.$$

Hence, $\{u_n\}$ is a Cauchy sequence in $L^2(I)$. It converges in $L^2(I)$ to a function $\tilde{u}$. Let

$$\tilde{\alpha}_k = (\tilde{u}, \bar{\varphi}_k), \quad k = 0, \pm 1, \pm 2, \dots$$

Then,

$$\tilde{\alpha}_k = (\lim_{n \to \infty} u_n, \bar{\varphi}_k) = \lim_{n \to \infty} \left( \sum_{|j| \leq n} \alpha_j \varphi_j, \bar{\varphi}_k \right) = \lim_{n \to \infty} \sum_{|j| \leq n} \alpha_j (\varphi_j, \bar{\varphi}_k) = \alpha_k.$$

Let $f = \tilde{u} - u$. Then $f$ satisfies (1.51). It follows from Theorem 1.19 that $f \equiv 0$. Thus $u = \tilde{u}$ a.e., and (1.45) holds. $\qquad\square$

It remains to give the proof of Lemma 1.17.

*Proof.* Assume that there is a point $x_0 \in I$ such that $f(x_0) > 0$. (If $f(x_0) < 0$, replace $f$ with $-f$). Thus, there are positive constants $\varepsilon, \delta$ such that

$$f(x) > \varepsilon, \quad |x - x_0| < \delta. \tag{1.55}$$

Let

$$w_n(x) = \frac{\sin^{2n}(x/2)}{\int_I \sin^{2n}(x/2)\, dx}. \tag{1.56}$$

We note that

$$w_n(x) \geq 0, \quad x \in I, \tag{1.57}$$

and

$$\int_I w_n(x)\, dx = 1. \tag{1.58}$$

Moreover,

$$w_n(x) = c_n(1 - \cos x)^n = c'_n(2 - e^{ix} - e^{-ix})^n.$$

Thus $w_n(x)$ is periodic in $x$ and is a linear combination of the functions $\varphi_k$, $|k| \leq n$. Consequently (1.51) implies

$$\int_I f(x) w_n(x + \pi - x_0)\, dx = 0. \tag{1.59}$$

Now, it follows from (1.56) that

$$w_n(x) \leq \frac{\sin^{2n}[(\pi - \delta)/2]}{\delta \sin^{2n}[(\pi - \delta/2)/2]}, \quad |x - \pi| > \delta,$$

since

$$\int_{\pi-(\delta/2)}^{\pi+(\delta/2)} \sin^{2n}(x/2)\,dx \le \int_I \sin^{2n}(x/2)\,dx.$$

Let

$$\theta = \frac{\sin^2[(\pi-\delta)/2]}{\sin^2[(\pi-\delta/2)/2]}.$$

Then $\theta < 1$, and

$$w_n(x) \le \theta^n/\delta, \quad |x - \pi| > \delta.$$

Hence,

$$w_n(x + \pi - x_0) \le \theta^n/\delta, \quad |x - x_0| > \delta. \tag{1.60}$$

Thus,

$$\int_I f(x)w_n(x + \pi - x_0)\,dx = \int_{x_0-\delta}^{x_0+\delta} + \int_{|x-x_0|>\delta}$$

$$\ge \varepsilon \int_{x_0-\delta}^{x_0+\delta} w_n(x + \pi - x_0)\,dx - 2\pi\theta^n M/\delta$$

$$\ge \varepsilon \int_I w_n(x + \pi - x_0)\,dx - 2\pi\theta^n(M + \varepsilon)/\delta$$

$$= \varepsilon - 2\pi\theta^n(M + \varepsilon)/\delta, \tag{1.61}$$

where

$$M = \max_I |f(x)|.$$

Since $\theta < 1$, we can take $n$ so large that the right-hand side of (1.61) is positive. This contradicts (1.59), and the proof follows. $\qquad\square$

## 1.5    Finding a functional

At this point we want to weaken the assumptions on the function $f(x,t)$. We may have to add to these assumptions from time to time, but such is life. To begin, we assume that $f(x,t)$ is a "Carathéodory" function on $I \times \mathbb{R}$. This means that $f(x,t)$ is a measurable function of $x$ in $I$ for each $t \in \mathbb{R}$, and it is a continuous function of $t$ in $\mathbb{R}$ for almost every $x \in I$. We assume that for each $R \in \mathbb{R}$ there is a constant $C_R$ such that

$$|f(x,t)| \le C_R, \quad x \in I, \ t \in \mathbb{R}, \ |t| \le R. \tag{1.62}$$

This assumption is used to carry out step (a) in our procedure. Based on our previous experience, we try

$$G(u) = \frac{1}{2}\|u\|_H^2 - \int_0^{2\pi} F(x, u(x))\, dx \qquad (1.63)$$

as our first candidate, where

$$F(x, t) = \int_0^t f(x, s)\, ds. \qquad (1.64)$$

First we must check that $G(u)$ is defined for each $u \in H$. For this purpose, we use Lemma 1.11. Note that (1.62) implies

$$|F(x, t)| \le RC_R, \quad t \in \mathbb{R}, \ |t| \le R. \qquad (1.65)$$

If $u \in H$, we have by (1.46) that

$$|F(x, u(x))| \le K\|u\|_H \le C'.$$

Note that $F(x, t)$ is also a "Carathéodory" function on $I \times \mathbb{R}$. We shall show that $F(x, u(x))$ is measurable on $I$ for each $u \in C(I)$ (Theorem B.38). Since it is majorized by a constant and $I$ is a bounded interval, we see that the integral in (1.63) exists (Theorem B.24). Thus, $G(u)$ is defined on $H$.

Next, we calculate the derivative of $G$. We have

$$G(u + v) - G(u) = \frac{1}{2}(\|u + v\|_H^2 - \|u\|_H^2)$$

$$- \int_0^{2\pi} [F(x, u + v) - F(x, u)]\, dx$$

$$= \frac{1}{2}(\|u\|_H^2 + 2(u, v)_H + \|v\|_H^2 - \|u\|_H^2)$$

$$- \int_0^{2\pi} \int_0^1 \frac{d}{d\theta} F(x, u + \theta v)\, d\theta\, dx$$

$$= (u, v)_H + \frac{1}{2}\|v\|_H^2 - \int_0^{2\pi} \int_0^1 f(x, u + \theta v)v\, d\theta\, dx.$$

Hence,

$$[G(u+v) - G(u)] - (u, v)_H + \int_0^{2\pi} f(x, u) v \, dx$$

$$= \frac{1}{2} \|v\|_H^2 - \int_0^{2\pi} \int_0^1 [f(x, u + \theta v) - f(x, u)] v \, d\theta \, dx.$$

The derivative will exist if we can show that the expression (1.34) converges to 0 together with $\|v\|_H$. If this were not true, there would be a sequence $\{v_k\} \subset H$ such that $\|v_k\|_H \to 0$ and

$$\int_I \int_0^1 |f(x, u + \theta v_k) - f(x, u)|^2 d\theta \, dx \geq \varepsilon > 0. \qquad (1.66)$$

By Lemma 1.11, $v_k(x) \to 0$ uniformly on $I$. Thus, $u + \theta v_k \to u$ uniformly on $I$. Consequently, the integrand of (1.66) converges to 0 a.e. The integrand of (1.66) is majorized by constants depending on the $H$ norms of $u + \theta v_k$ and $u$. Since these norms are bounded, the integrand is majorized by a constant. Since the integrand converges to 0 a.e., this implies that the integral converges to 0 (Theorem B.18). Consequently, the derivative of $G$ satisfies (1.37).

Next, we note that the derivative of $G$ is continuous on $H$. To see this, suppose that $\|u_k - u\|_H \to 0$. Then

$$(G'(u_k) - G'(u), v)_H = (u_k - u, v)_H - \int_I [f(x, u_k) - f(x, u)] v \, dx.$$

Thus,

$$|(G'(u_k) - G'(u), v)_H| \leq \|u_k - u\|_H \|v\|_H + \|v\|_H$$

$$\times \int_I |f(x, u_k) - f(x, u)| \, dx.$$

By Lemma 1.11, $u_k(x) \to u(x)$ uniformly in $I$. Consequently, $|f(x, u_k) - f(x, u)| \to 0$ a.e. in $I$. Moreover, it is majorized by constants depending on the norms $\|u_k\|_H, \|u\|_H$, which are bounded. Thus, the integral converges to 0. Now, we have by Corollary A.16

$$\|G'(u_k) - G'(u)\|_H = \sup_{v \in H} \{|(G'(u_k) - G'(u), v)_H| / \|v\|_H\}$$

$$\leq \|u_k - u\|_H + \int_I |f(x, u_k) - f(x, u)| \, dx \to 0.$$

This shows that $G'(u)$ is continuous on $H$. Thus we have proved

**Theorem 1.20.** *If $f(x,t)$ satisfies (1.62), then $G(u)$ given by (1.63) is continuously differentiable and satisfies (1.37).*

## 1.6     Finding a minimum, I

The next step is to find a $u \in H$ such that $G'(u) = 0$. The simplest situation is when $G(u)$ has an extremum. We now give a condition on $f(x,t)$ that will guarantee that $G(u)$ has a minimum on $H$. We assume that there is a function $W(x) \in L^1(I)$ such that

$$-W(x) \leq V(x,t) \equiv t^2 - 2F(x,t) \to \infty \quad \text{a.e. as } |t| \to \infty. \quad (1.67)$$

We shall need

**Lemma 1.21.** *If a sequence satisfies*

$$\rho_k = \|u_k\|_H \leq C, \quad (1.68)$$

*then there are a renamed subsequence (i.e., a subsequence for which we use the same notation) and a $u_0 \in H$ such that*

$$u_k \rightharpoonup u_0 \text{ in } H \quad (1.69)$$

*and*

$$u_k(x) \to u_0(x) \quad \text{uniformly in } I. \quad (1.70)$$

We shall prove Lemma 1.21 at the end of this section.

**Remark 1.22.** *The symbol "$\rightharpoonup$" signifies weak convergence (cf. Appendix A). In the case of (1.69) it means that*

$$(u_k - u_0, v)_H \to 0, \quad v \in H.$$

We let $N$ be the subspace of constant functions in $H$. It is of dimension one. Let $M$ be the subspace of those functions in $H$ which are orthogonal to $N$, that is, functions $w \in H$ which satisfy

$$(w, 1)_H = \int_I w(x)\, dx = 0.$$

We shall also need

$$\|w\| \leq \|w'\|, \quad w \in M, \quad (1.71)$$

and

**Lemma 1.23.** *If $w$ is in $M$ and $\|w\| = \|w'\|$, then*

$$w(x) = a\cos x + b\sin x \qquad (1.72)$$

*for some constants $a, b$.*

We shall also prove these at the end of the section. Now we show how they can be used to give

**Theorem 1.24.** *Under hypothesis (1.67) there is a $u$ in $H$ such that*

$$G(u) = \min_H G.$$

*Moreover, if $f(x, t)$ is continuous in both variables, any such minimum is a solution of (1.1),(1.2) in the usual sense.*

*Proof.* Let

$$\alpha = \inf_H G,$$

and let $\{u_k\}$ be a minimizing sequence, that is, a sequence satisfying

$$G(u_k) \searrow \alpha.$$

Assume first that

$$\rho_k = \|u_k\|_H \leq C.$$

Then, by Lemma 1.21, there is a renamed subsequence such that

$$u_k \rightharpoonup u_0 \text{ in } H \qquad (1.73)$$

and

$$u_k(x) \to u_0(x) \quad \text{uniformly in } I. \qquad (1.74)$$

Then

$$\int_I F(x, u_k)dx \to \int_I F(x, u_0)dx$$

by arguments given previously. Since

$$\|u_0\|_H^2 = \|u_k\|_H^2 - 2([u_k - u_0], u_0)_H - \|u_k - u_0\|_H^2,$$

we have

$$G(u_0) \leq \frac{1}{2}\|u_k\|_H^2 - ([u_k - u_0], u_0)_H - \int_I F(x, u_0)dx$$

$$= G(u_k) - ([u_k - u_0], u_0)_H + \int_I [F(x, u_k) - F(x, u_0)]dx$$

$$\to \alpha$$

Thus

$$G(u_0) \leq \alpha.$$

Since $u_0 \in H$, $\alpha \leq G(u_0)$. Consequently, $\alpha \leq G(u_0) \leq \alpha$, from which we conclude that $G(u_0) = \alpha$.

For each $u \in H$ we may write

$$u = w + v,$$

where $w \in M$ and $v \in N$. We have

$$2G(u) = \|w'\|^2 + \int_I V(x, u)\, dx \geq \frac{1}{2}\|w\|_H^2 - \int_I W(x)\, dx \qquad (1.75)$$

by (1.67) and (1.71). From this we see that if $\{u_k\}$ is a minimizing sequence for $G$, then we must have

$$|w_k(x)| \leq C\|w_k\|_H \leq C', \quad x \in I$$

by Lemma 1.11. But then we have

$$|u_k(x)| \geq |v_k| - |w_k(x)| \geq |v_k| - C', \quad x \in I.$$

Thus, the only way we can have $\|u_k\|_H \to \infty$ is if

$$|u_k(x)| \to \infty, \quad x \in I.$$

But then,

$$\int_I V(x, u_k(x))\, dx \to \infty \text{ as } k \to \infty$$

by (1.67), and this implies

$$G(u_k) \to \infty$$

by (1.75). But this is impossible for a minimizing sequence. Hence, a minimizing sequence must satisfy (1.68). Thus, the $\rho_k$ are bounded, and the proof of the first statement is complete.

To prove the second statement, note that $G'(u) = 0$. Consequently,

$$(u, v)_H - (f(\cdot, u), v) = 0, \quad v \in H.$$

Since $u \in H$, it is continuous in $I$ (Lemma 1.11). If $f(x, t)$ is continuous in both variables, $f(x, u(x))$ is continuous in $I$. Thus, $u'' = u - f(x, u)$ by Theorem 1.15, and consequently $u$ has a continuous second derivative and is a solution of (1.1),(1.2) in the usual sense. $\qquad \square$

It remains to prove Lemmas 1.21, 1.23 together with (1.71). Lemma 1.21 follows from a theorem in functional analysis (Theorem A.61) and the following two lemmas.

**Lemma 1.25.** *If $u \in H$, then*

$$|u(x) - u(x')| \leq |x - x'|^{1/2} \|u'\|, \quad x, x' \in I.$$

*Proof.* Assume first that $u \in C^1(I)$. Then

$$|u(x) - u(x')| = \left| \int_{x'}^{x} u'(y)\, dy \right| \leq \left| \int_{x'}^{x} dy \right|^{1/2} \left| \int_{x'}^{x} u'(y)^2 dy \right|^{1/2}$$
$$\leq |x - x'|^{1/2} \|u'\|.$$

If $u \in H$, there is a sequence $\{u_k\} \subset C^1(I)$ such that $\|u_k - u\|_H \to 0$. Thus,

$$|u_k(x) - u_k(x')| \leq |x - x'|^{1/2} \|u'_k\|.$$

The $u_k$ converge to $u$ uniformly (Lemma 1.11). Taking the limit, we obtain the desired inequality.  □

**Lemma 1.26.** *If $\{u_k\} \subset H$ and $\|u_k\|_H \leq C$, then there is a subsequence which converges uniformly on $I$.*

*Proof.* By Lemmas 1.11 and 1.25, we have

$$|u_k(x)| \leq K', \quad x \in I,$$

and

$$|u_k(x) - u_k(x')| \leq K''|x - x'|^{1/2}, \quad x, x' \in I.$$

Thus, the sequence $\{u_k\}$ is uniformly bounded on $I$ and equicontinuous there. The conclusion now follows from the Arzelà–Ascoli theorem (Theorem C.6).  □

Now we can prove Lemma 1.21.

*Proof.* From Lemma 1.26 we see that there is a renamed subsequence converging uniformly to a function $g \in C(I)$. Moreover, by a theorem in functional analysis (Theorem A.61), there is a renamed subsequence of this subsequence such that

$$(u_k, v) \to (u_0, v), \ (u'_k, v) \to (h, v), \quad v \in L^2(I),$$

where $u_0, h \in L^2(I)$. Now

$$(u_k, v') = -(u'_k, v), \quad v \in C^1(I).$$

Consequently,

$$(u_0, v) = -(h, v), \quad v \in C^1(I).$$

Thus, $h$ is the weak derivative of $u_0$, and $u_0 \in H$ (Definition 1.7). This means that

$$(u_k, v)_H \rightarrow (u_0, v)_H, \quad v \in H.$$

Moreover,

$$(g, v) \leftarrow (u_k, v) \rightarrow (u_0, v), \quad v \in L^2(I).$$

Consequently,

$$(u_0 - g, v) = 0, \quad v \in L^2(I).$$

This implies $\|u_0 - g\|^2 = 0$, showing that $u_0 \equiv g$ a.e. The proof is complete. $\qquad\square$

It now remains to prove (1.71) and Lemma 1.23. We prove them together.

*Proof.* By Lemma 1.9

$$\|u\|^2 = \lim_{n\to\infty} \| \sum_{|k|\le n} \alpha_k \varphi_k\|^2 = \lim_{n\to\infty} \sum_{|k|\le n} |\alpha_k|^2 = \sum_{k=-\infty}^{\infty} |\alpha_k|^2, \quad (1.76)$$

where the $\alpha_k, \varphi_k$ are given by (1.43) and (1.44), respectively. For the same reason,

$$\|u'\|^2 = \lim_{n\to\infty} \| \sum_{|k|\le n} \beta_k \varphi_k\|^2 = \sum_{k=-\infty}^{\infty} |\beta_k|^2 = \sum_{k=-\infty}^{\infty} k^2 |\alpha_k|^2, \quad (1.77)$$

where $\beta_k = (u', \bar\varphi_k) = ik\alpha_k$. If $u \in M$, then $\alpha_0 = 0$. Hence,

$$\|u\|^2 = \sum_{k\ne 0} |\alpha_k|^2 \le \sum k^2 |\alpha_k|^2 = \|u'\|^2,$$

which is (1.71). Moreover, if the two are equal, then

$$\|u'\|^2 - \|u\|^2 = \sum (k^2 - 1)|\alpha_k|^2 = 0.$$

Hence, $\alpha_k = 0$ if $|k| \ne 1$. This means that

$$u = (\alpha_1 e^{ix} + \alpha_{-1} e^{-ix})/\sqrt{2\pi} = a\cos x + b\sin x,$$

and the proof is complete. $\qquad\square$

## 1.7    Finding a minimum, II

It is still possible to show that $G(u)$ attains a minimum even when $F(x,t)$ does not satisfy (1.67). Now we assume that

$$|f(x,t)| \le C(|t|+1), \quad x \in I, \ t \in \mathbb{R}, \quad \text{and} \quad (1.78)$$

$$2F(x,t)/t^2 \to \beta(x) \ \text{a.e. as} \ |t| \to \infty,$$

where

$$\beta(x) \le 1, \ \beta(x) \not\equiv 1. \quad (1.79)$$

We now have

**Theorem 1.27.** *Under hypotheses (1.78), (1.79) there is a u in H such that*

$$G(u) = \min_H G.$$

*Moreover, if $f(x,t)$ is continuous in both variables, any such minimum is a solution of (1.1),(1.2) in the usual sense.*

*Proof.* First, we note that (1.78) implies (1.62). Thus, the functional given by (1.63) has a continuous derivative satisfying (1.37) (Theorem 1.20). Let

$$\alpha = \inf_H G.$$

(As far as we know now, we can have $\alpha = -\infty$.) Let $\{u_k\}$ be a minimizing sequence, that is, a sequence satisfying

$$G(u_k) \to \alpha.$$

Assume first that

$$\rho_k = \|u_k\|_H \le C.$$

Then we can conclude, as we did in the proof of Theorem 1.24, that $G(u_0) = \alpha$.

Next, assume that

$$\rho_k = \|u_k\|_H \to \infty,$$

and let $\tilde{u}_k = u_k/\rho_k$. Then $\|\tilde{u}_k\|_H = 1$, and there is a renamed subsequence such that

$$\tilde{u}_k \rightharpoonup \tilde{u} \ \text{in} \ H \quad (1.80)$$

and

$$\tilde{u}_k(x) \to \tilde{u}(x) \quad \text{uniformly in } I \tag{1.81}$$

(Lemma 1.21). Now,

$$2G(u_k)/\rho_k^2 = 1 - 2\int_I \frac{F(x, u_k)}{u_k^2} \cdot \tilde{u}_k^2 \, dx.$$

Let $\Omega_1$ be the set of points $x \in I$ such that $|u_k(x)| \to \infty$, and let $\Omega_2$ be the set of points $x \in I$ such that $|u_k(x)|$ is bounded. Let $\Omega_3 = I \setminus (\Omega_1 \cup \Omega_2)$. On $\Omega_1$ we have by (1.78)

$$\frac{2F(x, u_k(x))}{u_k(x)^2} \cdot \tilde{u}_k(x)^2 \to \beta(x)\tilde{u}(x)^2 \quad \text{a.e.} \tag{1.82}$$

On $\Omega_2$ we have $\tilde{u}_k(x) = u_k(x)/\rho_k \to 0$, and, consequently, $\tilde{u}(x) = 0$. In this case,

$$\frac{2F(x, u_k(x))}{u_k(x)^2} \cdot \tilde{u}_k(x)^2 = \frac{2F(x, u_k(x))}{\rho_k^2} \to 0 = \beta(x)\tilde{u}(x)^2 \quad \text{a.e.,}$$

and (1.82) holds as well. If $x \in \Omega_3$, there are subsequences for which $|u_k(x)| \to \infty$ and subsequences for which $|u_k(x)|$ is bounded. For the former (1.82) holds, and for the latter it holds as well since $\tilde{u}(x) = 0$. Hence, (1.82) holds a.e. on the whole of $I$. Moreover,

$$\frac{|F(x, u_k)|}{\rho_k^2} \le C\left(\frac{|u_k|^2}{\rho_k^2} + \frac{|u_k|}{\rho_k^2}\right) = C\left(|\tilde{u}_k|^2 + \frac{|\tilde{u}_k|}{\rho_k}\right) \le C \quad \text{a.e.} \tag{1.83}$$

by (1.78). Thus, by the Lebesgue dominated convergence theorem (Theorem B.18)

$$2\int_I F(x, u_k)/\rho_k^2 \, dx \to \int_I \beta(x)\tilde{u}(x)^2 \, dx. \tag{1.84}$$

Hence,

$$2G(u_k)/\rho_k^2 \to 1 - \int_I \beta(x)\tilde{u}(x)^2 \, dx$$

$$= (1 - \|\tilde{u}\|_H^2) + \|\tilde{u}'\|^2 + \int_I [1 - \beta(x)]\tilde{u}(x)^2 \, dx$$

$$= A + B + C.$$

Since $\|\tilde{u}\|_H \le 1$ and $\beta(x) \le 1$, the quantities $A, B, C$ are each $\ge 0$. The only way the sum can equal 0, is if each equals 0. If $B = 0$, then

$\tilde{u}'(x) \equiv 0$. Thus $\tilde{u}(x) \equiv$ constant and $\|\tilde{u}\|_H = \|\tilde{u}\|$. If $A = 0$, then $\|\tilde{u}\| = 1$. Thus, $\tilde{u}$ is a nonvanishing constant. If $C = 0$, then

$$\int_I [1 - \beta(x)]\, dx = 0.$$

Consequently,

$$G(u_k) \to \infty.$$

But this is impossible since $\{u_k\}$ is a minimizing sequence. Thus, the $\rho_k$ are bounded, and the proof of the first statement is complete. The second statement follows as in the proof of Theorem 1.24.     □

## 1.8     A slight improvement

We now give a slight improvement of Theorem 1.27. Now we assume that $f(x,t)$ is a Carathéodory function satisfying (1.62) and

$$F(x,t)/t^2 \leq W(x) \in L^1(I), \ |t| \geq 1, \ \text{and} \ \limsup_{|t| \to \infty} 2F(x,t)/t^2 \leq \beta(x) \text{ a.e.,}$$
$$(1.85)$$

where

$$\beta(x) \leq 1, \ \beta(x) \not\equiv 1. \tag{1.86}$$

We now have

**Theorem 1.28.** *Under hypotheses (1.85), (1.86) there is a u in H such that*

$$G(u) = \min_H G.$$

*Moreover, if $f(x,t)$ is continuous in both variables, any such minimum is a solution of (1.1),(1.2) in the usual sense.*

*Proof.* First, we note that the functional given by (1.63) has a continuous derivative satisfying (1.37) (Theorem 1.20). Let

$$\alpha = \inf_H G.$$

(As far as we know now, we can have $\alpha = -\infty$.) Let $\{u_k\}$ be a minimizing sequence, that is, a sequence satisfying

$$G(u_k) \to \alpha.$$

Assume first that

$$\rho_k = \|u_k\|_H \leq C.$$

Then we can conclude, as we did in the proof of Theorem 1.24, that $G(u_0) = \alpha$.

Next, assume that

$$\rho_k = \|u_k\|_H \to \infty,$$

and let $\tilde{u}_k = u_k/\rho_k$. Then $\|\tilde{u}_k\|_H = 1$, and there is a renamed subsequence such that

$$\tilde{u}_k \rightharpoonup \tilde{u} \quad \text{in } H \qquad (1.87)$$

and

$$\tilde{u}_k(x) \to \tilde{u}(x) \quad \text{uniformly in } I \qquad (1.88)$$

(Lemma 1.21). Now,

$$2G(u_k)/\rho_k^2 = 1 - 2\int_I \frac{F(x, u_k)}{u_k^2} \cdot \tilde{u}_k^2 \, dx.$$

Let $\Omega_1$ be the set of points $x \in I$ such that $|u_k(x)| \to \infty$, and let $\Omega_2$ be the set of points $x \in I$ such that $|u_k(x)|$ is bounded. Let $\Omega_3 = I \setminus (\Omega_1 \cup \Omega_2)$. On $\Omega_1$ we have by (1.85)

$$\limsup_{k \to \infty} \frac{2F(x, u_k(x))}{u_k(x)^2} \cdot \tilde{u}_k(x)^2 \leq \beta(x)\tilde{u}(x)^2. \qquad (1.89)$$

On $\Omega_2$ we have $\tilde{u}_k(x) = u_k(x)/\rho_k \to 0$, and, consequently, $\tilde{u}(x) = 0$. In this case,

$$\frac{2F(x, u_k(x))}{u_k(x)^2} \cdot \tilde{u}_k(x)^2 = \frac{2F(x, u_k(x))}{\rho_k^2} \to 0 = \beta(x)\tilde{u}(x)^2,$$

and (1.89) holds as well. If $x \in \Omega_3$, there are subsequences for which $|u_k(x)| \to \infty$ and subsequences for which $|u_k(x)|$ is bounded. For the former (1.89) holds, and for the latter it holds as well since $\tilde{u}(x) = 0$. Hence, (1.89) holds a.e. on the whole of $I$. Thus, by Theorem B.17

$$\limsup_{k \to \infty} 2\int_I F(x, u_k)/\rho_k^2 \, dx \leq \int_I \beta(x)\tilde{u}(x)^2 \, dx. \qquad (1.90)$$

Hence,

$$\liminf_{k\to\infty} 2G(u_k)/\rho_k^2 \geq 1 - \int_I \beta(x)\tilde{u}(x)^2 dx$$

$$= (1 - \|\tilde{u}\|_H^2) + \|\tilde{u}'\|^2 + \int_I [1 - \beta(x)]\tilde{u}(x)^2 dx$$

$$= A + B + C.$$

We now follow the proof of Theorem 1.27 to reach the desired conclusion. □

## 1.9 Finding a minimum, III

Here we show that one can obtain solutions of (1.1),(1.2) which minimize the functional (1.63) without making assumption (1.67) or assumptions (1.78) and (1.79), provided that the function $F(x,t)$ is concave in $t$. We call a function $g(t)$ on $\mathbb{R}$ **convex** if

$$g((1-\theta)s + \theta t) \leq (1-\theta)g(s) + \theta g(t), \quad s,t \in \mathbb{R}, \ 0 \leq \theta \leq 1. \quad (1.91)$$

It is called **concave** if $-g(t)$ is convex. We shall need

**Lemma 1.29.** *If $g(t) \in C^1(\mathbb{R})$ is convex on $\mathbb{R}$, then*

$$g(t) \geq g(s) + g'(s)(t - s). \quad (1.92)$$

*Proof.* From (1.91) we see that

$$\frac{g(s + \theta(t-s)) - g(s)}{\theta(t-s)}(t - s) \leq g(t) - g(s).$$

Letting $\theta \to 0$, we obtain

$$g'(s)(t - s) \leq g(t) - g(s).$$

□

We can now state

**Theorem 1.30.** *Assume that $f(x,t)$ satisfies (1.62) and that $F(x,t)$ is concave in $t$ for each $x \in I$. Then $G(u)$ given by (1.63) has a minimum on $H$.*

*Proof.* First we note that

$$F(x,t) \leq F(x,0) + f(x,0)t = f(x,0)t, \quad x \in I, t \in \mathbb{R}$$

by Lemma 1.29. Thus,

$$F(x,t) \leq C_0|t|, \quad t \in \mathbb{R}$$

by (1.62). This implies that

$$G(u) \geq \frac{1}{2}\|u\|_H^2 - \int_I f(x,0)u(x)\,dx$$

$$\geq \frac{1}{2}\|u\|_H^2 - c_1\|u\|, \quad u \in H \tag{1.93}$$

by Lemma 1.29. Let $\{u_k\} \subset H$ be a minimizing sequence for $G(u)$, that is, a sequence such that

$$G(u_k) \searrow \alpha = \inf_H G.$$

Since $G(u_k) \leq K$ and

$$G(u_k) \geq \|u_k\|_H^2 - c_1\|u_k\|,$$

we see that the $\|u_k\|_H$ are bounded. We now follow the proof of Theorem 1.24 to arrive at the desired conclusions. $\qquad\square$

## 1.10 The linear problem

You may be curious about the linear problem corresponding to (1.1), (1.2), namely

$$-u''(x) + u(x) = f(x), \quad x \in I = [0, 2\pi], \tag{1.94}$$

under the conditions

$$u(0) = u(2\pi), \quad u'(0) = u'(2\pi), \tag{1.95}$$

where the function $f(x)$ is continuous in $I$ and is periodic in $x$ with period $2\pi$. After a substantial calculation one finds that there is a unique solution given by

$$u(x) = Ae^x + Be^{-x} + \int_0^x \sinh(t - x)\, f(t)\, dt, \tag{1.96}$$

where

$$2A = \frac{e^{2\pi}}{e^{2\pi} - 1} \int_0^{2\pi} e^{-t} f(t)\, dt \tag{1.97}$$

and

$$2B = \frac{e^{-2\pi}}{1 - e^{-2\pi}} \int_0^{2\pi} e^{t} f(t)\, dt. \tag{1.98}$$

Can this solution be used to solve (1.1), (1.2)? It can be if $f(x,t)$ is bounded for all $x$ and $t$. For then we can define

$$Tu(x) = A(u)e^x + B(u)e^{-x} + \int_0^x \sinh(t-x)\, f(t, u(t))\, dt, \qquad (1.99)$$

where

$$2A(u) = \frac{e^{2\pi}}{e^{2\pi}-1} \int_0^{2\pi} e^{-t} f(t, u(t))\, dt \qquad (1.100)$$

and

$$2B(u) = \frac{e^{-2\pi}}{1-e^{-2\pi}} \int_0^{2\pi} e^t f(t, (u(t))\, dt. \qquad (1.101)$$

Then a solution of (1.1),(1.2) will exist if we can find a function $u(x)$ such that

$$Tu(x) = u(x), \quad x \in I. \qquad (1.102)$$

Such a function is called a **fixed point** of the operator $T$. In Chapter 6 we shall study techniques for obtaining fixed points of operators in various spaces. In the present case, one can show that there is indeed a fixed point for the operator $T$ when $f(x,t)$ is bounded.

It is also of interest to note that the linear problem (1.94),(1.95) can be solved easily by the Hilbert space techniques of this chapter. To see this note that

$$Fv = (v, f), \quad v \in H \qquad (1.103)$$

is a bounded linear functional on $H$ (see Appendix A). By the Riesz representation theorem (Theorem A.12), there is an element $u \in H$ such that

$$Fv = (v, u)_H, \quad v \in H.$$

Hence,

$$(u, v)_H = (f, v), \quad v \in H. \qquad (1.104)$$

Since $f$ is continuous, Theorem 1.15 tells us that $u''$ is continuous in $I$ and satisfies $u'' = u - f$.

Note that $u$ satisfies

$$u'(2\pi) = u'(0)$$

as well as $u(2\pi) = u(0)$. To see this, recall that (1.104) implies

$$(u, v) + (u', v') = (f, v), \quad v \in H,$$

which in turn says,

$$\int_0^{2\pi} [u'(x)v(x)]' dx = (u'', v) + (u', v')$$

$$= (u - f, v) + (f, v) - (u, v) = 0, \quad v \in H.$$

Thus

$$u'(2\pi)v(2\pi) - u'(0)v(0) = 0, \quad v \in H.$$

Since $v(2\pi) = v(0)$ for all $v \in H$, we have

$$[u'(2\pi) - u'(0)]v(0) = 0, \quad v \in H.$$

We now merely take $v(0) = u'(2\pi) - u'(0)$ to obtain the result.

All of this illustrates how much easier it is to be linear (straight).

## 1.11 Nontrivial solutions

Theorems 1.24, 1.27, and 1.30 guarantee us that a solution of (1.1),(1.2) exists, but as far as we know the solution may be identically 0. Such a solution is called "trivial" because it usually has no significance in applications. If $f(x, 0) \equiv 0$, we know that $u \equiv 0$ is a solution of (1.1),(1.2) and any method of solving it is an exercise in futility unless we know that the solution we get is not trivial. On the other hand, if $f(x, 0) \not\equiv 0$, then we do not have to worry about trivial solutions. We now consider the problem of insuring that the solutions provided by Theorems 1.24, 1.27, and 1.30 are indeed nontrivial even when $f(x, 0) \equiv 0$. We have

**Theorem 1.31.** *In addition to the hypotheses of Theorems 1.24, 1.27, or 1.30 assume that there is a $t_0 \in \mathbb{R}$ such that*

$$\int_I F(x, t_0) \, dx > \pi t_0^2. \tag{1.105}$$

*Then the solutions of (1.1),(1.2) provided by these theorems are nontrivial.*

*Proof.* We show that the minima $\alpha$ provided by Theorems 1.24, 1.27, and 1.30 are negative. If this is the case, then the solution $u_0$ satisfies

$G(u_0) < 0$. But $G(0) = 0$. This shows that $u_0 \neq 0$. To prove that $\alpha < 0$, let $v \equiv t_0$. Then

$$2G(v) = \|v\|_H^2 - 2 \int_I F(x,v)\, dx = 2\pi t_0^2 - 2 \int_I F(x,t_0)\, dx < 0$$

by (1.105). Hence $\alpha < 0$, and the proof is complete.                    □

Another question we can ask is if the solutions obtained by our theorems are constants. In answer to this we have

**Theorem 1.32.** *In addition to the hypotheses of Theorems 1.24, 1.27, or 1.30 assume that for each $t \in \mathbb{R}$*

$$f(x,t) \not\equiv \quad \text{constant.} \tag{1.106}$$

*Then the solutions of (1.1),(1.2) provided by these theorems are noncon-stant.*

*Proof.* If $u \in N$ and $v \in M$, then

$$(G'(u), v)/2 = (u, v)_H - \int_I f(x,u) v\, dx = - \int_I f(x,t) v\, dx,$$

where $u(x) \equiv t$. By (1.106), $f(x,t) \notin N$. Hence, there is a $v \in M$ such that

$$(G'(u), v)/2 = - \int_I f(x,t) v\, dx \neq 0.$$

Therefore, we cannot have $G'(u) = 0$ for $u \in N$.                    □

### 1.12    Approximate extrema

We now give a very useful method of finding points which are close to being extremum points even when no extremum exists. The following theorem is due to Ekeland.

**Theorem 1.33.** *Let $M$ be a complete metric space and let $G(u)$ be a lower semi-continuous (l.s.c.) functional on $M$. This means that $u_k \to u$ in $M$ implies*

$$G(u) \leq \liminf G(u_k).$$

*Assume that*

$$\inf_M G > -\infty.$$

*Then for every $\varepsilon > 0$, $C > 0$, $y \in M$ satisfying*

$$G(y) \le \inf_M G + \varepsilon, \tag{1.107}$$

*there is a $z \in M$ satisfying*

$$G(z) + Cd(z, y) \le G(y) \tag{1.108}$$

*and*

$$G(z) < G(u) + Cd(z, u), \quad u \in M, \ u \ne z. \tag{1.109}$$

*Proof.* Fix $C > 0$, and let

$$G_s(r) = G(r) + Cd(r, s), \quad r, s \in M,$$

and write $r \prec s$ if $G_s(r) \le G(s)$. Note that $r \prec r$ since $G_r(r) = G(r)$, and

$$r \prec s, \ s \prec t \Longrightarrow r \prec t,$$

since

$$G_s(r) \le G(s), \ G_t(s) \le G(t)$$

imply

$$G(r) + Cd(r, s) \le G(s), \ G(s) + Cd(s, t) \le G(t).$$

Hence,

$$\begin{aligned} G_t(r) = G(r) + Cd(r, t) &\le G(r) + Cd(r, s) + Cd(s, t) \\ &= G_s(r) + Cd(s, t) \le G(s) + Cd(s, t) \\ &= G_t(s) \le G(t). \end{aligned}$$

Let

$$I_s = \{r \in M : r \prec s\}.$$

Then $I_s$ is closed, since $r_k \in I_s$, $r_k \to r$ implies

$$G_r(s) = G(r) + Cd(r, s) \le \liminf\{G(r_k) + Cd(r_k, s)\} \le G(s).$$

Let $\varepsilon, y$ satisfy (1.107), and let $S_0 = I_y$. Pick $z_0 \in S_0$ such that

$$G(z_0) \le \inf_{S_0} G + 2.$$

Set $S_1 = I_{z_0}$, and pick $z_1 \in S_1$ such that

$$G(z_1) \le \inf_{S_1} G + 1.$$

Inductively, pick $z_n \in S_n = I_{z_{n-1}}$ such that

$$G(z_n) \leq \inf_{S_n} G + \frac{1}{n},$$

and set $S_{n+1} = I_{z_n}$. Since $z_n \in S_n = I_{z_{n-1}}$, we have $z_n \prec z_{n-1}$. If $u \in S_n$, then $u \prec z_{n-1} \prec z_{n-2}$. Thus, $u \in S_{n-1}$. This means that $S_n \subset S_{n-1}$. Hence, if $u \in S_{n+1}$, then

$$G_{z_n}(u) = G(u) + Cd(u, z_n) \leq G(z_n) \leq \inf_{S_n} G + \frac{1}{n} \leq G(u) + \frac{1}{n}.$$

In particular, this implies that

$$d(u, z_n) \leq \frac{1}{Cn}.$$

Since $u$ was any element of $S_{n+1}$, we see that the diameter of $S_{n+1}$ is $\leq 2/Cn \to 0$. We therefore have a nested sequence of closed sets whose diameters converge to 0. Consequently there is a unique

$$z \in S = \bigcap_{n=0}^{\infty} S_n$$

(Theorem C.5). Since $z \in S_0 = I_y$, we have $z \prec y$, that is, (1.108) holds. If $u \prec z$, then $u \prec z_n$ for every $n$, since $z \prec z_n$ for every $n$. Hence, $u \in S_n$ for every $n$. Thus, $u \in S$. But the only element in $S$ is $z$. Hence $u = z$. If $u \neq z$ then $u \not\prec z$. This means that $G_z(u) > G(z)$. This is the same as (1.109).                                                                      $\square$

**Corollary 1.34.** *Let $G \in C^1(H, \mathbb{R})$, where $H$ is a Hilbert space. Assume that*

$$\inf_H G > -\infty.$$

*Then for every $\varepsilon > 0$, $y \in H$ satisfying*

$$G(y) \leq \inf_H G + \varepsilon^2,$$

*there is a $u \in H$ such that*

$$G(u) \leq G(y), \quad \|u - y\| \leq \varepsilon, \quad \|G'(u)\| \leq \varepsilon. \tag{1.110}$$

*Proof.* By Theorem 1.33 there is a $u \in H$ such that

$$G(u) + \varepsilon\|u - y\| \leq G(y)$$

and

$$G(u) < G(w) + \varepsilon\|u - w\|, \quad w \neq u.$$

Thus, $G(u) \leq G(y)$ and

$$G(u) + \varepsilon\|u - y\| \leq G(y) \leq \inf_H G + \varepsilon^2 \leq G(u) + \varepsilon^2.$$

Thus, $\|u-y\| \leq \varepsilon$. To prove the last inequality, let $v$ be any fixed element of $H$, and take $w = u + tv$. Then we have

$$G(u) < G(u + tv) + \varepsilon\|tv\|.$$

Now, since the Fréchet derivative of $G$ exists, it satisfies

$$(G'(u), v) = \lim_{t \to 0}[G(u + tv) - G(u)]/t \geq -\varepsilon\|v\|.$$

Take $v = -G'(u)$. This gives

$$\|G'(u)\|^2 \leq \varepsilon\|G'(u)\|,$$

and the corollary is proved. □

We also have

**Corollary 1.35.** *Let $G \in C^1(H, \mathbb{R})$, where $H$ is a Hilbert space. Assume that*

$$\alpha = \inf_H G > -\infty.$$

*Then there is a sequence such that*

$$G(u_k) \to \alpha, \quad G'(u_k) \to 0. \tag{1.111}$$

*Proof.* For each positive integer $k$ there is an element $y_k \in H$ such that

$$G(y_k) \leq \alpha + \frac{1}{k^2}.$$

By Corollary 1.34, there is a $u_k \in H$ such that

$$G(u_k) \leq G(y_k), \quad \|G'(u_k)\| \leq \frac{1}{k}.$$

This gives the required sequence. □

In the next chapter, we shall give another proof of Corollary 1.35 based on Theorem 2.5. That proof will be simpler than the one given here.

**Remark 1.36.** *A sequence satisfying*

$$G(u_k) \to c, \quad G'(u_k) \to 0. \tag{1.112}$$

*is called a* **Palais–Smale sequence** *or* **PS sequence**. *Notice that Corollary 1.35 does not obtain a minimum for the functional $G$ on $H$.*

*However, if it turns out that the PS sequence has a convergent subsequence, then indeed we obtain a minimum for G and a solution of (1.1),(1.2).*

## 1.13    The Palais–Smale condition

Under the hypotheses of Corollary 1.35 we obtained the PS sequence (1.111). If a functional is such that every PS sequence has a convergent subsequence, we say that is satisfies the **Palais–Smale condition** or **PS condition.** In this section we allow $f(x,t)$ to satisfy (1.62), and give sufficient conditions which will guarantee that the PS condition holds for $G(u)$ given by (1.63). We have

**Theorem 1.37.** *If there are constants $\mu > 2$, $C$ such that*

$$H_\mu(x,t) := \mu F(x,t) - tf(x,t) \leq C(t^2 + 1) \qquad (1.113)$$

*and*

$$\limsup_{|t|\to\infty} H_\mu(x,t)/t^2 \leq 0, \qquad (1.114)$$

*then (1.112) implies that $\{u_k\}$ has a convergent subsequence which converges to a solution of (1.1),(1.2).*

*Proof.* If (1.112) holds, then

$$G(u_k) = \rho_k^2 - 2\int_I F(x,u_k)\,dx \to c \qquad (1.115)$$

and

$$(G'(u_k),u_k)_H = 2\rho_k^2 - 2(f(\cdot,u_k),u_k) = o(\rho_k), \qquad (1.116)$$

where $\rho_k = \|u_k\|_H$. Assume that $\rho_k \to \infty$, and let $\tilde{u}_k = u_k/\rho_k$. Since $\|\tilde{u}_k\|_H = 1$, there is a renamed subsequence such that (1.80) and (1.81) hold (Lemma 1.21). By (1.115) and (1.116)

$$\int_I \frac{2F(x,u_k)}{u_k^2}\tilde{u}_k^2\,dx \to 1$$

and

$$\int_I \frac{u_k f(x,u_k)}{u_k^2}\tilde{u}_k^2\,dx \to 1.$$

Thus,

$$\int_I \frac{H_\mu(x, u_k)}{u_k^2} \tilde{u}_k^2 \, dx = \int_I \frac{\mu F(x, u_k) - u_k f(x, u_k)}{u_k^2} \tilde{u}_k^2 \, dx \to \frac{1}{2}\mu - 1.$$

But by hypothesis (1.114),

$$\limsup \frac{H_\mu(x, u_k)}{u_k^2} \tilde{u}_k^2 = \limsup \frac{\mu F(x, u_k) - u_k f(x, u_k)}{u_k^2} \tilde{u}_k^2 \le 0.$$

This is obvious if $|u_k| \to \infty$. Otherwise, we use the fact that $\tilde{u}_k \to 0$. Moreover,

$$\frac{H_\mu(x, u_k)}{u_k^2} \tilde{u}_k^2 = \frac{\mu F(x, u_k) - u_k f(x, u_k)}{u_k^2} \tilde{u}_k^2 \le C\frac{u_k^2 + 1}{u_k^2} \tilde{u}_k^2 \le C'.$$

Note that if $\tilde{u}_k^2 \le 1$, then the right-hand side is

$$\le C\frac{u_k^2 + 1}{\rho_k^2} = C(\tilde{u}_k^2 + \frac{1}{\rho_k^2}) \le C',$$

while otherwise it is

$$\le C(1 + \frac{1}{u_k^2})\tilde{u}_k^2 \le 2C\tilde{u}_k^2 \le C'.$$

By Theorem B.17, this implies that $\frac{1}{2}\mu - 1 \le 0$, contrary to assumption. Hence, the $\rho_k$ are bounded. Consequently, we can conclude that there is a renamed subsequence satisfying (1.73),(1.74). Then

$$\int_I F(x, u_k) dx \to \int_I F(x, u_0) \, dx.$$

Therefore, (1.112) implies

$$(u_0, v)_H - (f(u_0), v) = 0, \quad v \in H,$$

which means that $u_0$ is a solution of (1.1),(1.2) and satisfies

$$\|u_0\|_H^2 - (f(u_0), u_0) = 0.$$

Moreover, by (1.112),

$$\|u_k\|_H^2 = (f(u_k), u_k) + o(1) \to (f(u_0), u_0) = \|u_0\|_H^2,$$

showing that $u_k$ converges to $u_0$ in $H$. Hence, the PS condition holds, and the proof is complete. $\qquad\square$

As a consequence we have

**Theorem 1.38.** *Under hypotheses (1.62),(1.113), and (1.114), if*

$$2F(x,t) \le t^2 + W(x), \quad x \in I,\ t \in \mathbb{R}, \tag{1.117}$$

*where $W(x) \in L^1(I)$, then the functional (1.63) has a minimum on $H$ providing a solution of (1.1),(1.2).*

*Proof.* We have

$$2\int_I F(x,u)\,dx \le \|u\|^2 + B,$$

where

$$B = \int_I W(x)\,dx.$$

Thus,

$$2G(u) \ge \|u\|_H^2 - \|u\|^2 - B = \|u'\|^2 - B \ge -B.$$

Consequently,

$$\alpha = \inf_H G > -\infty.$$

From Corollary 1.35 we conclude that there is a PS sequence satisfying (1.111). We can now apply Theorem 1.37 to reach the desired conclusion. ☐

### 1.14    Exercises

1. Show that

$$(u',v') + (u,v) = (-u'' + u, v) = (f(\cdot,u),v)$$

   holds for all $v \in C^1(I)$ satisfying (1.2).

2. Prove (1.15) and show that we cannot replace the denominator in (1.10) by $|h|$.

3. If $\langle g,h \rangle$ satisfies (1.18), show that for each $w \in H$ there is a unique $g \in H$ such that (1.19) holds.

4. If

$$G(u) = \|u'\|^2, \quad u \in H, \tag{1.118}$$

   show that

$$[G(u+v) - G(u) - 2(v',u')]/\|v\|_H = \|v'\|^2/\|v\|_H \to 0 \ \text{ as } \ \|v\|_H \to 0.$$

5. Prove Theorem 1.2.

6. Verify that (1.39) and (1.40) imply (1.41).

7. If $\alpha_k, \varphi_k$ are given by (1.43) and (1.44), let

$$\beta_k = (u', \bar{\varphi}_k), \quad k = 0, \pm 1, \pm 2, \ldots$$

Verify that

$$\beta_k = -(u, \bar{\varphi}'_k) = -(u, -ik\bar{\varphi}_k) = ik(u, \bar{\varphi}_k) = ik\alpha_k.$$

8. Show that there is a $x' \in I$ such that

$$2\pi u(x')^2 = \int_I u(y)^2 \, dy.$$

9. Prove

$$u(x)^2 \leq \frac{1}{2\pi} \|u\|^2 + \|u\|_H^2 \leq \left(1 + \frac{1}{2\pi}\right) \|u\|_H^2, \quad u \in H.$$

10. Why do we need $F(x)$ to be periodic in the proof of Theorem 1.19?

11. Show that $w_n(x)$ given by (1.56) satisfies

$$w_n(x) \leq \frac{\sin^{2n}[(\pi - \delta)/2]}{\delta \sin^{2n}[(\pi - \delta/2)/2]}, \quad |x - \pi| > \delta.$$

12. If

$$\theta = \frac{\sin^2[(\pi - \delta)/2]}{\sin^2[(\pi - \delta/2)/2]},$$

show that $\theta < 1$, and

$$w_n(x) \leq \theta^n/\delta, \quad |x - \pi| > \delta.$$

13. Prove (1.76) and (1.77).

14. It follows from the text that the Fréchet derivative of a functional $G$ on a Hilbert space $H$ at a point $u$ exists and equals $g \in H$ if and only if

$$[G(u + v) - G(u) - (v, g)_H]/\|v\|_H \to 0 \text{ as } \|v\|_H \to 0, \quad v \in H. \tag{1.119}$$

Why was this simple definition not given outright at the very beginning, rather than in a long drawn out discussion involving other expressions?

15. Show that the problem

$$-u''(x) + u(x) = f(x), \quad x \in I = [0, 2\pi],$$

under the conditions

$$u(0) = u(2\pi), \quad u'(0) = u'(2\pi), \tag{1.120}$$

where the function $f(x)$ is continuous in $I$ and is periodic in $x$ with period $2\pi$, has a unique solution given by

$$u(x) = Ae^x + Be^{-x} + \int_0^x \sinh(t - x) f(t) \, dt,$$

where

$$A = \frac{e^{2\pi}}{e^{2\pi} - 1} \int_0^{2\pi} e^{-t} f(t) \, dt$$

and

$$B = \frac{e^{-2\pi}}{1 - e^{-2\pi}} \int_0^{2\pi} e^t f(t) \, dt.$$

16. Derive these formulas.

17. Show that

$$Fv = (v, f), \quad v \in H$$

is a bounded linear functional on $H$ when $f \in L^2(I)$.

# 2
# Critical points

## 2.1 A simple problem

In order to obtain a minimum for the functional corresponding to our problem (1.1),(1.2), we required (1.67) or (1.78) and (1.79) or concavity. It is quite easy to show that no minimum exists if we assume that

$$|f(x,t)| \leq C(|t|+1), \quad t \in \mathbb{R}, \quad \text{and} \quad f(x,t)/t \to \beta(x) \text{ a.e. as } |t| \to \infty, \tag{2.1}$$

where

$$\int_I [1 - \beta(x)] \, dx < 0.$$

Let $u_k \equiv k$. Then

$$G(u_k)/k^2 = \frac{1}{2}\|1\|_H^2 - \int_I F(x,k)/k^2 \, dx$$

$$\to \pi - \frac{1}{2} \int_I \beta(x) \, dx = \frac{1}{2} \int_I [1 - \beta(x)] \, dx < 0.$$

Thus, $G(u_k) \to -\infty$. Can the problem (1.1),(1.2) be solved if (1.79) does not hold? Before we attempt to answer this question, let us consider a simple problem.

Suppose $f(x,y)$ is a $C^2$ function on $\mathbb{R}^2$ satisfying

$$m_0 = \inf_x f(x,0) \neq -\infty, \quad m_1 = \sup_y f(0,y) \neq \infty. \tag{2.2}$$

Does it follow that there is a point $p_0 = (x_0, y_0)$ such that

$$\nabla f(p_0) = 0? \tag{2.3}$$

The answer is negative. For instance, if we take $f(x,y) = e^x - e^y$, then $f(x,0) = e^x - 1$, while $f(0,y) = 1 - e^y$. Consequently,

$$\inf_x f(x,0) = -1, \quad \sup_y f(0,y) = 1.$$

But,

$$\nabla f(x,y) = (e^x, -e^{-y}) \neq 0, \quad (x,y) \in \mathbb{R}^2.$$

However, even though there is no point $p_0$ satisfying (2.3), there is a sequence $p_k = (x_k, y_k)$ such that

$$f(p_k) \to c, \; m_0 \leq c \leq m_1, \; \nabla f(p_k) \to 0. \tag{2.4}$$

In fact, if we take $x_k \to -\infty, \; y_k \to -\infty$, then

$$f(x_k, y_k) \to 0 \quad \text{and} \quad \nabla f(x_k, y_k) \to (0,0).$$

"What does this accomplish?" you ask. For the present example, nothing. But if we can produce a sequence satisfying (2.4) and this sequence has a convergent subsequence, then we do obtain a point $p_0$ satisfying

$$f(p_0) = c, \; m_0 \leq c \leq m_1, \; \nabla f(p_0) = 0. \tag{2.5}$$

It would therefore appear that a reasonable approach to solving (2.3) is to

(a) find a sequence satisfying (2.4) and
(b) show that this sequence has a convergent subsequence.

Actually, it would appear to be more prudent to check if (b) holds for a sequence satisfying (2.4) before attempting to find such a sequence. Indeed, this is the recommended approach.

## 2.2    A critical point

Even though the example given in the preceding section does not have a critical point, there does exist a sequence of points satisfying (2.4). It would then appear that the existence of such a sequence accomplishes nothing. However, if we change the example slightly and write

$$f(x,y) = e^x - e^y - xy,$$

then (2.2) also holds in this case. If we knew that (2.2) produced a sequence satisfying (2.4), then the situation would be different. For, in this case

$$\nabla f(x,y) = (e^x - y, -e^y - x).$$

If $p_k = (x_k, y_k)$ satisfies (2.4), then

$$e^{x_k} - e^{y_k} - x_k y_k \to c,$$
$$e^{x_k} - y_k \to 0, \tag{2.6}$$

and

$$-e^{y_k} - x_k \to 0. \tag{2.7}$$

If this is true, then the points $p_k$ must be bounded. For, $x_k$ cannot converge to $+\infty$ by (2.7). If it converges to $-\infty$, then we must have $y_k \to +\infty$ by the same expression. But that is precluded by (2.6). Also, $y_k$ cannot converge to $-\infty$ by (2.6), and if it converges to $+\infty$, then $x_k$ would have to do likewise by (2.6). But this is precluded by (2.7). Hence,

$$x_k^2 + y_k^2 \le C.$$

This implies that there is a renamed subsequence such that

$$x_k \to x_0, \quad y_k \to y_0.$$

But then

$$f(p_k) \to f(p_0)$$

and

$$\nabla f(p_k) \to \nabla f(p_0),$$

where

$$p_0 = (x_0, y_0).$$

Consequently, we would have

$$\nabla f(p_0) = 0.$$

This example shows that there are situations in which it is worth searching for a sequence satisfying (2.4).

## 2.3   Finding a Palais–Smale sequence

A sequence satisfying (2.4) is called a **Palais–Smale sequence**. We shall prove

**Theorem 2.1.** *Let $f(x, y)$ be a twice continuously differentiable function on $\mathbb{R}^2$ such that*

$$m_0 = \sup_y \inf_x f(x, y) \ne -\infty, \quad m_1 = \inf_x \sup_y f(x, y) \ne +\infty. \tag{2.8}$$

*Then there is a sequence $p_k = (x_k, y_k)$ such that (2.4) holds.*

To prove this, assume first that $m_0, m_1$ are given by (2.2). As usual when we do not know how to prove something, we assume that it is not so. This would mean that there is a $\delta > 0$ such that

$$|\nabla f(p)| \geq \delta \qquad (2.9)$$

when

$$m_0 - \delta \leq f(p) \leq m_1 + \delta. \qquad (2.10)$$

Otherwise, for every positive integer $k$ there would be a point $p_k$ such that

$$m_0 - \frac{1}{k} \leq f(p_k) \leq m_1 + \frac{1}{k}, \quad |\nabla f(p_k)| \leq \frac{1}{k},$$

and there would be a subsequence satisfying (2.4).

For each $x \in \mathbb{R}$, solve the differential equation

$$\sigma'(t) = \frac{\nabla f(\sigma(t))}{|\nabla f(\sigma(t))|}, \quad t \geq 0, \quad \sigma(0) = (x, 0), \qquad (2.11)$$

and call the solution $\sigma(t)x$. It is very easy to say, "Solve the equation," but it is not so easy to verify that it can be solved. Moreover, we want the solution to be unique. In order to apply Picard's theorem (cf. Theorem 2.13), we must verify that the right-hand side satisfies a local Lipschitz condition. The numerator in (2.11) does satisfy a local Lipschitz condition since $\nabla f$ has continuous derivatives (this is the reason for assuming that $f \in C^2$). Moreover, the same is true of the denominator, since the modulus of a function satisfying a Lipschitz condition also satisfies a Lipschitz condition. In addition, the quotient of two functions satisfying Lipschitz conditions satisfies a Lipschitz condition, provided that the denominator does not vanish.

I am not really concerned about the denominator vanishing, because if it did, we would have a critical point, and this whole discussion would be unnecessary. However, we can avoid the vanishing of the denominator by replacing (2.11) by

$$\sigma'(t) = \frac{\nabla f(\sigma(t))}{\max[|\nabla f(\sigma(t))|, \delta]}, \quad t \geq 0, \quad \sigma(0) = (x, 0). \qquad (2.12)$$

A more pressing concern is how far can we solve (2.12)? Picard's theorem only guarantees the solution in some neighborhood of 0. We shall need more than that. Fortunately, we can prove

**Theorem 2.2.** *Let $h(p)$ be a continuous map from $\mathbb{R}^2$ to $\mathbb{R}^2$ which satisfies a local Lipschitz condition in the neighborhood of each point. Assume also that $h(p)$ is uniformly bounded on $\mathbb{R}^2$. Then for each $p \in \mathbb{R}^2$ there is a unique solution of*

$$\sigma'(t) = h(\sigma(t)), \ t \in \mathbb{R}, \quad \sigma(0) = p. \tag{2.13}$$

*If $\sigma(t)p$ is the solution of (2.13), then it is a continuous map from $\mathbb{R} \times \mathbb{R}^2$ to $\mathbb{R}^2$.*

We shall prove Theorem 2.2 later (cf. Theorem 2.9). Now we use it to continue the proof of Theorem 2.1.

*Proof.* By Theorem 2.2 we can solve (2.12) as far as we like for each $x \in \mathbb{R}$. Note that

$$|\sigma'(t)x| \leq 1, \quad t \geq 0, \ x \in \mathbb{R}.$$

Consequently,

$$|\sigma(t_1)x - \sigma(t_2)x| \leq |t_1 - t_2|, \quad t_1, t_2 \geq 0, \ x \in \mathbb{R}. \tag{2.14}$$

Now,

$$\frac{d}{dt} f(\sigma(t)x) = \nabla f(\sigma(t)x) \cdot \sigma'(t)x = \frac{|\nabla f(\sigma(t)x)|^2}{\max[|\nabla f(\sigma(t)x)|, \delta]} \geq 0. \tag{2.15}$$

Thus, $f(\sigma(t)x) \geq m_0$ for all $t, x$ by (2.2). If $f(\sigma(t)x) \leq m_1 + \delta$ as well, then the right-hand side of (2.15) is $\geq \delta$ by (2.9). As long as this is true, we have

$$f(\sigma(t)x) \geq f(x, 0) + \delta t \geq m_0 + \delta t. \tag{2.16}$$

Let

$$T = (m_1 - m_0 + 2\delta)/\delta. \tag{2.17}$$

Then $f(\sigma(t)x)$ will reach the value $m_1 + \delta$ before $t$ reaches the value $T$. Thus, if $f(x, 0) \leq m_1 + \delta$, there is a smallest number $T_x < T$ such that

$$f(\sigma(T_x)x) = m_1 + \delta. \tag{2.18}$$

By Theorem 2.2, $T_x$ depends continuously on $x$. If $f(x, 0) > m_1 + \delta$, we take $T_x = 0$. It results that $T_x$ is a continuous function of $x$. We know that this is true if $f(x, 0) < m_1 + \delta$ or if $f(x, 0) > m_1 + \delta$. Suppose $f(x_0, 0) = m_1 + \delta$. Then $T_{x_0} = 0$. If $x_k \to x_0$ and $f(x_k, 0) \geq m_1 + \delta$, then $T_{x_k} = 0$, and $T_{x_k} \to 0 = T_{x_0}$. If $x_k \to x_0$ and $f(x_k, 0) < m_1 + \delta$, then there is a renamed subsequence such that $T_{x_k} \to \tilde{T}$. Now $\sigma(T_{x_k})x_k \to$

$\sigma(\tilde{T})x_0$ by Theorem 2.2. Consequently, $f(\sigma(T_{x_k})x_k) \to f(\sigma(\tilde{T})x_0)$. Since $f(\sigma(T_{x_k})x_k) = m_1 + \delta$, we must have $f(\sigma(\tilde{T})x_0) = m_1 + \delta$. This implies that $\tilde{T} = 0$. Otherwise, we would have

$$f(\sigma(\tilde{T})x_0) - f(x_0, 0) = \int_0^{\tilde{T}} \frac{d}{dt} f(\sigma(t)x_0)\, dt$$

$$= \int_0^{\tilde{T}} \frac{|\nabla f(\sigma(t)x_0)|^2}{\max[|\nabla f(\sigma(t)x_0)|, \delta]}\, dt$$

$$= 0.$$

Consequently,

$$\nabla f(\sigma(t)x_0) = 0, \quad 0 \le t \le \tilde{T},$$

which implies

$$\frac{d}{dt} f(\sigma(t)x_0)\, dt = 0, \quad 0 \le t \le \tilde{T},$$

giving

$$f(\sigma(t)x_0) = m_1 + \delta, \quad 0 \le t \le \tilde{T},$$

implying

$$|\nabla f(\sigma(t)x_0)| \ge \delta, \quad 0 \le t \le \tilde{T},$$

an obvious contradiction. Since this is true for each subsequence we see that $T_{x_k} \to 0$ for the entire sequence.

Let

$$h(x) = \sigma(T_x)x. \tag{2.19}$$

Then,

$$|h(x) - (x, 0)| = |\sigma(T_x)x - \sigma(0)x| \le T_x < T \tag{2.20}$$

by (2.14), and

$$f(\sigma(T_x)x) = m_1 + \delta \tag{2.21}$$

by the definition of $T_x$. It follows that $h(x)$ is a continuous function of $x$.

Suppose $x > T$. Then I claim that $h(x)$ is to the right of the $y$-axis, that is, if $h(x) = (h_1(x), h_2(x))$, then $h_1(x) > 0$. To see this, note that

$$x = x - h_1(x) + h_1(x) \le |(x, 0) - h(x)| + h_1(x) \le T + h_1(x).$$

Consequently,

$$0 < x - T \leq h_1(x).$$

Similarly, if $x < -T$, then $h_1(x) < 0$. For then we have

$$h_1(x) = h_1(x) - x + x \leq |h(x) - (x, 0)| + x \leq T + x < 0.$$

Hence we have

$$h_1(x) > 0, \quad x > T,$$

and

$$h_1(x) < 0, \quad x < -T.$$

Since $h_1(x)$ is a continuous function of $x$, there must be an $\hat{x} \in \mathbb{R}$ such that $h_1(\hat{x}) = 0$. This means that $h(\hat{x}) = (0, h_2(\hat{x}))$ is on the $y$-axis, while

$$f(h(\hat{x})) = f(\sigma(T_{\hat{x}})\hat{x}) \geq m_1 + \delta$$

by the definition of $T_x$. But this contradicts (2.2).

This argument proves Theorem 2.1 when (2.2) holds. To prove it when (2.8) holds, let $\varepsilon > 0$ be given. Then there is a point $p_0 = (x_0, y_0) \in \mathbb{R}$ such that

$$m_0 < \inf_x f(x, y_0) + \varepsilon, \quad \sup_y f(x_0, y) < m_1 + \varepsilon.$$

Let

$$g(p) = f(p + p_0), \quad p \in \mathbb{R}^2.$$

Then

$$m_0 < \inf_x g(x, 0) + \varepsilon, \quad \sup_y g(0, y) < m_1 + \varepsilon.$$

By what we have already proved, there is a sequence $\{\tilde{p}_k\}$ such that

$$g(\tilde{p}_k) \to c, \ m_0 - \varepsilon \leq c \leq m_1 + \varepsilon, \ \nabla g(\tilde{p}_k) \to 0.$$

Consequently,

$$f(\tilde{p}_k + p_0) \to c, \ m_0 - \varepsilon \leq c \leq m_1 + \varepsilon, \ \nabla f(\tilde{p}_k + p_0) \to 0.$$

This shows us that there is a point $p_\varepsilon \in \mathbb{R}^2$ such that

$$m_0 - 2\varepsilon \leq f(p_\varepsilon) \leq m_1 + 2\varepsilon, \ |\nabla f(p_\varepsilon)| < \varepsilon.$$

By taking a sequence of $\varepsilon \to 0$ we find a Palais–Smale sequence satisfying (2.4). This completes the proof of Theorem 2.1. $\square$

## 2.4    Pseudo-gradients

In Theorem 2.1 we were required to assume that $f(x, y)$ was twice continuously differentiable on $\mathbb{R}^2$. This was done in order to apply Theorem 2.2 to (2.11). For that purpose we needed the right-hand side of (2.11) or (2.12) to satisfy a local Lipschitz condition. On the other hand, the conclusion of Theorem 2.1 only involves the gradient of $f(x, y)$, that is, the first order derivatives of $f$. It makes one suspicious that the requirement on the second derivatives might not be necessary. Indeed, we are going to show that Theorem 2.1 remains true even when $f$ is only known to have continuous first order derivatives.

How can this be accomplished? Let $Q$ denote the set (2.10). Suppose we can find a function $V$ from $Q$ to $\mathbb{R}^2$ which is locally Lipschitz continuous and satisfies

$$|V(p)| \leq 1, \quad p \in Q, \tag{2.22}$$

and

$$V(p) \cdot \nabla f(p) \geq \frac{1}{2}\delta, \quad p \in Q. \tag{2.23}$$

Instead of solving (2.11), we solve

$$\sigma'(t) = V(\sigma(t)), \quad t > 0, \ \sigma(0) = (x, 0). \tag{2.24}$$

This can be done by Theorem 2.2. The solution exists for $0 \leq t < \infty$. Then

$$|\sigma'(t)| \leq 1, \quad |\sigma(t) - (x, 0)| \leq t,$$

and

$$\frac{df(\sigma(t))}{dt} = \nabla f(\sigma(t)) \cdot \sigma'(t) = \nabla f(\sigma) \cdot V(\sigma) \geq \frac{\delta}{2}$$

as long as $\sigma(t) \in Q$. Therefore,

$$f(\sigma(t)) - f(x, 0) = \int_0^t \left[ \frac{df(\sigma(s))}{ds} \right] ds \geq t\frac{\delta}{2}$$

as long as $\sigma(s) \in Q$ for $0 \leq s \leq t$. We can now follow the proof of Theorem 2.1 to come to the same conclusion.

Of course, all of this depends on our finding such a function $V$. In carrying out our construction we shall make use of the following simple but useful lemma.

**Lemma 2.3.** *If $A$ is any set in $\mathbb{R}^2$ and*

$$g(p) = d(p, A) = \inf_{q \in A} |p - q|,$$

*then*

$$|g(p) - g(p')| \leq |p - p'|.$$

*Proof.* If $q \in A$, then $p - q = p - p' + p' - q$, and

$$d(p, A) \leq |p - q| \leq |p - p'| + |p' - q|.$$

Thus,

$$d(p, A) \leq |p - p'| + d(p', A).$$

Consequently,

$$d(p, A) - d(p', A) \leq |p - p'|.$$

Interchanging $p$ and $p'$ gives

$$d(p', A) - d(p, A) \leq |p' - p|,$$

which produces the desired inequality. $\qquad \square$

Now to the construction of $V(p)$. For each $p \in Q$, let

$$\gamma(p) = \nabla f(p)/|\nabla f(p)|.$$

This is finite by (2.9). Since

$$\nabla f(p) \cdot \gamma(p) = |\nabla f(p)| \geq \delta,$$

there is a neighborhood $N(p)$ of $p$ such that

$$\nabla f(q) \cdot \gamma(p) \geq \frac{\delta}{2}, \quad q \in N(p).$$

(We may assume that the diameter of $N(p)$ is less than one.) Thus,

$$Q \subset \bigcup_{p \in Q} N(p).$$

Let

$$B_R = \{p \in \mathbb{R}^2 : |p| < R\}.$$

Then for each integer $J \geq 0$, the set

$$Q \cap \overline{[B_{J+1} \backslash B_J]}$$

can be covered by a finite number of these neighborhoods by the Heine–Borel theorem. Consequently, there are integers $K_J$, $L_J$ and points $p_k$ in this set such that

$$Q \cap \overline{[B_{J+1} \backslash B_J]} \subset \bigcup_{k=K_J}^{L_J} N(p_k).$$

Then,

$$Q \subset \bigcup_{k=1}^{\infty} N(p_k).$$

Let

$$g_k(p) = d(p, \mathbb{R}^2 \backslash N(p_k)) = \inf_{q \notin N(p_k)} |p - q|.$$

Then,

$$g_k(p) \equiv 0, \quad p \notin N(p_k),$$

and $g_k(p)$ is Lipschitz continuous by Lemma 2.3. Define

$$\psi_k(p) = \frac{g_k(p)}{\sum\limits_{j=1}^{\infty} g_j(p)}, \quad p \in Q, \quad k = 1, 2, \ldots$$

The denominator is positive and finite for each $p \in Q$. The reason for this is that each $p \in Q$ is contained in at least one, but not more than a finite number of, $N(p_k)$, and $g_j(p) = 0$ when $p$ is not in $N(p_j)$.

Let $\tilde{N}(p)$ be a small neighborhood of $p$. Then

$$\tilde{N}(p) \cap N(p_k) \neq \phi$$

for only a finite number of $p_k$. Thus, there is only a finite number of $g_j$ which do not vanish on $\tilde{N}(p)$. Since each $g_j$ is locally Lipschitz continuous, the same is true of the denominator of each $\psi_k$. Thus the same is true of each $\psi_k$ itself. We take

$$V(p) = \sum_{k=1}^{\infty} \psi_k(p) \gamma(p_k).$$

Now, each $\psi_k(p)$ is locally Lipschitz continuous in $N(p_k)$, and $\gamma(p_k)$ is constant there. Moreover, $\psi_k \equiv 0$ outside $N(p_k)$. For each $\tilde{N}(q)$ sufficiently small, only a finite number of functions $\psi_k(p) \gamma(p_k) \neq 0$ in $\tilde{N}(q)$. Consequently, $V(p)$ is locally Lipschitz continuous.

Now,

$$|V(p)| \leq \sum_{k=1}^{\infty} \psi_k(p)|\gamma(p_k)| = \sum_{k=1}^{\infty} \psi_k(p) = 1, \quad p \in Q.$$

Also,

$$V(p) \cdot \nabla f(p) = \sum_{k=1}^{\infty} \psi_k(p)\gamma(p_k) \cdot \nabla f(p) \geq \left(\frac{\delta}{2}\right) \sum_{k=1}^{\infty} \psi_k(p) = \frac{\delta}{2}.$$

The reason for this is that $\psi_k(p) = 0$ if $p \notin N(p_k)$ and $\gamma(p_k) \cdot \nabla f(p) \geq \delta/2$ if $p \in N(p_k)$. Thus, $V(p)$ satisfies (2.22) and (2.23).

We can now state

**Theorem 2.4.** *If $f(x,y)$ is only once continuously differentiable on $\mathbb{R}^2$ but otherwise satisfies the hypotheses of Theorem 2.1, then the conclusions of that theorem hold.*

We are going to show in Appendix $D$ (Theorem D.3) that a pseudo-gradient can be constructed even in a Banach space of infinite dimensions. This will be needed in future work.

## 2.5    A sandwich theorem

In order to apply the ideas of the preceding section to the problem (1.1),(1.2) that we have been studying, we must generalize Theorem 2.1 to a Hilbert space setting. In this case it reads as follows.

**Theorem 2.5.** *Let $M$,$N$ be closed subspaces of a Hilbert space $E$ such that $M = N^{\perp}$. Assume that at least one of these subspaces is finite dimensional. Let $G$ be a differentiable functional on $E$ such that $G'$ is locally Lipschitz continuous and satisfies*

$$m_0 = \sup_{v \in N} \inf_{w \in M} G(v + w) \neq -\infty \qquad (2.25)$$

*and*

$$m_1 = \inf_{w \in M} \sup_{v \in N} G(v + w) \neq \infty. \qquad (2.26)$$

*Then there is a sequence $\{u_k\} \subset E$ such that*

$$G(u_k) \rightarrow c, \; m_0 \leq c \leq m_1, \; G'(u_k) \rightarrow 0. \qquad (2.27)$$

We note that Corollary 1.35 is a simple consequence of Theorem 2.5. To show this, all we need do is consider the case $M = E$, $N = \{0\}$. Then

$$m_0 = m_1 = \alpha = \inf_E G.$$

The result now follows immediately from Theorem 2.5.

We shall begin the proof of Theorem 2.5 in the next section and complete it in Chapter 6. A stronger version will be given in Appendix $D$ (Theorem D.5), where we remove the requirement that $G'$ be locally Lipschitz continuous and replace it with mere continuity. Now we show how it can be used to solve (1.1),(1.2) when (1.79) does not hold. Of course, then we cannot expect to obtain a minimum for $G$ as we showed before. However, Theorem 2.5 allows us to obtain a Palais–Smale sequence if we can find subspaces of $H$ such that (2.25) and (2.26) hold. An example of this is given by

**Theorem 2.6.** *Assume that (1.78) holds with*

$$1 \leq \beta(x) \leq 2, \quad \beta(x) \not\equiv 1, \ \beta(x) \not\equiv 2 \ \text{a.e.} \tag{2.28}$$

*If $G(u)$ is given by (1.63) and $G'$ is locally Lipschitz continuous, then there is a $u_0 \in H$ such that*

$$G'(u_0) = 0. \tag{2.29}$$

*In particular, if $f(x,t)$ is continuous in both variables, then $u_0$ is a solution of (1.1),(1.2) in the usual sense.*

*Proof.* We let $N$ be the subspace of constant functions in $H$. It is of dimension one. Let $M$ be the subspace of those functions in $H$ which are orthogonal to $N$, that is, functions $w \in H$ which satisfy

$$(w, 1)_H = \int_I w(x)\, dx = 0.$$

I claim that

$$m_0 = \inf_M G > -\infty, \quad m_1 = \sup_N G < \infty.$$

For suppose $\{w_k\} \subset M$ and $G(w_k) \searrow m_0$. If $\rho_k = \|w_k\|_H \leq C$, then by Lemma 1.11, (1.63) and (1.65) imply that $m_0 > -\infty$. If $\rho_k \to \infty$, let $\tilde{w}_k = w_k/\rho_k$. Then $\|\tilde{w}_k\|_H = 1$. Consequently, there is a renamed subsequence such that

$$\tilde{w}_k \rightharpoonup \tilde{w} \ \text{in} \ H \tag{2.30}$$

and

$$\tilde{w}_k \to \tilde{w} \quad \text{uniformly in } I \tag{2.31}$$

(Lemma 1.21). Thus,

$$2G(w_k)/\rho_k^2 = 1 - 2 \int_I \frac{F(x, w_k)}{w_k^2} \tilde{w}_k^2 \, dx \tag{2.32}$$

$$\to 1 - \int_I \beta(x)\tilde{w}^2(x) \, dx$$

$$= (1 - \|\tilde{w}\|_H^2)$$

$$+ (\|\tilde{w}\|_H^2 - 2\|\tilde{w}\|^2)$$

$$+ \int_I [2 - \beta(x)]\tilde{w}^2(x) \, dx$$

$$= A + B + C,$$

as we saw before. Now, I claim that $A, B, C \geq 0$. Since $\|\tilde{w}_k\|_H = 1$, we see that $A \geq 0$. We also note that (1.71) implies that $B = \|\tilde{w}'\|^2 - \|\tilde{w}\|^2 \geq 0$. The only way the right-hand side of (2.32) can vanish is if $A = B = C = 0$. If $A = 0$, we see that $\tilde{w} \not\equiv 0$. If $B = 0$, then $\tilde{w}$ is of the form (1.72). If $\tilde{w} \not\equiv 0$, then $a$ and $b$ cannot both vanish. In such a case, $\tilde{w}(x)$ can vanish at only a finite number of points. Finally, for such a function, if $C = 0$, then we must have $\beta(x) \equiv 2$ a.e. But this is excluded by hypothesis. Hence, $A, B, C$ cannot all vanish. This means that the right-hand side of (2.32) is positive. But this implies that $m_0 = \infty$, an impossibility. Thus, the $\rho_k$ must be bounded, and $m_0 > -\infty$.

To prove that $m_1 < \infty$, let $\{u_k\}$ be a sequence $u_k \equiv c_k$, where $|c_k| \to \infty$. Then

$$2G(u_k)/c_k^2 = 2\pi - 2 \int_I F(x, u_k)/c_k^2 \, dx \to 2\pi - \int_I \beta(x) \, dx$$

$$= \int_I [1 - \beta(x)] \, dx < 0$$

by hypothesis. Thus,

$$G(u_k) \to -\infty \quad \text{as } |k| \to \infty.$$

Since $G$ is continuous, we see that $m_1 < \infty$.

We can now apply Theorem 2.5 to conclude that there is a sequence $\{u_k\}$ satisfying (2.27). By (1.37),

$$(G'(u_k), v)_H = (u_k, v)_H - (f(\cdot, u_k), v) = o(\|v\|_H), \quad \|v\|_H \to 0.$$

Assume first that

$$\rho_k = \|u_k\|_H \to \infty. \tag{2.33}$$

Set $\tilde{u}_k = u_k/\rho_k$. Then $\|\tilde{u}_k\|_H = 1$, and consequently, by Lemma 1.21, there is a renamed subsequence such that

$$\tilde{u}_k \rightharpoonup \tilde{u} \text{ in } H, \quad \tilde{u}_k \to \tilde{u} \text{ uniformly on } I. \tag{2.34}$$

Thus,

$$(\tilde{u}_k, v)_H - (f(x, u_k)/\rho_k, v) \to 0, \quad v \in H.$$

As we saw before, this implies in the limit that

$$(\tilde{u}, v)_H = (\beta\tilde{u}, v), \quad v \in H.$$

Take

$$\tilde{u} = \tilde{w} + \gamma, \ v = \tilde{w} - \gamma, \quad \text{where } \tilde{w} \in M, \ \gamma \in N.$$

Then

$$([\tilde{w} + \gamma], [\tilde{w} - \gamma])_H = (\beta[\tilde{w} + \gamma], \tilde{w} - \gamma).$$

This gives

$$\|\tilde{w}\|_H^2 - 2\pi\gamma^2 = (\beta\tilde{w}, \tilde{w}) - \gamma^2(\beta, 1).$$

We write this as

$$(\|\tilde{w}'\|^2 - \|\tilde{w}\|^2) + \int_0^{2\pi} [2 - \beta(x)]\tilde{w}(x)^2 \, dx + \gamma^2 \int_0^{2\pi} [\beta(x) - 1] \, dx$$
$$= A + B + C = 0.$$

Note that $A, B, C$ are all nonnegative. Since their sum is 0, they must each vanish. If $A = 0$, then we must have, in view of Lemma 1.23,

$$\tilde{w} = a\cos x + b\sin x.$$

If $a, b$ are not both 0, then $\tilde{w} \neq 0$ a.e. If $B = 0$, then $[2 - \beta(x)]\tilde{w}(x)^2 \equiv 0$ a.e., and since $\beta(x) \not\equiv 2$, we must have $a = b = 0$. Hence, $\tilde{w}(x) \equiv 0$. If $C = 0$, we must have $\gamma = 0$. Hence, $\tilde{u}(x) \equiv 0$.

On the other hand, we also have

$$2G(u_k)/\rho_k^2 = 1 - 2\int_I F(x, u_k)/\rho_k^2 \, dx \to 0.$$

Thus,

$$1 - \int_I \beta(x)\tilde{u}^2 \, dx = 0.$$

This cannot happen if $\tilde{u} \equiv 0$. Thus (2.33) cannot hold, and the $\rho_k$ are bounded. Consequently, there is a renamed subsequence such that

$$u_k \rightharpoonup u \text{ in } H, \quad u_k \to u \text{ uniformly in } I \qquad (2.35)$$

(Lemma 1.21). By (2.27),

$$(G'(u_k), v) = (u_k, v)_H - (f(x, u_k), v) \to 0, \quad v \in H,$$

and we have in the limit

$$(u, v)_H - (f(x, u), v) = 0, \quad v \in H.$$

Thus, (2.29) holds with $u_0 = u$. Since $u \in H$, it is continuous in $I$. If $f(x, t)$ is continuous in both variables, then $f(x, u(x))$ is continuous in $I$. Thus, $u'' = u - f(x, u)$ in the usual sense by Theorem 1.15. Hence, $u$ is a solution of (1.1),(1.2). $\square$

In order to complete the program, we must find conditions on $G$ which will imply local Lipschitz continuity of $G'$. A simple choice is

$$|f(x, t) - f(x, s)| \le K|t - s|, \quad x \in I, \ s, t \in \mathbb{R}. \qquad (2.36)$$

For then we have

$$\int_I |f(x, u + h) - f(x, u)| \cdot |v| \, dx \le K \int_I |h| \cdot |v| \, dx \le K \|h\| \cdot \|v\|.$$

Hence,

$$|(G'(u + h) - G'(u), v)_H| \le \|h\|_H \|v\|_H + K \|h\| \cdot \|v\| \le C \|h\|_H \|v\|_H,$$

yielding

$$\|G'(u + h) - G'(u)\|_H \le C \|h\|_H, \quad h \in H$$

(cf. (A.5)).

Can we do better? The answer is yes. In fact, we have

**Lemma 2.7.** *If $q < \infty$, and $f(x, t)$ satisfies*

$$|f(x, t) - f(x, s)| \le C(|t|^q + |s|^q + 1)|t - s|, \quad x \in I, \ s, t \in \mathbb{R}, \qquad (2.37)$$

*then $G'(u)$ satisfies*

$$\|G'(u) - G'(v)\|_H \leq C(\|u\|_H^q + \|v\|_H^q + 1)\|u - v\|_H, \quad u, v \in H. \tag{2.38}$$

*Proof.* It follows from (2.37) that

$$\int_I |f(x, u) - f(x, v)| \cdot |h| \, dx \leq C \int_I (|u|^q + |v|^q + 1)|u - v| \cdot |h| \, dx$$

$$\leq C'(\|u\|_H^q + \|v\|_H^q + 1)\|u - v\|_H \|h\|_H. \tag{2.39}$$

This implies

$$|(G'(u) - G'(v), h)_H| \leq \|(u - v)\|_H \|h\|_H$$
$$+ C'(\|u\|_H^q + \|v\|_H^q + 1)\|u - v\|_H \|h\|_H,$$

which implies (2.38).                                           □


Thus we have

**Theorem 2.8.** *Under hypotheses (1.78), (2.28), and (2.37), if $f(x, t)$ is continuous in both variables, then there is at least one solution of (1.1),(1.2).*


## 2.6   A saddle point

In this section we begin the proof of Theorem 2.5. First, we consider the case needed in the proof of Theorem 2.6. The general case will be proved later in Chapter 6. We take $E = H$, $M = V$, $N = W$ and follow the proof of Theorem 2.1. Thus, we have the following situation. Space $H$ splits into the sum of orthogonal subspaces: $H = V \oplus W$, where $V$ represents the constant functions. The functional $G$ is bounded above on $V$ and below on $W$. Let

$$m_0 = \inf_W G, \quad m_1 = \sup_V G, \tag{2.40}$$

and suppose that there is no sequence satisfying (2.27). Then there is a $\delta > 0$ such that

$$\|G'(u)\|_H \geq \delta \tag{2.41}$$

whenever

$$m_0 - \delta \leq G(u) \leq m_1 + \delta. \tag{2.42}$$

Once we have this, we try to draw a "curve" from each point (function) in $V$ along which $G$ decreases at a rate of at least $\delta$. Thus, if $\sigma(t)v$ is the curve emanating from $v \in V$, then we have

$$G(\sigma(t)v) \leq G(v) - t\delta \leq m_1 - t\delta. \tag{2.43}$$

Hence, if

$$T > 1 + (m_1 - m_0)/\delta, \tag{2.44}$$

we have

$$G(\sigma(T)v) \leq m_0 - \delta, \quad v \in V. \tag{2.45}$$

If $\sigma(T)v$ intersects $W$, we have a contradiction of (2.40). This implies that (2.41) cannot hold in the interval (2.42). We can now conclude that there is a sequence satisfying (2.27).

However, there are quite a few obstacles that have to be overcome before we can reach this conclusion. First of all, how do we obtain a curve such as $\sigma(t)v$? Following the reasoning we used in the case of $\mathbb{R}^2$, we look for a solution of the differential equation

$$\frac{d\sigma(t)v}{dt} = -\frac{G'(\sigma(t)v)}{\|G'(\sigma(t)v)\|_H}, \quad t > 0, \quad \sigma(0)v = v. \tag{2.46}$$

If we can solve this for $0 \leq t \leq T$, we will have

$$dG(\sigma(t)v)/dt = (G'(\sigma(t)v), \sigma'(t)v) = -\|G'(\sigma(t)v)\|_H \leq -\delta. \tag{2.47}$$

This will give (2.43), and (2.45) will hold for $T$ satisfying (2.44). It is not obvious that we can solve (2.46) for $0 \leq t \leq T$. As before, we have to be concerned with the denominator vanishing before we get to $t = T$. This is not too much of a concern, since the vanishing of the denominator produces a solution of $G'(u) = 0$. Moreover, the vanishing of the denominator at some point in the interval $[0, T]$ means that $G(\sigma(t)v) \leq m_0 - \delta$ at some earlier point. Actually, we can always "adjust" the right-hand side of (2.46) to avoid this problem. For instance, we can replace (2.46) with

$$\frac{d\sigma(t)v}{dt} = -\frac{G'(\sigma(t)v)}{\max[\|G'(\sigma(t))\|_H, \delta]}, \tag{2.48}$$

which will imply (2.45) in any case. However, we have bigger problems.

We do not know how small $\delta$ is, and consequently, $T$ may be very large. Therefore, we must be able to solve (2.46) for all $t \geq 0$. We shall need the following extension of the well known Picard theorem.

**Theorem 2.9.** *Let $h(t, u)$ be a continuous map from $\mathbb{R} \times H$ to $H$, where $H$ is a Banach space. Assume that for each point $(t_0, u_0) \in \mathbb{R} \times H$, there are constants $K, b > 0$ such that*

$$\|h(t, u) - h(t, v)\| \leq K \|u - v\|, \quad |t - t_0| < b, \ \|u - u_0\| < b, \ \|v - u_0\| < b. \tag{2.49}$$

*Assume also that there is a constant $M$ such that*

$$\|h(t, u)\| \leq M, \quad t \in \mathbb{R}, \ u \in H. \tag{2.50}$$

*Then for each $u \in H$ there is a unique solution $\sigma(t)u$ of the equation*

$$\frac{d\sigma(t)u}{dt} = h(t, \sigma(t)u), \quad t \in \mathbb{R}, \quad \sigma(0)u = u. \tag{2.51}$$

*Moreover, $\sigma(t)u$ is a continuous map from $\mathbb{R} \times H$ to $H$.*

We shall give the proof of Theorem 2.9 later in Section 2.12. Now we shall see how it can be used to help us solve our problem. We want to solve (2.48) for $t > 0$. If we take

$$h(u) = -\frac{G'(u)}{\max[\|G'(u)\|_H, \delta]}, \tag{2.52}$$

we see that (2.50) holds with $M = 1$. However, we also need a local Lipschitz condition of the form

$$\|h(u) - h(v)\|_H \leq K \|u - v\|_H, \quad \|u - u_0\|_H < b, \ \|v - u_0\|_H < b \tag{2.53}$$

in order to satisfy (2.49). If we can show that

$$\|G'(u) - G'(v)\|_H \leq K \|u - v\|_H, \quad \|u - u_0\|_H < b, \ \|v - u_0\|_H < b, \tag{2.54}$$

then the numerator in (2.52) will satisfy a local Lipschitz condition. As we noted in the $\mathbb{R}^2$ case, the same will be true of the entire fraction because

  (a) the norm of a mapping satisfying a Lipschitz condition satisfies the same condition,
  (b) the maximum of two functions satisfying Lipschitz conditions satisfies a Lipschitz condition and

(c) the ratio of two functions satisfying Lipschitz conditions satisfies a Lipschitz condition provided the denominator does not vanish.

Suppose we can prove (2.54). Then Theorem 2.9 tells us that (2.48) can be solved for all $t \in \mathbb{R}$. Moreover,

$$dG(\sigma(t)v)/dt = -\|G'(\sigma(t)v)\|_H^2 / \max[\|G'(\sigma(t)v)\|_H, \delta] \leq 0. \qquad (2.55)$$

Let $T$ satisfy (2.44). If there is a $t_1 \leq T$ such that

$$G(\sigma(t_1)v) \leq m_0 - \delta,$$

we have

$$G(\sigma(T)v) \leq G(\sigma(t_1)v) \leq m_0 - \delta, \qquad (2.56)$$

and (2.45) holds. On the other hand, if

$$G(\sigma(t)v) > m_0 - \delta, \quad 0 \leq t \leq T,$$

then (2.40), (2.41) and (2.55) imply (2.43), and (2.45) holds in this case as well. Once we know that (2.45) holds, we would like to conclude that the set $S = \{\sigma(T)v : v \in V\}$ intersects $W$. Theorem 2.9 tells us that $S$ is a continuous curve in $H$. Let $P$ be the projection of $H$ onto $V$. We shall say that the point $\sigma(T)v$ lies "above" $W$ if $P\sigma(T)v > 0$, and we shall say that it lies "below" $W$ if $P\sigma(T)v < 0$. There are points in $S$ which lie above $W$. For if $v \in V$, then

$$\|\sigma(T)v - v\|_H \leq \int_0^T \|\sigma'(s)v\|_H \, ds \leq T.$$

Consequently,

$$|P\sigma(T)v - v| \leq T,$$

and

$$P\sigma(T)v \geq v - T, \quad P\sigma(T)v \leq v + T.$$

Thus if $v > T$, we must have $P\sigma(T)v > 0$. Similarly, if $v < -T$, we must have $P\sigma(T)v < 0$. Hence, there are points in $S$ which lie above $W$ and points which lie below. Since $S$ is a continuous curve, there must be at least one point $\sigma(T)v_1 \in S$ such that $P\sigma(T)v_1 = 0$. This means that $\sigma(T)v_1 \in W$. In view of (2.40), this gives $G(\sigma(T)v_1) \geq m_0$, contradicting (2.45).

**Remark 2.10.** *It remains, among other things, to prove Theorem 2.9. This will be done in the next few sections. It also remains to complete the proof of Theorem 2.5. We did start it and finished what was necessary for the proof of Theorem 2.6. The general case will be considered in Chapter 6. In Appendix D we shall show that the requirement of Lipschitz continuity of $G'$ can be replaced by mere continuity in Theorem 2.5. This allows us to remove this requirement from Theorem 2.6 and remove the hypothesis (2.37) from Theorem 2.8. We will deal with this and other matters and complete the proof of Theorem 2.5 in Chapter 6 and Appendix D.*

## 2.7    The chain rule

In the previous section we used the formula

$$\frac{dG(\sigma(t)v)}{dt} = (G'(\sigma(t)v), \sigma'(t)v).$$

This is the counterpart of the chain rule from elementary calculus. We present it formally here.

**Lemma 2.11.** *Let $H$ be a Hilbert space, and let $\sigma(t)$ be a mapping of $H \times \mathbb{R}$ to $H$ which has a derivative with respect to $t$ and such that $\sigma(t)$ and $\sigma'(t)$ are continuous from $H \times \mathbb{R}$ to $H$. Let $G$ be a $C^1$ functional from $H$ to $\mathbb{R}$. Then*

$$g(t) = G(\sigma(t))$$

*is a continuously differentiable function from $\mathbb{R}$ to $\mathbb{R}$ and satisfies*

$$g'(t) = (G'(\sigma(t)), \sigma'(t)), \quad t \in \mathbb{R}.$$

*Proof.* Since $\sigma(t)$ is differentiable, we have

$$\frac{[\sigma(t+h) - \sigma(t) - h\sigma'(t)]}{h} \to 0 \text{ as } h \to 0$$

for each fixed $t \in \mathbb{R}$. Fix $t$. Then

$$\sigma(t+h) = \sigma(t) + h\sigma'(t) + o(h),$$

where $o(h)/h \to 0$ as $h \to 0$. Since $G \in C^1(H, \mathbb{R})$, we have

$$\frac{[G(u+v) - G(u) - (G'(u), v)]}{\|v\|} \to 0 \text{ as } \|v\| \to 0.$$

Thus,

$$G(u+v) = G(u) + (G'(u), v) + o(\|v\|),$$

where $o(\|v\|)/\|v\| \to 0$ as $\|v\| \to 0$. Consequently,

$$G(\sigma(t+h)) = G(\sigma(t) + h\sigma'(t) + o(h))$$
$$= G(\sigma(t)) + (G'(\sigma(t)), h\sigma'(t) + o(h))$$
$$+ o(\|h\sigma'(t) + o(h)\|).$$

This means that

$$\frac{[g(t+h) - g(t)]}{h} - (G'(\sigma(t)), \sigma'(t)) = \frac{o(h)}{h} \to 0 \text{ as } h \to 0.$$

This gives the desired result. $\qquad\qquad\qquad\qquad\qquad\qquad\qquad \Box$

## 2.8    The Banach fixed point theorem

In proving Theorem 2.9 we shall make use of several important results from functional analysis concerning mappings on Banach spaces. They will allow us to solve differential equations in such spaces.

Let $X$ be a Banach space, let $M \neq \phi$ be a closed subset, and let $f(x)$ be a map from $M$ to itself such that

$$\|f(x) - f(y)\| \leq \theta\|x - y\|, \quad x, y \in M, \qquad (2.57)$$

for some $\theta < 1$. We have

**Theorem 2.12.** *Under the above hypotheses there is a unique $x_0 \in M$ such that*

$$f(x_0) = x_0. \qquad (2.58)$$

*Proof.* Let $z_0$ be any point in $M$. Define

$$z_{k+1} = f(z_k), \quad k = 0, 1, 2, \ldots$$

Then $z_k \in M$ for every $k$, and

$$z_{k+1} - z_k = f(z_k) - f(z_{k-1}).$$

Hence

$$\|z_{k+1} - z_k\| = \|f(z_k) - f(z_{k-1})\| \leq \theta \|z_k - z_{k-1}\|$$
$$\leq \theta^2 \|z_{k-1} - z_{k-2}\|$$
$$\leq \theta^3 \|z_{k-2} - z_{k-3}\|$$
$$\leq \cdots \leq \theta^k \|z_1 - z_0\|.$$

Therefore, if $m < n$, then

$$\|z_n - z_m\| \leq \|z_n - z_{n-1}\| + \|z_{n-1} - z_{n-2}\| + \cdots + \|z_{m+1} - z_m\|$$
$$\leq (\theta^{n-1} + \theta^{n-2} + \cdots + \theta^m)\|z_1 - z_0\|$$
$$\leq \sum_{k=m}^{\infty} \theta^k \|z_1 - z_0\| = \frac{\theta^m}{1-\theta}\|z_1 - z_0\| \to 0 \ \text{ as } \ m, n \to \infty.$$

By the completeness of $X$, $z_k$ converges to some limit $x_0$ in $X$. By (2.57)

$$\|f(z_k) - f(x_0)\| \leq \theta \|z_k - x_0\| \to 0.$$

Hence

$$f(x_0) \leftarrow f(z_k) = z_{k+1} \to x_0,$$

showing that $x_0$ is a solution of (2.58). If $y_0$ is another solution, we have

$$\|y_0 - x_0\| = \|f(y_0) - f(x_0)\| \leq \theta \|y_0 - x_0\|,$$

which can happen only if $y_0 = x_0$. This proves the theorem. $\qquad\square$

Theorem 2.12 is known as the **Banach fixed point theorem** or the **contraction mapping principle**.

## 2.9    Picard's theorem

Banach's fixed point theorem can be used to prove the following theorem of Picard.

**Theorem 2.13.** *Let $X$ be a Banach space, and let*

$$B_0 = \{x \in X : \|x - x_0\| \leq R_0\}$$

*and*

$$I_0 = \{t \in \mathbb{R} : |t - t_0| \leq T_0\}.$$

*Assume that $g(t,x)$ is a continuous map of $I_0 \times B_0$ into $X$ such that*

$$\|g(t, x) - g(t, y)\| \leq K_0 \|x - y\|, \quad x, y \in B_0, \ t \in I_0 \qquad (2.59)$$

*and*

$$\|g(t, x)\| \leq M_0, \quad x \in B_0, \ t \in I_0. \qquad (2.60)$$

*Let $T_1$ be such that*

$$T_1 \leq \min(T_0, R_0/M_0), \quad K_0 T_1 < 1. \qquad (2.61)$$

*Then there is a unique solution x(t) of*

$$\frac{dx(t)}{dt} = g(t, x(t)), \quad |t - t_0| \leq T_1, \quad x(t_0) = x_0. \tag{2.62}$$

**Remark 2.14.** *We deal with mappings from $\mathbb{R}$ to $X$ in the same way that we deal with mappings from $\mathbb{R}$ to itself. In particular, continuity, differentiability, and Riemann integrability are defined in the same way, and the same theorems hold. In particular, continuous functions are Riemann integrable, and the fundamental theorems of differential and integral calculus are valid.*

We now give the proof of Theorem 2.13.

*Proof.* First we note that $x(t)$ is a solution of (2.62) iff it is a solution of

$$x(t) = x_0 + \int_{t_0}^{t} g(s, x(s)) \, ds, \quad t \in I_1 = \{t \in \mathbb{R} : |t - t_0| \leq T_1\}. \tag{2.63}$$

For if $x(t)$ is a solution of (2.62), we can integrate to obtain (2.63). Note that $x(t)$ will be in $B$ as long as $t$ is in $I_1$ by (2.61). Conversely, if $x(t)$ satisfies (2.63), it is continuous in $t$ since

$$x(t + h) - x(t) = \int_{t}^{t+h} g(s, x(s)) \, ds,$$

and consequently,

$$\|x(t + h) - x(t)\| \leq M_0|h| \to 0 \text{ as } h \to 0.$$

It is also differentiable since $g(s, x(s))$ is continuous, and

$$[x(t + h) - x(t)]/h = \frac{1}{h} \int_{t}^{t+h} g(s, x(s)) \, ds \to g(t, x(t)) \text{ as } h \to 0.$$

Let $Y$ be the Banach space of all continuous functions $x(t)$ from $I_1$ to $X$ with norm

$$|||x||| = \max_{t \in I_1} \|x(t)\|. \tag{2.64}$$

For $x(t) \in Y$, let $f(x(t))$ be the right-hand side of (2.63), and let

$$Q = \{x(t) \in Y : |||x - \hat{x}_0||| \leq R_0\},$$

where $\hat{x}_0(t) \equiv x_0, \ t \in I_1$. If $x(t)$ is in $Q$, then $f(x(t))$ satisfies

$$|||f(x) - \hat{x}_0||| \leq \int_{t_0}^{t_0+T_1} M_0 \, ds \leq R_0.$$

Thus $f$ maps $Q$ into $Q$. Also

$$|||f(x) - f(y)||| = \max_{I_1} \|\int_{t_0}^t [g(s, x(s)) - g(s, y(s))] ds\|$$

$$\leq K_0 \int_{t_0}^{t_0+T_1} \|x(s) - y(s)\| \, ds \leq K_0 T_1 |||x - y|||.$$

Since $K_0 T_1 < 1$ by (2.61), we can use Theorem 2.12 to conclude that there is a unique $x(t) \in Q$ such that $x(t) = f(x(t))$. Hence, $x(t)$ satisfies (2.63), and therefore (2.62). Since every solution of (2.62) is a solution of (2.63), the uniqueness follows from Theorem 2.12.                    □

## 2.10    Continuous dependence of solutions

We now show that the unique solution of (2.62) provided by Theorem 2.13 depends continuously on both $t$ and $x_0$. We denote this solution by $x(t, x_0)$. Thus,

$$x(t, x_0) = x_0 + \int_{t_0}^t g(s, x(s, x_0)) \, dx,$$

and

$$x(t, x_1) = x_1 + \int_{t_0}^t g(s, x(s, x_1)) \, ds,$$

for $x_1 \in B_0$. Hence,

$$\|x(t, x_0) - x(t, x_1)\| \leq \|x_0 - x_1\| + \int_{t_0}^t \|g(s, x(s, x_0)) - g(s, x(s, x_1))\| \, ds$$

$$\leq \|x_0 - x_1\| + K_0 \int_{t_0}^t \|x(s, x_0) - x(s, x_1)\| \, ds.$$

Let

$$w(t) = \|x(t, x_0) - x(t, x_1)\|.$$

Then

$$w(t) \leq \|x_0 - x_1\| + K_0 \int_{t_0}^t w(s) \, ds.$$

This implies

$$\frac{d}{dt}\left[e^{-K_0 t} \int_{t_0}^t w(s) \, ds\right] = e^{-K_0 t}\left[w(t) - K_0 \int_{t_0}^t w(s) \, ds\right] \leq e^{-K_0 t}\|x_0 - x_1\|$$

and

$$e^{-K_0 T} \int_{t_0}^{T} w(s)\, ds \leq \|x_0 - x_1\| \int_{t_0}^{T} e^{-K_0 t}\, dt$$

$$= \|x_0 - x_1\| \, [e^{-K_0 t_0} - e^{-K_0 T}]/K_0,$$

where $T$ satisfies

$$T \leq \min(T_0, (R_0 - \|x_1\|)/M_0), \quad K_0 T < 1.$$

It follows that

$$\int_{t_0}^{T} w(s)\, ds \leq \|x_0 - x_1\| \, [e^{K_0(T - t_0)} - 1]/K_0,$$

and

$$w(t) \leq e^{K_0(T - t_0)} \|x_0 - x_1\|.$$

Thus we have

**Theorem 2.15.** *The solution of (2.62) obtained in Theorem 2.13 is continuous in t and $x_0$.*


## 2.11  Continuation of solutions

We now discuss the question concerning the interval in which (2.62) can be solved. First we note

**Lemma 2.16.** *Assume that $g(t, x)$ is continuous in $\mathbb{R} \times X$, and that $x(t)$ is a solution of*

$$\frac{dx(t)}{dt} = g(t, x(t)), \quad a \leq t < b,$$

*while $y(t)$ is a solution of*

$$\frac{dy(t)}{dt} = g(t, y(t)), \quad b < t \leq c,$$

*where $a < b < c$. Assume also that*

$$\lim_{t \nearrow b} x(t) = \lim_{t \searrow b} y(t) = z_0.$$

*Let*

$$z(t) = x(t), \quad a \leq t < b,$$
$$z(b) = z_0,$$
$$z(t) = y(t), \quad b < t \leq c.$$

*Then $z(t)$ is a solution of*

$$\frac{dz(t)}{dt} = g(t, z(t)), \quad a \le t \le c. \tag{2.65}$$

*Proof.* It is quite clear that $z(t)$ is a solution in the intervals $a < b,\ b < c$ and that $z(t)$ is continuous in $[a, c]$. Moreover,

$$\lim_{t \nearrow b} \frac{dx(t)}{dt} = \lim_{t \nearrow b} g(t, x(t)) = g(b, z_0),$$

and

$$\lim_{t \searrow b} \frac{dy(t)}{dt} = \lim_{t \searrow b} g(t, y(t)) = g(b, z_0),$$

showing that $z(t)$ is continuously differentiable at $b$ and satisfies (2.65). □

We also have

**Corollary 2.17.** *Assume that $g(t, x)$ is continuous in $\mathbb{R} \times X$ and that $x(t)$ is a solution of*

$$\frac{dx(t)}{dt} = g(t, x(t)), \quad a \le t < b,$$

*satisfying*

$$\lim_{t \nearrow b} x(t) = z_0.$$

*Assume also that there is a unique solution $z(t)$ of*

$$\frac{dz(t)}{dt} = g(t, z(t)), \quad a \le t \le c, \quad z(b) = z_0,$$

*where $c > b$. Then $x(t) \equiv z(t)$ in $a \le t \le b$.*

*Proof.* Let

$$w(t) = x(t), \quad a \le t < b,$$
$$w(b) = z_0,$$
$$w(t) = z(t), \quad b < t \le c.$$

By Lemma 2.16, $w(t)$ is a solution of

$$\frac{dw(t)}{dt} = g(t, w(t)), \quad a \le t \le c,$$

and $w(b) = z_0$. By hypothesis, $z(t)$ is the only solution of this equation equal to $z_0$ at $b$. Hence, $w(t) \equiv z(t)$ in $[a, c]$. Since $w(t) \equiv x(t)$ in $[a, b)$, the result follows. □

## 2.12    Extending solutions

We are now able to give the proof of Theorem 2.9.

*Proof.* By Theorems 2.13 and 2.15 there is an interval $|t| < r$ in which a unique solution of (2.51) exists and is continuous with respect to both $t$ and $u$ (with $r$ depending on $u$). Let $T$ be the supremum of all numbers $m$ such that (2.51) has a unique solution in $[0, m]$. Let $m_k$ be a sequence of such $m$ converging to $T$. If $m_j < m_k$, then the solution in $[0, m_j]$ coincides with that of $[0, m_k]$ since such solutions are unique. Thus a unique solution of (2.51) exists for $0 \leq t < T$. The values $\sigma(m_k)u$ are uniquely defined for each $u \in H$.

Assume $T < \infty$. Since

$$\sigma(m_k)u - \sigma(m_j)u = \int_{m_j}^{m_k} h(t, \sigma(t)u)\, dt,$$

we have by (2.50)

$$\|\sigma(m_k)u - \sigma(m_j)u\| \leq M|m_k - m_j|.$$

Thus $\{\sigma(m_k)u\}$ is a Cauchy sequence in $H$. Since $H$ is complete, $\sigma(m_k)u$ converges to an element $w \in H$. Moreover, we note that

$$\sigma(t)u \to w \text{ as } t \to T.$$

To see this, let $\varepsilon > 0$ be given. Then there is a $k$ such that

$$\|\sigma(m_k)u - w\| < \varepsilon, \quad M(T - m_k) < \varepsilon.$$

Then for $m_k \leq t < T$,

$$\|\sigma(t)u - w\| \leq \|\sigma(t)u - \sigma(m_k)u\| + \|\sigma(m_k)u - w\|$$
$$\leq M|t - m_k| + \|\sigma(m_k) - w\| < 2\varepsilon.$$

We define $\sigma(T)u = w$. Then, we have a solution of (2.51) in $[0, T]$. By Theorem 2.13, there is a unique solution of

$$\frac{d\sigma(t)u}{dt} = h(t, \sigma(t)u), \quad \sigma(T)u = w \tag{2.66}$$

in some interval $|t - T| < \delta$. By the uniqueness, the solution of (2.66) coincides with the solution of (2.51) in the interval $(T - \delta, T]$. This gives a solution of (2.51) in the interval $[0, T + \delta)$, contradicting the definition of $T$. Hence, $T = \infty$. Similar reasoning applies to the interval $(-\infty, 0]$. $\qquad\square$

## 2.13     Resonance

In Theorem 1.27 (cf. (1.79)) we did not allow $\beta(x) \equiv 1$ for a good reason. Consider, for instance the case

$$2F(x,t) = t^2 - t^{3/2}, \quad t > 0,$$
$$2F(x,t) = t^2 + |t|^{3/2}, \quad t < 0.$$

For this function

$$G(c) \to \pm\infty \text{ as } c \to \pm\infty.$$

Consequently, neither method used so far will work because the functional $G$ is unbounded on the constants from both above and below. Why the difficulty? If we try to solve the linear problem

$$-u'' + u = \lambda u,$$

for functions satisfying (1.2), we note that the only solution is $u \equiv 0$ unless

$$\lambda = n^2 + 1, \quad n = 0, \pm 1, \pm 2, \ldots$$

For these values of $\lambda$, called **eigenvalues**, there are more solutions, 0, $\cos nx$, $\sin nx$. Thus, the number of solutions jumps from one to three when $\lambda$ hits an eigenvalue. In other words, the number of solutions is unstable in the neighborhood of an eigenvalue. This instability causes the nonlinear problem which asymptotically equals a linear problem at an eigenvalue to become much more delicate than otherwise. We call such a situation "resonance." So let us assume that $f$ satisfies (1.78) with $\beta(x) \equiv 1$. Is there any way of attacking the problem? We shall try the following. Assume that there are constants $\theta, \sigma$ such that $2\theta < \sigma < \theta + 1$, and

$$|f(x,t) - t| \le C(|t|^\theta + 1), \quad x \in I, \ t \in \mathbb{R}, \tag{2.67}$$

$$|F(x,t) - \frac{1}{2}t^2| \le C(|t|^\sigma + 1), \quad x \in I, \ t \in \mathbb{R}, \tag{2.68}$$

$$[2F(x,t) - t^2]/|t|^\sigma \to F_\pm(x) \text{ as } t \to \pm\infty, \tag{2.69}$$

$$\int_I F_\pm(x)\, dx > 0. \tag{2.70}$$

We have

**Theorem 2.18.** *In addition to the above hypotheses, assume that (1.78) holds with $\beta \equiv 1$. Then (1.1),(1.2) has at least one solution.*

*Proof.* First we note that $G$ is bounded from above on $V$. In fact, we have

$$2G(c)/|c|^\sigma = \int_I [c^2 - 2F(x,c)]\,dx/|c|^\sigma \to -\int_I F_\pm(x)\,dx < 0. \quad (2.71)$$

Hence,

$$G(c) \to -\infty \text{ as } |c| \to \infty.$$

On the other hand, $G(c)$ is continuous and bounded on any finite interval. Thus, $G$ is bounded from above on $V$. In contrast, it is bounded from below on $W$. For if $w_k \in W$ and $\rho_k = \|w_k\|_H \to \infty$, let $\tilde{w}_k = w_k/\rho_k$. Then

$$2G(w_k)/\rho_k^2 = \|\tilde{w}_k\|_H^2 - 2\int_I F(x,w_k)\,dx/\rho_k^2$$

$$= 1 - \|\tilde{w}_k\|^2 + \int_I [w_k^2 - 2F(x,w_k)]\,dx/\rho_k^2 \to 1 - \|\tilde{w}\|^2,$$

where we took a renamed subsequence such that $\tilde{w}_k \to \tilde{w}$ weakly in $H$, and uniformly in $I$ (Lemma 1.21). But

$$1 - \|\tilde{w}\|^2 = (1 - \|\tilde{w}\|_H^2) + (\|\tilde{w}'\|^2 - \|\tilde{w}\|^2) + \|\tilde{w}\|^2,$$

which is nonnegative (cf. (1.71)). The only way it can vanish is if

$$\|\tilde{w}\|_H = 1,$$

and

$$\|\tilde{w}\| = 0.$$

The obvious contradiction shows that

$$G(w) \to \infty \text{ as } \|w\| \to \infty, \quad w \in W.$$

From this it follows that $G$ is bounded from below on $W$. We can now follow the procedure in the proof of Theorem 2.8. Let $m_0, m_1$ be given by (2.40). Then there is a sequence $\{u_k\} \subset H$ such that

$$G(u_k) \to a, \quad G'(u_k) \to 0, \quad (2.72)$$

where $a \in [m_0, m_1]$. At this point we must depart from the proof of Theorem 2.8 because (1.78) will not imply that $\tilde{u} \equiv 0$ in this case. We must use a different argument. We write $u_k = v_k + w_k$, where $v_k \in V$, $w_k \in W$. By (2.72)

$$(G'(u_k), w_k)_H = \|w_k'\|^2 - (f(\cdot, u_k) - u_k, w_k) = o(\|w_k\|_H).$$

Now by (2.67)

$$\int_I [f(x, u_k) - u_k] w_k \, dx \le C \left( \int_I (|u_k|^\theta + 1)^2 \, dx \right)^{1/2} \|w_k\|$$

$$\le C' \left( \int_I (|u_k|^{2\theta} + 1) \, dx \right)^{1/2} \|w_k\|$$

$$\le C''(\|u_k\|^\theta + 1)\|w_k\|,$$

since

$$\int_I |u_k|^{2\theta} \, dx \le \left( \int_I |u_k|^2 dx \right)^\theta \left( \int_I dx \right)^{1-\theta}.$$

Consequently,

$$\|w_k'\|^2 \le C(\|u_k\|^\theta + 1)\|w_k\| + o(\|w_k\|_H).$$

In view of (1.71), the same bound holds for $\|w_k\|^2$. Thus

$$\|w_k\|_H^2 \le C(\|u_k\|^\theta + 1)\|w_k\| + o(\|w_k\|_H). \qquad (2.73)$$

Assume that $\rho_k = \|u_k\|_H \to \infty$, and let $\tilde{u}_k = u_k/\rho_k$, $\tilde{v}_k = v_k/\rho_k$, $\tilde{w}_k = w_k/\rho_k$. Then $\|\tilde{u}_k\|_H = 1$, and there is a renamed subsequence such that $\tilde{u}_k \to \tilde{u}$ weakly in $H$ and uniformly in $I$ (Lemma 1.21). Now (2.73) implies

$$\|w_k\|_H \le C(\rho_k^\theta + 1) + o(1),$$

and hence, $\tilde{w}_k \to 0$ in $H$. This means that $\tilde{u}_k \to \tilde{v}$ in $H$, and $\|\tilde{v}\|_H = 1$. But

$$\int_I \left[ F(x, u_k) - \frac{1}{2} u_k^2 \right] dx/\rho_k^\sigma = \int_I \left[ \left( F(x, u_k) - \frac{1}{2} u_k^2 \right) /|u_k|^\sigma \right] |\tilde{u}_k|^\sigma \, dx$$

$$\to \int_{\tilde{v}>0} F_+(x)|\tilde{v}|^\sigma dx + \int_{\tilde{v}<0} F_-(x)|\tilde{v}|^\sigma \, dx$$

$$\to \begin{cases} |\tilde{v}|^\sigma \int_I F_+(x) \, dx, & \tilde{v} > 0, \\ |\tilde{v}|^\sigma \int_I F_-(x) \, dx, & \tilde{v} < 0, \end{cases}$$

by (2.68) and (2.69), since $\tilde{v}$ is a nonzero constant. In view of (2.72),

$$2G(u_k)/\rho_k^\sigma = \|w_k'\|^2/\rho_k^\sigma - 2\int_I \left[ F(x, u_k) - \frac{1}{2} u_k^2 \right] dx/\rho_k^\sigma \to 0.$$

But this implies

$$\int_I F_+(x)\,dx = 0 \ \text{ or } \ \int_I F_-(x)\,dx = 0,$$

contradicting (2.70). Therefore, we must have $\rho_k \leq C$. Once we know that the $u_k$ are bounded in $H$, we can now follow the proof of Theorem 2.8 to reach the desired conclusion. □

## 2.14     The question of nontriviality

In the case of Theorems 2.8 and 2.18, we have the same problem that faced us when we proved Theorem 1.27, namely if $f(x,0) \equiv 0$, what guarantee do we have that the solution provided is not $u \equiv 0$. Of course, if $f(x,0) \not\equiv 0$, then we can relax because 0 cannot be a solution of (1.1),(1.2). But if $f(x,0) \equiv 0$, then we have no such guarantee. In the case of Theorem 1.27 we were able to find a criterion (namely (1.105)) which provides such a guarantee (cf. Theorem 1.31). This was accomplished by showing that the minimum obtained by Theorem 1.27 was less than 0. Since $G(0) = 0$, this shows that the solution obtained was not 0. However, this does not work in the case of Theorem 2.8 since $m_0 \leq G(0) \leq m_1$ (cf. (2.40)). Thus, in order to guarantee that our solutions are not $\equiv 0$, we must devise another means of attack. One plan is as follows. Suppose, in addition to the hypotheses of Theorem 2.8, we have assumptions which will provide positive constants $\varepsilon, \rho$ so that

$$G(u) \geq \varepsilon, \quad \|u\|_H = \rho, \ u \in H.$$

Since $G(0) = 0$, this creates the image of 0 being in a valley surrounded by mountains of minimum height $\varepsilon$. Now, we maintain that there is a sequence $\{u_k\} \subset H$ such that

$$G(u_k) \to c, \ \varepsilon \leq c \leq m_1, \ G'(u_k) \to 0. \qquad (2.74)$$

With this added help, there is a situation in which we can be sure that (1.1),(1.2) has a nontrivial solution. In fact, we have

**Lemma 2.19.** *In addition to the hypotheses of Theorem 2.8, assume that there are positive constants $\varepsilon$, $\rho$ such that*

$$G(u) \geq \varepsilon \qquad (2.75)$$

*when*

$$\|u\|_H = \rho. \qquad (2.76)$$

*Then there is a solution $u$ of (1.1),(1.2) satisfying (2.75).*

*Proof.* We proceed as in the proof of Theorem 2.8. However, this time we suppose that (2.41) holds for all $u \in H$ satisfying

$$\varepsilon - \delta \le G(u) \le m_1 + \delta. \tag{2.77}$$

If we now take

$$T > 1 + (m_1 - \varepsilon)/\delta, \tag{2.78}$$

we find that

$$G(\sigma(T)v) < \varepsilon - \delta, \quad v \in V. \tag{2.79}$$

Next, we note that

$$P\sigma(T)v > v - T$$

as before. Thus, if $v > T + \rho$, then $\sigma(T)v$ must be outside the sphere (2.76). However, if $v$ is close to 0, then $G(v) < \varepsilon$ (since $G(0) = 0$). By (2.55), $G(\sigma(t)v)$ cannot increase as $t$ increases. Hence, for such $v \in V$, $\sigma(t)v$ must remain inside the sphere (2.76) for ever. In particular, $\sigma(T)v$ is inside. Since $S = \{\sigma(T)v : v \in V\}$ is a continuous curve, there must be at least one point $\sigma(T)v_1 \in S$ such that $\|\sigma(T)v_1\|_H = \rho$. This means that

$$G(\sigma(T)v_1) \ge \varepsilon$$

by (2.75) and (2.76). But this contradicts (2.79) and completes the proof. $\square$

Similarly, we have

**Lemma 2.20.** *In addition to the hypotheses of Theorem 2.18 assume that (2.75) holds when u is on the sphere (2.76). Then (1.1),(1.2) has a solution satisfying (2.75).*

Since the surrounding of the origin by mountains is very helpful, we would like to find criteria which will ensure that this is the case. We take up this matter in the next section.

## 2.15    The mountain pass method

We now want to give sufficient conditions on $F(x, t)$ which will imply that the origin is surrounded by mountains. This can be done as follows.

**Theorem 2.21.** *Assume that (1.62) holds and that there is a $\delta > 0$ such that*

$$2F(x,t) \leq t^2, \quad |t| \leq \delta. \tag{2.80}$$

*Then for each positive $\rho \leq \delta/2$, we have either*

**(a)** *there is an $\varepsilon > 0$ such that*

$$G(u) \geq \varepsilon, \quad \|u\|_H = \rho, \tag{2.81}$$

*or*

**(b)** *there is a constant $t \in \mathbb{R}$ such that $|t| = \rho/(2\pi)^{\frac{1}{2}} \leq \delta/2$, and*

$$f(x,t) \equiv t. \tag{2.82}$$

*Moreover, the constant function $v \in V$ satisfying $v \equiv t$ is a solution of (1.1) and (1.2).*

*Proof.* For each $u \in H$ write $u = v + w$, where $v \in V$, $w \in W$. Then

$$2G(u) = \|u\|_H^2 - 2 \int_I F(x,u)\,dx$$

$$= \|u'\|^2 - \int_I [2F(x,u) - u^2]\,dx$$

$$\geq \|w'\|^2 - \int_{|u|>\delta} [2F(x,u) - u^2]\,dx.$$

Now

$$\|u\|_H \leq \rho \implies \|v\|^2 + \|w\|_H^2 \leq \rho^2 \implies (2\pi)^{\frac{1}{2}}|v| \leq \rho.$$

Thus, if $\rho \leq \delta/2$, then $|v| \leq \delta/2$. Hence, if

$$\|u\|_H \leq \rho, \quad |u(x)| \geq \delta,$$

then

$$\delta \leq |u(x)| \leq |v| + |w(x)| \leq \delta/2 + |w(x)|.$$

Consequently,

$$\delta \leq |u(x)| \leq 2|w(x)|.$$

Then

$$2G(u) \geq \|w'\|^2 - C \int_{|u|>\delta} (|u|^{q+1} + u^2 + |u|) \, dx$$

$$\geq \|w'\|^2 - C(1 + \delta^{1-q} + \delta^{-q}) \int_{|u|>\delta} |u|^{q+1} dx$$

$$\geq \frac{1}{2}\|w\|_H^2 - C' \int_{2|w|>\delta} |w|^{q+1} \, dx$$

$$\geq \frac{1}{2}\|w\|_H^2 - C'' \int_I \|w\|_H^{q+1} \, dx$$

$$\geq \frac{1}{2}\|w\|_H^2 - C'''\|w\|_H^{q+1}$$

$$= \left( \frac{1}{2} - C'''\|w\|_H^{q-1} \right) \|w\|_H^2$$

by Lemma 1.11 and (1.71). Hence,

$$G(u) \geq \frac{1}{5}\|w\|_H^2, \quad \|u\|_H \leq \rho, \tag{2.83}$$

for $\rho > 0$ sufficiently small. For such $\rho$ assume that there is no $\varepsilon > 0$ for which (2.81) holds. Then there is a $\{u_k\} \subset H$ such that $\|u_k\|_H = \rho$ and $G(u_k) \to 0$. Write $u_k = v_k + w_k$, where $v_k \in V$, $w_k \in W$. Then $\|w_k\|_H \to 0$ by (2.83). This means that $\|v_k\|_H \to \rho$. Thus, $|v_k| \to \rho/(2\pi)^{\frac{1}{2}}$. Since the constants $\{v_k\}$ are bounded, there is a renamed subsequence such that $v_k \to v_0$. Clearly $|v_0| = \rho/(2\pi)^{\frac{1}{2}} \leq \delta/2$, and

$$\int_I [v_0^2 - 2F(x, v_0)] \, dx = G(v_0) = 0.$$

In view of (2.80), $F(x, v_0) \equiv v_0^2$, and $v_0$ is a minimum point of $t^2 - 2F(x, t)$ in $|t| < \delta$. Hence, the derivative of $t^2 - 2F(x, t)$ with respect to $t$ must vanish at $t = v_0$. This gives $f(x, v_0) \equiv v_0$. Since $\rho$ was any sufficiently small constant, we see that (b) holds. This completes the proof.                                                                        □

We note that (b) implies that every constant function $v \in V$ satisfying (2.82) is a solution of $G'(v) = 0$. We therefore have

**Corollary 2.22.** *Under the hypotheses of Theorem 2.21, either (a) holds for all $\rho > 0$ sufficiently small, or (1.1),(1.2) has an infinite number of solutions.*

We are now able to prove

**Theorem 2.23.** *Assume that (1.78), (2.28), (2.37), and (2.80) hold. Then there is a nontrivial solution of (1.1) and (1.2).*

*Proof.* By Corollary 2.22, either (2.81) holds for some positive constants $\varepsilon$, $\rho$, or (1.1),(1.2) has an infinite number of solutions. Thus, we may assume that (2.81) holds. Then by Lemma 2.19 there is a solution of (1.1),(1.2) satisfying (2.75). But such a solution cannot be trivial since $G(0) = 0$.  □

## 2.16     Other intervals for asymptotic limits

Suppose $f(x,t)$ satisfies (1.78), but $\beta(x)$ does not satisfy (2.28). Are there other intervals $(a,b)$ such that a solution of (1.1),(1.2) can be found when $a \leq \beta(x) \leq b$? We are going to show that this is indeed the case. In fact we have

**Theorem 2.24.** *Let $n$ be an integer $\geq 0$. Assume that (1.78) holds with $\beta(x)$ satisfying*

$$1 + n^2 \leq \beta(x) \leq 1 + (n+1)^2 \quad 1 + n^2 \not\equiv \beta(x) \not\equiv 1 + (n+1)^2. \quad (2.84)$$

*If (2.37) holds, then (1.1),(1.2) has a solution.*

*Proof.* First, we note by Lemma 1.9, (1.76), and (1.77), that

$$\|u\|_H^2 = \sum (1 + k^2)|\alpha_k|^2, \quad u \in H, \quad (2.85)$$

where the $\alpha_k$ are given by (1.43) and (1.44). Let

$$N = \{u \in H : \alpha_k = 0 \ for \ |k| > n\}.$$

Thus,

$$\|u\|_H^2 = \sum_{|k| \leq n} (1 + k^2)|\alpha_k|^2 \leq (1 + n^2)\|u\|^2, \quad u \in N. \quad (2.86)$$

Let

$$M = \{u \in H : \alpha_k = 0 \ for \ |k| \leq n\}.$$

In this case,

$$\|u\|_H^2 = \sum_{|k| \geq n+1} (1 + k^2)|\alpha_k|^2 \geq (1 + (n+1)^2)\|u\|^2, \quad u \in M. \quad (2.87)$$

Note that $M, N$ are closed subspaces of $H$ and that $M = N^\perp$. Note also that $N$ is finite dimensional. Next, we consider the functional (1.63) and show that

$$G(v) \to -\infty \text{ as } \|v\|_H \to \infty, \quad v \in N, \tag{2.88}$$

and

$$G(w) \to \infty \text{ as } \|w\|_H \to \infty, \quad w \in M. \tag{2.89}$$

Assuming these for the moment, we note that they imply

$$\inf_M G > -\infty; \quad \sup_N G < \infty. \tag{2.90}$$

This is easily seen from the fact that (2.89) implies that there is an $R > 0$ such that

$$G(w) > 0, \quad \|w\|_H > R, \; w \in M.$$

Consequently, if the first statement in (2.90) were false, there would be a sequence satisfying

$$G(w_k) \to -\infty, \quad \|w_k\|_H \le R, \; w_k \in H.$$

But this would imply that there is a renamed subsequence converging uniformly to a limit $w_0$ in $I$. Thus,

$$G(w_k) \ge - \int_I F(x, w_k) dx \to - \int_I F(x, w_0) \, dx > -\infty.$$

This contradiction verifies the first statement in (2.90). The second is verified similarly by (2.88).

We are now in a position to apply Theorem 2.5. This produces a sequence in $H$ satisfying

$$G(u_k) \to c, \quad G'(u_k) \to 0, \tag{2.91}$$

where $c$ is finite. In particular, this implies

$$(G'(u_k), v)_H = (u_k, v)_H - (f(\cdot, u_k), v) = o(\|v\|_H), \quad \|v\|_H \to 0. \tag{2.92}$$

Assume first that

$$\rho_k = \|u_k\|_H \to \infty. \tag{2.93}$$

Set $\tilde{u}_k = u_k/\rho_k$. Then $\|\tilde{u}_k\|_H = 1$, and consequently, by Lemma 1.21, there is a renamed subsequence satisfying (2.34). Thus

$$(\tilde{u}_k, v)_H - (f(\cdot, u_k)/\rho_k, v) \to 0, \quad v \in H. \tag{2.94}$$

As we saw before, this implies in the limit that

$$(\tilde{u}, v)_H = (\beta\tilde{u}, v), \quad v \in H. \tag{2.95}$$

Let

$$\tilde{u} = \tilde{w} + \tilde{v}, \quad \hat{u} = \tilde{w} - \tilde{v}. \tag{2.96}$$

Then

$$(\tilde{u}, \hat{u})_H = (\beta\tilde{u}, \hat{u}).$$

This implies

$$\|\tilde{w}\|_H^2 - \|\tilde{v}\|_H^2 = (\beta[\tilde{w} + \tilde{v}], \tilde{w} - \tilde{v}) = (\beta\tilde{w}, \tilde{w}) - (\beta\tilde{v}, \tilde{v}),$$

since

$$(\beta\tilde{v}, \tilde{w}) = (\beta\tilde{w}, \tilde{v}) = \int_I \beta(x)\tilde{v}(x)\tilde{w}(x)\,dx.$$

Thus,

$$\|\tilde{w}\|_H^2 - (\beta\tilde{w}, \tilde{w}) = \|\tilde{v}\|_H^2 - (\beta\tilde{v}, \tilde{v}).$$

This becomes

$$\left(\|\tilde{w}\|_H^2 - (1 + (n+1)^2)\|\tilde{w}\|^2\right) + \int_I [1 + (n+1)^2 - \beta(x)]\tilde{w}^2\,dx$$

$$= \left(\|\tilde{v}\|_H^2 - (1 + n^2)\|\tilde{v}\|^2\right) + \int_I [1 + n^2 - \beta(x)]\tilde{v}^2\,dx.$$

We write this as $A + B = C + D$. In view of (2.84), (2.86) and (2.87), $A \geq 0$, $B \geq 0$, $C \leq 0$, $D \leq 0$. But this implies $A = B = C = D = 0$. If

$$\tilde{u} = \sum \tilde{\alpha}_k \varphi_k,$$

then in view of (2.87) the only way $A$ can vanish is if

$$\tilde{w} = a \, \cos(n+1)x + b \, \sin(n+1)x.$$

If $a$ and $b$ are not both 0, then this function can vanish only at a finite number of points. But then, $B$ cannot vanish in view of (2.84). Hence, $\tilde{w} \equiv 0$. Similar reasoning shows that $C = D = 0$ implies that $\tilde{v} \equiv 0$. On the other hand, (2.91) implies

$$2G(u_k)/\rho_k^2 = \|\tilde{u}_k\|_H^2 - 2\int_I F(x, \tilde{u}_k)dx/\rho_k^2 \to 1 - 2\int_I \beta(x)\tilde{u}^2 dx = 0,$$

from which we conclude that $\tilde{u} \not\equiv 0$. This contradiction shows that the assumption (2.93) is incorrect. Once this is known, we can conclude that

there is a renamed subsequence such that (2.35) holds (Lemma 1.21). It
then follows from (2.92) that

$$(u, v)_H - (f(\cdot, u), v) = 0, \quad v \in H. \tag{2.97}$$

It remains to prove (2.88) and (2.89). Let $\{w_k\} \subset M$ be any sequence
such that $\rho_k = \|w_k\| \to \infty$. Let $\tilde{w}_k = w_k/\rho_k$. Then $\|\tilde{w}_k\|_H = 1$. Thus,
there is a renamed subsequence such that (2.30) and (2.31) hold. This
implies

$$2G(w_k)/\rho_k^2 = 1 - 2 \int_I \frac{f(x, w_k)}{w_k^2} \tilde{w}_k^2 \, dx \to 1 - \int_I \beta(x)\tilde{w}^2(x) dx$$

$$\geq (1 - \|\tilde{w}\|_H^2) + (\|\tilde{w}\|_H^2 - (1 + (n+1)^2)\|\tilde{w}\|^2)$$

$$+ \int_I [1 + (n+1)^2 - \beta(x)]\tilde{w}^2(x) \, dx$$

$$= A + B + C.$$

As before, we note that $A \geq 0$, $B \geq 0$, $C \geq 0$. The only way $G(w_k)$
can fail to become infinite is if $A = B = C = 0$. As before, $B = C = 0$
implies that $\tilde{w} \equiv 0$. But this contradicts the fact that $A = 0$. Thus,
$G(w_k) \to \infty$ for each such sequence. This proves (2.89). The limit (2.88)
is proved in a similar fashion. This completes the proof of Theorem 2.24.
□

We also have

**Theorem 2.25.** *If, in addition, (2.80) holds, then (1.1),(1.2) has a
nontrivial solution.*

*Proof.* Follow the proof of Theorem 2.23.                                    □

## 2.17    Super-linear problems

When $f(x, t)$ satisfies (1.35), we refer to problem (1.1),(1.2) as sub-linear.
If $f(x, t)$ does not satisfy (1.35), we call problem (1.1),(1.2) super-linear.
As we saw in Theorem 1.20, assumption (1.62) will make the functional
$G(u)$ given by (1.63) continuous and have a continuous derivative on $H$.

We now come to a situation which causes a serious departure from the
sub-linear case. In that case we assumed (1.78) with $\beta(x)$ having certain
properties. From these properties we were able to infer that either $G$ is

bounded from below (Theorem 1.27), or (2.40) holds with $m_0, m_1$ finite. If (1.78) does not hold, these configurations are not true. We must look for another "geometry." The simplest is the one we used in connection with nontriviality. If we can show that 0 is in a "valley" surrounded by "mountains" and that there are villages beyond the mountains, then we can adapt to this situation the splitting subspaces method that we used before.

Suppose we can show that

$$G(u) \geq \varepsilon, \quad \|u\|_H = \rho, \tag{2.98}$$

holds for some positive $\varepsilon, \rho$, and that $G$ is bounded from above on $V_1 = \{v \in V : v > 0\}$. Now, $G(0) = 0$. Hence, 0 would be in a valley surrounded by mountains. If we can draw a curve of bounded length from each $v \in V_1$ along which $G$ decreases and such that (a) the endpoint of each curve depends continuously on the beginning point and (b) $G$ is less than $\varepsilon$ (the height of the mountains) at the endpoint of each curve, then we will have the desired contradiction.

The reason is simple. Since $G$ decreases along the curves and $G(0) = 0$, curves emanating from points in $V_1$ near 0 will be trapped inside the mountain sphere $\|u\|_H = \rho$. Moreover, there will be points in $V_1$ so far away from the origin that the curves emanating from them will remain outside the sphere $\|u\|_H = \rho$. As before, the continuity of the endpoint curve will imply that there is an endpoint on the sphere, providing the contradiction.

We now need three sets of hypotheses; (a) those that will imply that (2.98) holds, (b) those that imply that $G$ is bounded from above on $V_1$, and (c) those that imply that for each $v \in V_1$, there is a curve of bounded length emanating from $v$ such that the endpoint depends continuously on $v$ and $G < \varepsilon$ at the endpoint. In the next section we shall give the details.

## 2.18   A general mountain pass theorem

In proving our theorems we have been using ideas which can be combined into a general theorem. We have

**Theorem 2.26.** *Let $G$ be a continuously differentiable functional on a Hilbert space $H$ such that $G'(u)$ satisfies a local Lipschitz condition.*

*Assume that $G(0) = 0$ and that there are positive numbers $\varepsilon, \rho$ such that*

$$G(u) \geq \varepsilon \quad \text{when} \quad \|u\| = \rho. \tag{2.99}$$

*Assume also that there is a nonzero element $\varphi_0 \in H$ such that*

$$G(r\varphi_0) \leq C_0, \quad r > 0, \tag{2.100}$$

*for some constant $C_0$. Then there is a sequence $\{u_k\} \subset H$ such that*

$$G(u_k) \to c, \quad \varepsilon \leq c \leq C_0, \quad G'(u_k) \to 0. \tag{2.101}$$

*Proof.* We have essentially proved this theorem already, so we will merely sketch the main points. If the conclusion (2.101) were not true, there would be a positive number $\delta$ such that

$$\|G'(u)\| \geq \delta \quad \text{when} \quad \varepsilon - \delta \leq G(u) \leq C_0 + \delta. \tag{2.102}$$

From each $v = r\varphi_0, r > 0$, construct a curve $\sigma(t)v$ such that $\sigma(0)v = v$ and

$$dG(\sigma(t)v)/dt \leq -\delta \quad \text{when} \quad \varepsilon - \delta \leq G(\sigma(t)v) \leq C_0 + \delta,$$

and

$$dG(\sigma(t)v)/dt \leq 0$$

otherwise. We do this by applying Theorem 2.9 to the equation (2.48) using the fact that $G'(u)$ is locally Lipschitz continuous. If we take

$$T > 1 + (C_0 - \varepsilon)/\delta,$$

we see that

$$G(\sigma(T)v) \leq \varepsilon - \delta$$

for every $v$. As we showed before, the endpoints $\sigma(T)v$ depend continuously on $v$ and form a continuous curve. Since $G$ decreases along each curve, the endpoints of those $v$ near 0 cannot escape from the ball if the radius is $\rho$, while there are points $v$ so far away that their end points never reach the ball. Hence there must be an endpoint $\sigma(T)v$ such that

$$\|\sigma(T)v\| = \rho.$$

But this is impossible, since

$$G(\sigma(T)v) \leq \varepsilon - \delta$$

and

$$G(u) \geq \varepsilon \quad \text{when} \quad \|u\| = \rho.$$

This contradiction proves the theorem.                                    □

## 2.19  The Palais–Smale condition

In solving the problem (1.1),(1.2), our approach has been to find a sequence $\{u_k\}$ such that (2.27) holds and then show that this implies that $\{u_k\}$ has a convergent subsequence (i.e., the functional (1.63) satisfies the PS condition). So far we have shown this when $f(x,t)$ satisfies (1.78) or the hypotheses of Theorem 1.37.

We can now combine Theorems 2.21 and 1.37 to solve a super-linear problem. We have

**Theorem 2.27.** *Assume that either*

$$t^2 - 2F(x,t) \le W(x) \in L^1(\Omega), \quad t > 0 \tag{2.103}$$

*or*

$$t^2 - 2F(x,t) \le W(x) \in L^1(\Omega), \quad t < 0. \tag{2.104}$$

*Then under the hypotheses of Theorems 2.21, 1.37, and Lemma 2.7, problem (1.1),(1.2) has at least one nontrivial solution.*

*Proof.* Use Theorem 2.26. We must show that (2.100) holds. To show this, let $\varphi(x) \equiv \pm 1$, depending on whether (2.103) or (2.104) holds. Then

$$2G(k\varphi) = \int_I [k^2 - 2F(x, \pm k)]\, dx \le \int_I W(x)\, dx < \infty.$$

This completes the proof. □

## 2.20  Exercises

1. Why does

$$x_k^2 + y_k^2 \le C$$

   imply that there is a renamed subsequence such that

$$x_k \to x_0, \quad y_k \to y_0?$$

2. Show that $\nabla f$ satisfies a local Lipschitz condition if $f \in C^2(\mathbb{R}^n, \mathbb{R})$.

3. Show that the modulus of a function satisfying a local Lipschitz condition satisfies a local Lipschitz condition.

4. Show that the ratio of two functions on $\mathbb{R}^n$ satisfying local Lipschitz conditions satisfies a local Lipschitz condition provided that the denominator does not vanish.

5. Prove (2.15).

6. Why is there a renamed subsequence such that $T_{x_k} \to \tilde{T}$ in the proof of Theorem 2.1?

7. Follow the proof of Theorem 2.1 replacing (2.11) with (2.24).

8. Prove: If $f \in C^1(\mathbb{R}^n, \mathbb{R})$ and

$$\nabla f(p) \cdot \gamma(p) = |\nabla f(p)| \geq \delta,$$

there is a neighborhood $N(p)$ of $p$ such that

$$\nabla f(q) \cdot \gamma(p) \geq \frac{\delta}{2}, \quad q \in N(p).$$

9. Why was it assumed that the diameter of each neighborhood $N(p)$ was less than one in the construction of $V$ following Lemma 2.3?

10. Show that if $\tilde{N}(p)$ is a small neighborhood of $p$, then

$$\tilde{N}(p) \cap N(p_k) \neq \phi$$

for only a finite number of $p_k$. Why do we need $\tilde{N}(p)$ to be small?

11. Show that $\psi_k$ is locally Lipschitz continuous.

12. Show that (2.39) implies (2.38).

13. Show that the set $Y$ of all continuous functions $x(t)$ from $I_1$ to $X$ with norm

$$|||x||| = \max_{t \in I_1} ||x(t)|| \qquad (2.105)$$

is a Banach space.

14. Prove the statements in the remark following Theorem 2.13.

15. If $T < \infty$, why do we obtain a solution of (2.51) in $[0, T]$ in the proof of Theorem 2.9 in Section 2.12?

16. Show that

$$\int_I [w_k^2 - 2F(x, w_k)] \, dx / \rho_k^2 \to 0 \text{ as } k \to \infty$$

in the proof of Theorem 2.18.

17. Show that Theorem 2.18 applies to problem (1.1),(1.2) when

$$2F(x, t) = t^2 + |t|^{3/2}, \quad t \in \mathbb{R}.$$

# 3

# Boundary value problems

## 3.1 Introduction

Until now we have studied problems for periodic functions. However, there are many problems arising from applications which search for functions satisfying boundary conditions. These functions are not required to be periodic, but their values are prescribed on the boundary of the region under consideration. In dealing with such problems, we are unable to use some of the tools which helped us in the periodic case (e.g., Fourier series). Consequently, we must search for other means of solving such problems. The spaces we shall use will be different to adjust to the new situations. The methods will change as well as the outcomes.

## 3.2 The Dirichlet problem

Let us change problem (1.1), (1.2) slightly. Suppose we want to solve

$$-u'' + u = f(x, u(x)), \quad x \in \Omega = (a, b), \tag{3.1}$$

under the condition

$$u(a) = u(b) = 0. \tag{3.2}$$

We assume that $f(x, t)$ satisfies (1.62). We must make some changes from the periodic case. We shall use the set

$$C_0^1(\Omega) = \{u \in C^1(\bar{\Omega}) : u(a) = u(b) = 0\}.$$

Here, $C^1(\bar{\Omega})$ denotes the set of continuously differentiable functions on $\bar{\Omega}$; no periodicity is required. We let $H_0^1 = H_0^1(\Omega)$ be the completion of $C_0^1(\Omega)$ with respect to the norm given by

$$\|u\|_H^2 = \|u\|^2 + \|u'\|^2.$$

The corresponding scalar product is given by

$$(u, v)_H = (u, v) + (u', v'),$$

where

$$(u, v) = \int_a^b u(x)v(x)dx.$$

The question of completing a normed vector space arose before with respect to the Hilbert space $H$. We will have to address it here as well. As we saw in the case of periodic functions, we shall need the concept of a weak derivative. Here it will differ slightly from that for periodic functions. Before we proceed we shall need some of the results of the next two sections.

### 3.3  Mollifiers

Let $j(x)$ be an infinitely differentiable function on $\mathbb{R}^n$ such that $j(x) > 0$ for $|x| < 1$, $j(x) = 0$ for $|x| \geq 1$, and

$$\int j(x)dx = 1. \tag{3.3}$$

An example of such a function is

$$j(x) = \begin{cases} ae^{-1/(1-|x|^2)}, & |x| < 1, \\ 0, & |x| \geq 1, \end{cases}$$

where the constant $a$ is suitably chosen. For $u \in L^p = L^p(\mathbb{R}^n)$, $p \geq 1$, we define

$$J_\varepsilon u(x) = \int j_\varepsilon(x - y)u(y)\, dy = \int j(z)u(x - \varepsilon z)\, dz, \tag{3.4}$$

where

$$j_\varepsilon(x) = \varepsilon^{-n} j(x/\varepsilon).$$

We shall prove

$$|J_\varepsilon u|_p \leq |u|_p, \quad u \in L^p, \tag{3.5}$$

and

$$|J_\varepsilon u - u|_p \to 0, \quad u \in L^p. \tag{3.6}$$

Recall that

$$|u|_p = \left( \int |u(x)|^p dx \right)^{1/p}.$$

In proving these statements, note that

$$\int j_\varepsilon(x)\, dx = 1.$$

By (3.4) and Hölder's inequality (Theorem B.23),

$$|J_\varepsilon u(x)| \leq \int [j_\varepsilon(x-y)^{1/p}|u(y)|]\, j_\varepsilon(x-y)^{1/p'}\, dy$$

$$\leq \left( \int j_\varepsilon(x-y)|u(y)|^p dy \right)^{1/p} \left( \int j_\varepsilon(x-y)\, dy \right)^{1/p'},$$

where

$$\frac{1}{p} + \frac{1}{p'} = 1.$$

Hence

$$|J_\varepsilon u(x)|^p \leq \int j_\varepsilon(x-y)|u(y)|^p dy.$$

Integrating with respect to $x$ and using Fubini's theorem (Theorem B.26), we obtain

$$\int |J_\varepsilon u(x)|^p dx \leq \int \left[ \int j_\varepsilon(x-y)|u(y)|^p dy \right] dx$$

$$= \int \left[ \int j_\varepsilon(x-y)\, dx \right] |u(y)|^p dy$$

$$= |u|_p^p.$$

This gives (3.5). In proving (3.6), assume first that $u$ is continuous. Let $\delta > 0$ be given, and take $R$ so large that

$$\int_{|x|>R-1} |u(x)|^p dx < \delta^p.$$

On the set $|x| \leq R+1$, the function $u(x)$ is uniformly continuous. Hence, there is an $\varepsilon > 0$ such that $\varepsilon < 1$ and

$$|u(x+y) - u(x)| < \delta/R^{n/p}, \quad |x| < R, \ |y| < \varepsilon.$$

Consequently,

$$
\begin{aligned}
|J_\varepsilon u(x) - u(x)|^p &= \left| \int j(z)^{1/p} [u(x - \varepsilon z) - u(x)] j(z)^{1/p'} dz \right|^p \\
&\le \int j(z) |u(x - \varepsilon z) - u(x)|^p dz \\
&\le \begin{cases} \int j(z) \delta^p / R^n dz, & |x| < R, \\ \int j(z)(|u(x - \varepsilon z)| + |u(x)|)^p dz. & |x| > R \end{cases}
\end{aligned}
$$

Hence,

$$
\begin{aligned}
\int |J_\varepsilon u(x) - u(x)|^p dx &\le \int_{|x|<R} \int j(z) \delta^p / R^n dz\, dx \\
&\quad + \int \int_{|x|>R} j(z)(|u(x - \varepsilon z)| + |u(x)|)^p dx\, dz \\
&\le 2^n \delta^p + 2^{p+1} \int_{|x|>R-1} |u(x)|^p dx \\
&\le (2^n + 2^{p+1}) \delta^p.
\end{aligned}
$$

Since $\delta$ was arbitrary, we see that (3.6) holds for continuous functions in $L^p(\mathbb{R}^n)$. If $u \in L^p(\mathbb{R}^n)$, then for each $\delta > 0$ there is a continuous function $v \in L^p(\mathbb{R}^n)$ such that $|u - v|_p < \delta$. Thus,

$$
|J_\varepsilon u - u|_p \le |J_\varepsilon(u - v)|_p + |J_\varepsilon v - v|_p + |v - u|_p \le \delta + \delta + \delta,
$$

for $\varepsilon$ sufficiently small. This proves (3.6).

## 3.4    Test functions

We let $C_0^\infty$ denote the set of infinitely differentiable functions with compact supports in $\mathbb{R}^n$. This means that they vanish for $|x|$ large. The function $j(x)$ described in the previous section is such a function. Thus, $C_0^\infty$ is not empty. In fact, we shall prove

**Theorem 3.1.** *For* $1 \le p < \infty$, $C_0^\infty$ *is dense in* $L^p = L^p(\mathbb{R}^n)$.

*Proof.* Let $u$ be any function in $L^p$, and take $R$ so large that $|u - u_R|_p < \delta$, where $\delta > 0$ is given and

$$
u_R(x) = \begin{cases} u(x), & |x| \le R \\ 0, & |x| > R. \end{cases}
$$

Then $J_\varepsilon u_R \in C_0^\infty$ by (3.4). Moreover, by (3.6) we can take $\varepsilon > 0$ so small that

$$|J_\varepsilon u_R - u_R|_p < \delta.$$

Thus

$$|J_\varepsilon u_R - u|_p \le |J_\varepsilon u_R - u_R|_p + |u_R - u|_p < 2\delta.$$

This proves the theorem. $\qquad\qquad\qquad\qquad\qquad\qquad\qquad\qquad\square$

For any open set $\Omega \subset \mathbb{R}^n$, we let $C_0^\infty(\Omega)$ denote the set of infinitely differentiable functions with compact supports in $\Omega$, that is, functions in $C_0^\infty$ which vanish outside $\Omega$ and near $\partial\Omega$. We shall show that $C_0^\infty(\Omega)$ is not empty for any open set $\Omega$. In fact, we have

**Theorem 3.2.** *For $1 \le p < \infty$, $C_0^\infty(\Omega)$ is dense in $L^p(\Omega)$.*

*Proof.* Let $u$ be any function in $L^p(\Omega)$. For $x \in \mathbb{R}^n$, let

$$d(x, \partial\Omega) = \min_{y \in \partial\Omega} |x - y|$$

be the distance from $x$ to $\partial\Omega$. Let

$$u_R(x) = \begin{cases} u(x), & |x| \le R, \ x \in \Omega, \ d(x, \partial\Omega) \ge 1/R, \\ 0, & \text{otherwise.} \end{cases}$$

We note that

$$|u - u_R|_p \to 0 \quad \text{as} \quad R \to \infty.$$

To see this, note that

$$|u(x) - u_R(x)| \le |u(x)|, \quad x \in \Omega,$$

and

$$u_R(x) \to u(x), \quad x \in \Omega.$$

Apply the Lebesgue dominated convergence theorem (Theorem B.18). Let $\delta > 0$ be given. Take $R$ so large that $|u - u_R|_p < \delta$. For $\varepsilon > 0$ sufficiently small, $J_\varepsilon u_R \in C_0^\infty(\Omega)$, and we can take it so small that $|J_\varepsilon u_R - u_R|_p < \delta$. Thus

$$|J_\varepsilon u_R - u|_p \le |J_\varepsilon u_R - u_R|_p + |u_R - u|_p < 2\delta.$$

This completes the proof. $\qquad\qquad\qquad\qquad\qquad\qquad\qquad\qquad\square$

### 3.5     Differentiability

We return to the one dimensional case $\Omega = (a, b) \in \mathbb{R}$. A corollary of Theorem 3.2 is

**Lemma 3.3.** *The set* $C_0^1(\Omega)$ *is dense in* $L^2(\Omega)$. *Hence, if* $h \in L^2(\Omega)$ *and*

$$\int h(x)\varphi(x)\, dx = 0, \quad \varphi \in C_0^\infty(\Omega),$$

*then* $h(x) = 0$ *a.e.*

This will be important in defining a weak derivative. We shall say that $h \in L^2(\Omega)$ is a weak derivative of $u \in L^2(\Omega)$ if

$$(u, \varphi') = -(h, \varphi), \quad \varphi \in C_0^1(\Omega). \tag{3.7}$$

Lemma 3.3 tells us that weak derivatives are unique. For if $h_1$ were another, we would have

$$(h - h_1, \varphi) = 0, \quad \varphi \in C_0^1(\Omega).$$

The density of $C_0^1(\Omega)$ would then show that $h = h_1$ a.e.

The proof of Lemma 3.3 is easily given.

*Proof.* By Theorem 3.2, there is a sequence $\{\varphi_k(x)\} \subset C_0^\infty(\Omega)$ converging to $h(x)$ in $L^2(\Omega)$. Then,

$$\int h(x)\varphi_k(x)\, dx = 0, \quad k = 1, 2, \dots$$

In the limit we have

$$\int |h(x)|^2\, dx = 0,$$

which implies our conclusion.                                              □

We want $H_0^1(\Omega)$ to be the smallest Hilbert space containing $C_0^1(\Omega)$ using the norm $\|u\|_H$. To this end, we let $H_0^1 = H_0^1(\Omega)$ be the set of those $u \in L^2(\Omega)$ such that there is an $h \in L^2(\Omega)$ and a sequence $\{u_k\} \subset C_0^1(\Omega)$ such that

$$u_k \to u, \ u_k' \to h \text{ in } L^2(\Omega). \tag{3.8}$$

Not every function in $L^2(\Omega)$ has a weak derivative, but we have

**Lemma 3.4.** *Every function in* $H_0^1(\Omega)$ *has a weak derivative.*

*Proof.* If $u \in H_0^1(\Omega)$, then there is a sequence $\{u_k\} \subset C_0^1(\Omega)$ such that
$$\|u_k - u\|_H \to 0.$$
Thus $u_k \to u$ in $L^2(\Omega)$ and $\{u_k'\}$ is a Cauchy sequence in $L^2(\Omega)$. Hence, there is an $h \in L^2(\Omega)$ such that $u_k' \to h$ in $L^2(\Omega)$. Then,
$$(u_k, \varphi') = -(u_k', \varphi), \quad \varphi \in C_0^1(\Omega).$$
In the limit this gives (3.7). □

It follows from Lemma 3.4 that the function $h$ in the definition (3.8) is the weak derivative of $u$; it will be denoted by $u'$. We shall need

**Theorem 3.5.** *The space $H_0^1(\Omega)$ is complete.*

*Proof.* Suppose $\{u_k\}$ is a Cauchy sequence in $H_0^1(\Omega)$. Then there are functions $u, h \in L^2(\Omega)$ such that (3.8) holds. Since each $u_k \in H_0^1(\Omega)$, there is a $v_k \in C_0^1(\Omega)$ such that
$$\|v_k - u_k\| < \frac{1}{k}, \quad \|v_k' - u_k'\| < \frac{1}{k}.$$
Thus,
$$\|v_k - u\| \le \|v_k - u_k\| + \|u_k - u\| \to 0, \quad \|v_k' - h\| \le \|v_k' - u_k'\| + \|u_k' - h\| \to 0.$$
Hence, $u \in H_0^1(\Omega)$ and $h = u'$ a.e. □

Next we have

**Theorem 3.6.** *The space $H_0^1 \subset C(\bar{\Omega})$ and*
$$|u(x)| \le \min(|x - a|^{\frac{1}{2}}, |b - x|^{\frac{1}{2}})\|u'\|, \quad u \in H_0^1. \tag{3.9}$$

*Proof.* First assume that $u \in C_0^1(\Omega)$. Then
$$|u(x)| = \left| \int_a^x u'(t) dt \right| \le (x - a)^{\frac{1}{2}} \|u'\|,$$
and
$$|u(x)| = \left| \int_x^b u'(t) \, dt \right| \le (b - x)^{\frac{1}{2}} \|u'\|.$$
This gives (3.9). Now if $u \in H_0^1$, there is a sequence $\{u_k\} \subset C_0^1(\Omega)$ converging to $u$ in $H_0^1$ (definition). By inequality (3.9), $u_k$ converges uniformly to a limit in $C(\bar{\Omega})$ which must coincide with $u(x)$ a.e. Moreover, $u_k'$ converges to $u'$ in $L^2(\Omega)$. By inequality (3.9),
$$|u_k(x)| \le \min(|x - a|^{\frac{1}{2}}, |b - x|^{\frac{1}{2}})\|u_k'\|,$$
and in the limit the inequality applies to $u(x)$ itself. □

Next we have

**Theorem 3.7.** *If $u \in L^2(\Omega)$ has a weak derivative $h \in L^2(\Omega)$, then $u \in C(\Omega)$. If $h \in C(\Omega)$, then $u \in C^1(\Omega)$, and $h$ is its derivative in the usual sense.*

*Proof.* Let $\eta > 0$ be given, and let $(a', b') = \Omega' \subset \Omega$ be an open interval such that $d(\Omega', \partial\Omega) = \min(b - b', a' - a) > \eta$. For $\varepsilon < \eta$ and $x \in \Omega'$, we note that $j_\varepsilon(x - y)$ as a function of $y$ is in $C_0^\infty(\Omega)$. Thus

$$D_x(J_\varepsilon u) = \int D_x j_\varepsilon(x - y)u(y)\, dy$$

$$= -\int D_y j_\varepsilon(x - y)u(y)\, dy$$

$$= \int j_\varepsilon(x - y)h(y)\, dy$$

$$= J_\varepsilon h(x) \to h(x) \quad \text{in } L^2(\Omega'),$$

where $D_x = \partial/\partial x$. Since $u \in L^2(\Omega)$, we know that

$$J_\varepsilon u(x) \to u(x) \quad \text{in } L^2(\Omega).$$

This implies that there is a sequence $\{\varepsilon_k\}$ such that $\varepsilon_k \to 0$ and

$$J_{\varepsilon_k} u(x) \to u(x) \quad \text{a.e.,} \quad x \in \Omega$$

(Theorem B.25). Let $M \subset \Omega'$ be the set where

$$J_{\varepsilon_k} u(x) \to u(x), \quad x \in M.$$

Then the set $\Omega' \setminus M$ has measure zero. Thus,

$$J_{\varepsilon_k} u(x) - J_{\varepsilon_k} u(x') = \int_{x'}^x D_t(J_{\varepsilon_k} u(t))\, dt \to \int_{x'}^x h(t)\, dt, \quad x, x' \in M.$$

Consequently,

$$u(x) - u(x') = \int_{x'}^x h(t)\, dt, \quad x, x' \in M. \tag{3.10}$$

Fix $x' \in M$ and define

$$\tilde{u}(x) = u(x') + \int_{x'}^x h(t)\, dt, \quad x \in \Omega'.$$

The right-hand side is continuous on $\Omega'$. Thus, $u(x)$ is a.e. equal to a continuous function $\tilde{u}(x)$ on $\Omega'$. By correcting it on a set of measure 0, we can make $u(x)$ continuous there. Since $\eta$ was arbitrary, we see that $u(x)$ is continuous on the whole of $\Omega$.

If $h(t) \in C(\Omega)$, then $u(x)$ is differentiable and

$$u'(x) = h(x), \quad x \in \Omega'.$$

Since $\eta$ was arbitrary, the result follows. □

As a result we have

**Theorem 3.8.** *If $u \in H_0^1$, $f \in L^2(\Omega)$ and*

$$(u,v)_H = (f,v), \quad v \in C_0^1(\Omega), \tag{3.11}$$

*then $u'$ is continuous in $\Omega$ and is the derivative of $u$ in the usual sense. If $f \in C(\Omega)$, then $u'$ is differentiable in the usual sense and satisfies $u'' = u - f$ in $\Omega$.*

*Proof.* We see from (3.11) that

$$(u',v') = -(u - f, v), \quad v \in C_0^1(\Omega).$$

Therefore, $u'$ has a weak derivative equal to $u - f$. By Theorem 3.7, $u'$ is continuous in $\Omega$ and is the derivative of $u$ in the usual sense. If $f \in C(\Omega)$ then $u - f \in C(\Omega)$, and $u'$ is differentiable in the usual sense and satisfies $u'' = u - f$ in $\Omega$. □

Corresponding to Lemma 1.9 we have

**Lemma 3.9.** *If $I = [0, \pi]$ and $u(x) \in L^2(I)$, then*

$$\left\| u - \sum_{k=1}^{n} b_k \sin kx \right\| \to 0 \quad \text{as } n \to \infty \tag{3.12}$$

*and*

$$\|u\|^2 = \frac{\pi}{2} \sum_{k=1}^{\infty} b_k^2, \tag{3.13}$$

*where*

$$b_k = \frac{2}{\pi} \int_0^{\pi} u(x) \sin kx \, dx, \quad k = 1, 2, \ldots \tag{3.14}$$

*Proof.* Define

$$\tilde{u} = \begin{cases} u(x), & 0 \le x \le \pi, \\ -u(-x), & -\pi \le x < 0. \end{cases}$$

Then $\tilde{u}$ is an odd function in $\tilde{I} = [-\pi, \pi]$:

$$\tilde{u}(-x) = -\tilde{u}(x), \quad x \in \tilde{I}.$$

By Lemma 1.9,

$$\left\| \tilde{u} - \sum_{|k|\le n} \tilde{\alpha}_k \varphi_k \right\| \to 0,$$

where $\varphi_k$ is given by (1.44),

$$\tilde{\alpha}_k = (\tilde{u}, \bar{\varphi}_k) = \frac{1}{\sqrt{2\pi}} \int_{-\pi}^{\pi} \tilde{u}(t) e^{-ikt}\, dt,$$

and the norm is that of $L^2(-\pi, \pi)$. Since $\tilde{u}$ is odd,

$$\int_{-\pi}^{0} \tilde{u}(t) e^{-ikt}\, dt = -\int_{0}^{\pi} u(s) e^{iks}\, ds.$$

Hence,

$$\tilde{\alpha}_k = \frac{1}{\sqrt{2\pi}} \int_{0}^{\pi} u(t) [e^{-ikt} - e^{ikt}]\, dt = -\frac{2i}{\sqrt{2\pi}} \int_{0}^{\pi} u(t) \sin kt\, dt.$$

Consequently,

$$\begin{aligned}
\tilde{\alpha}_k \varphi_k + \tilde{\alpha}_{-k}\varphi_{-k} &= -\frac{2i}{\sqrt{2\pi}} \int_{0}^{\pi} u(t) \sin kt\, dt\, \varphi_k \\
&\quad + \frac{2i}{\sqrt{2\pi}} \int_{0}^{\pi} u(t) \sin kt\, dt\, \varphi_{-k} \\
&= -\frac{2i}{2\pi} \int_{0}^{\pi} u(t) \sin kt\, dt\, (i \sin kt) \\
&\quad + \frac{2i}{2\pi} \int_{0}^{\pi} u(t) \sin kt\, dt\, (-i \sin kt) \\
&= \frac{4}{2\pi} \int_{0}^{\pi} u(t) \sin kt\, dt\, \sin kx \\
&= b_k \sin kx.
\end{aligned}$$

Thus,

$$\sum_{|k|\le n} \tilde{\alpha}_k \varphi_k = \sum_{k=1}^{n} b_k \sin kx,$$

and

$$\left\| \tilde{u} - \sum_{k=1}^{n} b_k \sin kx \right\| \to 0 \quad \text{as } n \to \infty,$$

where the norm is that of $L^2(-\pi, \pi)$. This implies (3.12) with the norm of $L^2(0, \pi)$ and completes the proof. Inequality (3.13) follows from (3.12) and the fact that the functions $\sin kx$ are orthonormal in $L^2(\Omega)$. $\qquad\square$

**Lemma 3.10.** *If $u \in C(\mathbb{R})$ has a weak derivative $u' \in L^2(\Omega)$ and satisfies*

$$u(x) \equiv 0, \quad x \leq a + \varepsilon, \ x \geq b - \varepsilon$$

*for some $\varepsilon > 0$, then $u \in H_0^1(\Omega)$.*

*Proof.* If $\delta < \varepsilon$, then

$$J_\delta u(x) \in C_0^\infty(\Omega) \subset C_0^1(\Omega).$$

This follows from the fact that

$$J_\delta u(x) = 0, \quad x \leq a + \varepsilon - \delta, \ x \geq b - \varepsilon + \delta.$$

By (3.6),

$$J_\delta u \to u \text{ in } L^2(\Omega).$$

Moreover, if $a + \varepsilon \leq x \leq b - \varepsilon$, then $j_\delta(x - y) \in C_0^\infty(\Omega)$ as a function of $y$. Thus

$$
\begin{aligned}
J_\delta' u(x) &= \int \frac{\partial}{\partial x} j_\delta(x - y) u(y) \, dy \\
&= -\int \frac{\partial}{\partial y} j_\delta(x - y) u(y) \, dy \\
&= \int j_\delta(x - y) u'(y) \, dy \\
&= J_\delta u'(x) \to u'(x) \text{ in } L^2(\Omega).
\end{aligned}
$$

Therefore, $u \in H_0^1(\Omega)$ and

$$J_\delta u \to u \text{ in } H_0^1(\Omega).$$

$\square$

**Lemma 3.11.** *If $u \in C(\mathbb{R})$ has a weak derivative $u' \in L^2(\Omega)$ and satisfies*

$$u(x) \equiv 0, \quad x \leq a, \ x \geq b,$$

*then $u \in H_0^1(\Omega)$.*

*Proof.* We may assume that $a = 0$. For $0 < \varepsilon < b/2$, let

$$
x_\varepsilon = \begin{cases}
0, & 0 \leq x \leq \varepsilon, \\
b(x - \varepsilon)/(b - 2\varepsilon), & \varepsilon \leq x \leq b - \varepsilon, \\
b, & b - \varepsilon \leq x \leq b.
\end{cases}
$$

Then

$$x_\varepsilon - x \to 0 \text{ as } \varepsilon \to 0, \text{ uniformly in } \bar{\Omega}.$$

If $h \in C(\bar{\Omega})$, then

$$h_\varepsilon(x) = h(x_\varepsilon) \to h(x) \text{ as } \varepsilon \to 0, \text{ uniformly in } \bar{\Omega}.$$

If $h \in L^2(\Omega)$, then

$$\begin{aligned}
\|h_\varepsilon\|^2 &= \int_\varepsilon^{b-\varepsilon} |h(x_\varepsilon)|^2 dx \\
&= \frac{b - 2\varepsilon}{b} \int_0^b |h(x_\varepsilon)|^2 dx_\varepsilon \\
&= \frac{b - 2\varepsilon}{b} \int_0^b |h(y)|^2 dy \\
&\to \|h\|^2 \text{ as } \varepsilon \to 0.
\end{aligned}$$

Moreover,

$$h_\varepsilon(x) \to h(x) \text{ in } L^2(\Omega) \text{ as } \varepsilon \to 0.$$

To see this, note that for any $\delta > 0$, there is a $g \in C(\bar{\Omega})$ such that

$$\|g - h\| < \delta.$$

Take $\varepsilon$ so small that

$$\|g_\varepsilon - g\| < \delta.$$

Then

$$\|h_\varepsilon - h\| \le \|h_\varepsilon - g_\varepsilon\| + \|g_\varepsilon - g\| + \|g - h\| < \delta + \delta + \delta$$

for $\varepsilon$ sufficiently small.

If $u$ satisfies the hypotheses of the lemma, the $u_\varepsilon \in H_0^1(\Omega)$ for each $\varepsilon > 0$ by Lemma 3.10. Moreover, $u_\varepsilon \to u$, $u_\varepsilon' \to u'$ in $L^2(\Omega)$ as $\varepsilon \to 0$. Therefore, there is a sequence $\{u_k\}$ of functions in $C_0^1(\Omega)$ such that $u_k \to u$, $u_k' \to u'$ in $L^2(\Omega)$. Thus, $u \in H_0^1(\Omega)$, and the proof is complete.  $\square$

We now have

**Corollary 3.12.** *A function $u$ is in $H_0^1(\Omega)$ if, and only if, it is in $C(\bar{\Omega})$, satisfies (3.2), and has a weak derivative in $L^2(\Omega)$.*

*Proof.* If $u \in H_0^1(\Omega)$, then it is in $C(\bar{\Omega})$ and satisfies (3.2) by Theorem 3.6. It has a weak derivative by Lemma 3.4. Conversely, if it has these properties, then it can be extended to be continuous in the whole of $\mathbb{R}$ by defining it to be 0 outside $\bar{\Omega}$. We can then apply Lemma 3.11 to conclude that $u \in H_0^1(\Omega)$. □

## 3.6    The functional

We can now proceed to tackle problem (3.1), (3.2). Actually, all we need to do is follow the methods used to solve (1.1),(1.2). First, we must choose a space and a functional. This is not difficult. We merely use $H_0^1 = H_0^1(\Omega)$ in place of $H$ and replace $G(u)$ with

$$G_0(u) = \frac{1}{2}\|u\|_H^2 - \int_a^b F(x, u(x))\, dx, \quad u \in H_0^1, \tag{3.15}$$

where $F(x,t)$ is given by (1.64). Replacing (1.46) with (3.9), we follow the proof of Theorem 1.20 to conclude

**Theorem 3.13.** *If $f(x,t)$ satisfies (1.62), the functional $G_0(u)$ is continuously differentiable on $H_0$ and satisfies*

$$(G_0'(u), v)_H = (u, v)_H - (f(\cdot, u), v), \quad u, v \in H_0^1. \tag{3.16}$$

It therefore follows that if we can find a function $u_0 \in H_0^1$ satisfying

$$G_0'(u_0) = 0, \tag{3.17}$$

then it will satisfy

$$(u_0, v)_H = (f(\cdot, u_0), v), \quad v \in H_0^1. \tag{3.18}$$

Since $u_0 \in H_0^1$, it is in $C(\bar{\Omega})$ by Theorem 3.6 and satisfies (3.2) by (3.9). This means that $f(x, u_0(x))$ is in $C(\Omega)$. Thus, by Theorem 3.8, $u_0$ has continuous second derivatives and satisfies

$$u_0''(x) = u_0 - f(x, u_0(x)), \quad x \in \Omega,$$

which is (3.1).

Now we are ready to search for solutions of (3.17). The first attempt to find such a solution is to look for a minimum as we did in the case of problem (1.1),(1.2). In order to keep the same numbers, we take $\Omega = (0, \pi)$. We shall need the counterpart of Lemma 1.21.

**Lemma 3.14.** *If $\{u_k\} \subset H_0^1$ satisfies*

$$\|u_k\|_H \leq C, \tag{3.19}$$

*then there is a renamed subsequence and a $u_0 \in H_0^1$ such that*

$$u_k \rightharpoonup u_0 \text{ in } H_0^1 \tag{3.20}$$

*and*

$$u_k(x) \to u_0(x) \text{ uniformly in } \bar{\Omega}. \tag{3.21}$$

In proving Lemma 3.14 we shall use

**Lemma 3.15.** *If $u \in L^2(\Omega)$ has a weak derivative $h \in L^2(\Omega)$, then*

$$(u, v') = -(h, v), \quad v \in H_0^1. \tag{3.22}$$

*Proof.* By definition,

$$(u, \varphi') = -(h, \varphi), \quad \varphi \in C_0^1(\Omega).$$

If $v \in H_0^1$, then there is a sequence $\{\varphi_k\} \subset C_0^1(\Omega)$ converging to $v$ in $H_0^1$. Since

$$(u, \varphi_k') = -(h, \varphi_k)$$

we obtain (3.22) in the limit. □

We can now give the proof of Lemma 3.14.

*Proof.* We note that Lemmas 1.25 and 1.26 remain valid when we replace $H$ by $H_0^1$. From Lemma 1.26 we see that there is a renamed subsequence converging to a function $w \in C(\bar{\Omega})$. Moreover, by a theorem in functional analysis (Theorem A.61), there is a renamed subsequence of this subsequence which converges weakly in $H_0^1$ to a function $u_0 \in H_0^1$. Let $v$ be a function in $H_0^1$ such that $v'$ has a weak derivative $v''$. Then

$$(u_k, v - v'') = (u_k, v)_H \to (u_0, v)_H = (u, v - v'').$$

But

$$(u_k, v - v'') \to (w, v - v'').$$

Thus

$$(u_0 - w, v - v'') = 0.$$

Let $h$ be any function in $L^2(\Omega)$. Define

$$\Phi(\varphi) = (\varphi, h), \quad \varphi \in H_0^1.$$

Then $\Phi(\varphi)$ is a linear functional on $H_0$. It is also bounded, since

$$|\Phi(\varphi)| \leq \|\varphi\| \cdot \|h\| \leq C\|\varphi\|_{H_0^1}.$$

By the Riesz representation theorem (Theorem A.12), there is a $v \in H_0^1$ such that

$$\Phi(\varphi) = (\varphi, v)_H, \quad \varphi \in H_0^1.$$

Thus

$$(\varphi, h) = (\varphi, v) + (\varphi', v'), \quad \varphi \in H_0^1.$$

This means that $v'$ has a weak derivative

$$v'' = v - h.$$

Since $h$ was arbitrary, we see that

$$(u_0 - w, h) = 0, \quad h \in L^2(\Omega).$$

If we take $h = u_0 - w$, we see that $u_0 \equiv w$. This completes the proof. $\square$

## 3.7    Finding a minimum

The next step is to find a $u \in H_0^1$ such that $G_0'(u) = 0$. The simplest situation is when $G_0(u)$ has an extremum. We now give a condition on $f(x,t)$ that will guarantee that $G_0(u)$ has a minimum. We assume that there is a function $W(x) \in L^1(I)$ such that

$$-W(x) \leq V(x,t) \equiv t^2 - F(x,t) \to \infty \text{ a.e. as } |t| \to \infty. \tag{3.23}$$

**Remark 3.16.** *Note that this assumption is weaker than hypothesis (1.67) which was used in Theorem 1.24. Recall that we are now assuming that $\Omega = (0, \pi)$.*

We shall also make use of

$$\|u\| \leq \|u'\|, \quad u \in H_0^1, \tag{3.24}$$

and

**Lemma 3.17.** *If $u$ is in $H_0^1$, and $\|u\| = \|u'\|$, then*

$$u(x) = b \sin x \tag{3.25}$$

*for some constant b. If*

$$\int_0^\pi u(x) \sin x \, dx = 0, \tag{3.26}$$

*then*

$$2\|u\| \le \|u'\|. \tag{3.27}$$

*If u satisfies (3.26) and*

$$2\|u\| = \|u'\|,$$

*then*

$$u(x) = c \sin 2x.$$

We shall prove these at the end of the section. Now we show how they can be used to give

**Theorem 3.18.** *Under hypothesis (3.23) there is a u in $H_0^1$ such that*

$$G_0(u) = \min_{H_0^1} G_0.$$

*Moreover, if $f(x,t)$ is continuous in both variables, any such minimum is a solution of (3.1),(3.2) in the usual sense.*

*Proof.* Let

$$\alpha = \inf_{H_0^1} G_0.$$

Let $\{u_k\}$ be a minimizing sequence, that is, a sequence in $H_0^1$ satisfying

$$G_0(u_k) \searrow \alpha.$$

I claim that

$$\rho_k = \|u_k\|_H \le C.$$

To see this, note that for each $u \in H_0^1$ we may write

$$u = w + v,$$

where $w$ satisfies (3.26) (and consequently (3.27)) and $v$ is of the form (3.25). We have

$$2G_0(u) \ge \frac{3}{4}\|w'\|^2 + 2\int_\Omega V(x,u)\,dx \ge \frac{3}{5}\|w\|_H^2 - 2\int_\Omega W(x)\,dx \tag{3.28}$$

by (3.24) and (3.27). In fact these latter inequalities imply

$$4\|w\|^2 \le \|w'\|^2, \quad \|v\| = \|v'\|,$$

and consequently

$$\|u\|_H^2 \geq \frac{3}{4}\|w'\|^2 + 2\|u\|^2 \geq \frac{3}{5}\|w\|_H^2 + 2\|u\|^2.$$

From this we see that if $\{u_k\}$ is a minimizing sequence for $G_0$, then we must have

$$|w_k(x)| \leq C\|w_k\|_H \leq C', \quad x \in \Omega$$

by Theorem 3.6. The only way we can have

$$\rho_k \to \infty$$

is if

$$\|v_k\|_H \to \infty.$$

Since

$$v_k(x) = c_k \sin x,$$

this means that

$$\|v_k\|_H^2 = \pi c_k^2 \to \infty.$$

Whence,

$$|v_k(x)| = |c_k| \sin x \to \infty, \quad x \in \Omega.$$

But then we have

$$|u_k(x)| \geq |v_k(x)| - |w_k(x)| \geq |v_k(x)| - C' \to \infty, \quad x \in \Omega.$$

Thus, the only way we can have $\|u_k\|_H \to \infty$ is if

$$|u_k(x)| \to \infty, \quad x \in \Omega.$$

But then,

$$\int_\Omega V(x, u_k(x))\, dx \to \infty \text{ as } k \to \infty$$

by (3.23), and this implies

$$G_0(u_k) \to \infty$$

by (3.28). Thus, the $\rho_k$ are bounded. Then, by Lemma 3.14, there is a renamed subsequence such that

$$u_k \rightharpoonup u_0 \text{ in } H_0^1 \tag{3.29}$$

and

$$u_k(x) \to u_0(x) \quad \text{uniformly in } \Omega. \tag{3.30}$$

Then

$$\int_\Omega F(x, u_k)\, dx \to \int_\Omega F(x, u_0)\, dx$$

by arguments given before. Since

$$\|u_0\|_H^2 = \|u_k\|_H^2 - 2([u_k - u_0], u_0)_H - \|u_k - u_0\|_H^2,$$

we have

$$G_0(u_0) \le \frac{1}{2}\|u_k\|_H^2 - ([u_k - u_0], u_0)_H - \int_\Omega F(x, u_0)\, dx$$

$$= G(u_k) - ([u_k - u_0], u_0)_H + \int_\Omega [F(x, u_k) - F(x, u_0)]\, dx$$

$$\to \alpha.$$

Thus

$$G(u_0) \le \alpha.$$

Since $u_0 \in H_0^1$, $\alpha \le G(u_0)$. Consequently, $\alpha \le G(u_0) \le \alpha$, from which we conclude that $G_0(u_0) = \alpha$. Thus, the proof of the first statement is complete.

To prove the second statement, note that $G_0'(u_0) = 0$. Consequently,

$$(u_0, v)_H - (f(x, u_0), v) = 0, \quad v \in H_0^1.$$

Since $u_0 \in H_0^1$, it is continuous in $\bar{\Omega}$. If $f(x, t)$ is continuous in both variables, $f(x, u_0(x))$ is continuous in $\bar{\Omega}$. Thus, $u_0'' = u_0 - f(x, u_0)$ by Theorem 3.8. Consequently, $u_0$ has a continuous second derivative and is a solution of (3.1),(3.2) in the usual sense. $\qquad\square$

It remains to prove (3.24) and Lemma 3.17. We prove them together.

*Proof.* We follow the proof of Lemma 1.23. (By now, you should know it by heart.) By Lemma 1.9

$$\|u\|^2 = \lim_{n\to\infty} \|\sum_{|k|\le n} \alpha_k \varphi_k\|^2 = \lim_{n\to\infty} \sum_{|k|\le n} |\alpha_k|^2 = \sum_{k=-\infty}^{\infty} |\alpha_k|^2, \quad (3.31)$$

where the $\alpha_k, \varphi_k$ are given by (1.43) and (1.44). For the same reason,

$$\|u'\|^2 = \lim_{n\to\infty} \|\sum_{|k|\le n} \beta_k \varphi_k\|^2 = \sum_{k=-\infty}^{\infty} |\beta_k|^2 = \sum_{k=-\infty}^{\infty} k^2 |\alpha_k|^2, \quad (3.32)$$

since $\beta_k = ik\alpha_k$. If $u \in H_0^1$, then $\alpha_0 = 0$. Hence,

$$\|u\|^2 = \sum_{k \neq 0} |\alpha_k|^2 \leq \sum k^2 |\alpha_k|^2 = \|u'\|^2,$$

which is (3.24). Moreover, if the two are equal, then

$$\|u'\|^2 - \|u\|^2 = \sum (k^2 - 1)|\alpha_k|^2 = 0.$$

Hence, $\alpha_k = 0$ if $|k| \neq 1$. This means that

$$u = (\alpha_1 e^{ix} + \alpha_{-1} e^{-ix})/\sqrt{2\pi} = a \cos x + b \sin x.$$

Since $u \in H_0^1$, we see that $a = 0$. This completes the proof. □

Next, we have the counterpart of Theorem 1.27.

**Theorem 3.19.** *Assume that $f(x,t)$ satisfies (1.78) for some $\beta$ satisfying*

$$\beta(x) \leq 2, \ \beta(x) \not\equiv 2 \ \text{a.e.} \tag{3.33}$$

*If $\Omega = (0, \pi)$, then there is a $u \in H_0^1$ such that*

$$G_0(u) = \min_{H_0^1} G_0. \tag{3.34}$$

*Moreover, if $f(x,t)$ is continuous in both variables, then any such minimum is a solution of (3.1),(3.2) in the usual sense.*

The proof of Theorem 3.19 is almost identical to that of Theorem 1.27 if we replace $I, H$ and $G$ with $\Omega, H_0^1$ and $G_0$, Lemma 1.21 with Lemma 3.14 and Theorem 1.15 with Theorem 3.8. This includes Lemmas 1.25 and 1.26. We follow that proof until we reach the conclusion

$$2G_0(u_k)/\rho_k^2 \to 1 - \int_\Omega \beta(x)\tilde{u}(x)^2 \, dx$$

$$= (1 - \|\tilde{u}\|_H^2) + (\|\tilde{u}'\|^2 - \|\tilde{u}\|^2) + \int_\Omega [2 - \beta(x)]\tilde{u}(x)^2 \, dx$$

$$= A + B + C.$$

Since $\|\tilde{u}\|_H \leq 1$ and $\beta(x) \leq 2$, the quantities $A, B, C$ are each $\geq 0$. The only way the sum can equal 0, is if each equals 0. If $B = 0$, then $\tilde{u}(x) = b \sin x$ by Lemma 3.17. If $C = 0$, then

$$b^2 \int_\Omega [2 - \beta(x)] \sin^2 x \, dx = 0.$$

By hypothesis (3.33), the only way this can happen is if $b = 0$. Then

$\tilde{u}(x) \equiv 0$, causing $A$ to be positive. Thus, the $\rho_k$ are bounded, and the proof of the first statement is complete. The second statement follows as in the proof of Theorem 3.18.

We also have a counterpart of Theorem 1.30.

**Theorem 3.20.** *Assume that $f(x,t)$ satisfies (1.62) and that $F(x,t)$ is concave in $t$ for each $x \in \Omega$. Then $G_0(u)$ given by (3.15) has a minimum on $H_0^1$.*

*Proof.* Reasoning as in the proof of Theorem 1.30, we first note that

$$F(x,t) \leq F(x,0) + f(x,0)t = f(x,0)t, \quad x \in \Omega, \, t \in \mathbb{R}$$

by Lemma 1.29. Thus,

$$F(x,t) \leq C_0|t|, \quad t \in \mathbb{R}$$

by (1.62). This implies that

$$G_0(u) \geq \frac{1}{2}\|u\|_H^2 - \int_\Omega f(x,0)u(x)\,dx$$

$$\geq \frac{1}{2}\|u\|_H^2 - c_1\|u\|, \quad u \in H_0^1 \qquad (3.35)$$

by Lemma 1.29. Let $\{u_k\} \subset H_0^1$ be a minimizing sequence for $G_0(u)$, that is, a sequence such that

$$G_0(u_k) \searrow \alpha = \inf_{H_0^1} G_0.$$

Since $G_0(u_k) \leq K$ and

$$G_0(u_k) \geq \frac{1}{2}\|u_k\|_H^2 - c_1\|u_k\|,$$

we see that the $\|u_k\|_H$ are bounded. We now follow the proofs of Theorems 3.18 and 3.19 to arrive at the desired conclusions. $\qquad\square$

We also have a counterpart of Theorem 1.31.

**Theorem 3.21.** *In addition to the hypotheses of Theorem 3.18, 3.19, or 3.20, assume that there is a $t_0 \in \mathbb{R}$ such that*

$$\int_\Omega F(x, t_0 \sin x)\,dx > \frac{\pi}{2}t_0^2. \qquad (3.36)$$

*Then the solutions of problem (3.1),(3.2) given by these theorems are nontrivial.*

*Proof.* Let $\psi(x) = t_0 \sin x$. Then

$$G_0(\psi) = \frac{\pi}{2} t_0^2 - \int_\Omega F(x, t_0 \sin x) dx < 0.$$

Thus, the minimum obtained in Theorem 3.19 is negative. Since $G_0(0) = 0$, the theorem follows. $\qquad\square$

## 3.8  Finding saddle points

We also want to obtain theorems for the situations in which no extrema exist for the functional. As in the periodic case, we can obtain such results. Recall that we are now assuming that $\Omega = (0, \pi)$. For instance, as a counterpart of Theorem 2.6, we have

**Theorem 3.22.** *Assume that (1.78) holds with*

$$2 \le \beta(x) \le 5, \quad \beta(x) \not\equiv 2, \ \beta(x) \not\equiv 5 \ \text{a.e.} \tag{3.37}$$

*If $G_0(u)$ is given by (3.15) and $G_0'(u)$ is locally Lipschitz continuous, then there is a $u_0 \in H_0^1$ such that*

$$G_0'(u_0) = 0. \tag{3.38}$$

*In particular, if $f(x, t)$ is continuous in both variables, then $u_0$ is a solution of (3.1),(3.2) in the usual sense.*

*Proof.* We let $N$ be the subspace of functions in $H_0^1$ which are multiples of $\varphi_0 = \sin x$. It is of dimension one. Let $M$ be the subspace of those functions in $H_0^1$ which are orthogonal to $N$, that is, functions $w \in H_0^1$ which satisfy

$$\int_\Omega w(x) \sin x \, dx = 0.$$

Note that this implies

$$(w', \varphi_0') = 0,$$

and consequently that

$$(w, \varphi)_H = 0.$$

I claim that

$$m_0 = \inf_M G_0 > -\infty, \quad m_1 = \sup_N G_0 < \infty.$$

For suppose $\{w_k\} \subset M$ and $G_0(w_k) \searrow m_0$. If $\rho_k = \|w_k\|_H \le C$, then Theorem 3.6 and Lemma 3.14, imply that $m_0 > -\infty$. If $\rho_k \to \infty$, let $\tilde{w}_k = w_k/\rho_k$. Then $\|\tilde{w}_k\|_H = 1$. Consequently, there is a renamed subsequence such that

$$\tilde{w}_k \rightharpoonup \tilde{w} \text{ in } H_0^1 \tag{3.39}$$

and

$$\tilde{w}_k \to \tilde{w} \text{ uniformly in } \bar{\Omega} \tag{3.40}$$

(Lemma 3.14). Thus,

$$2G_0(w_k)/\rho_k^2 = 1 - 2\int_\Omega \frac{F(x, w_k)}{w_k^2}\tilde{w}_k^2 dx \to 1 - \int_\Omega \beta(x)\tilde{w}^2(x)\,dx$$

$$= (1 - \|\tilde{w}\|_H^2) + (\|\tilde{w}\|_H^2 - 5\|\tilde{w}\|^2) + \int_\Omega [5 - \beta(x)]\tilde{w}^2(x)dx$$

$$= A + B + C, \tag{3.41}$$

as we saw before. Now, I claim that $A, B, C \geq 0$. We note that Lemma 3.17 implies that $B = \|\tilde{w}'\|^2 - 4\|\tilde{w}\|^2 \geq 0$. The only way the right-hand side of (3.41) can vanish is if $A = B = C = 0$. If $A = 0$, we see that $\tilde{w} \not\equiv 0$. If $B = 0$, then $\tilde{w}$ is of the form

$$\tilde{w}(x) = c \sin 2x.$$

If $\tilde{w} \not\equiv 0$, then $c \neq 0$. Finally, for such a function, if $C = 0$, then we must have $\beta(x) \equiv 5$ a.e. But this is excluded by hypothesis. Hence, $A, B, C$ cannot all vanish. This means that the right-hand side of (3.41) is positive. But this implies that $m_0 = \infty$, an impossibility. Thus, the $\rho_k$ must be bounded, and $m_0 > -\infty$.

To prove that $m_1 < \infty$, let $\{u_k\}$ be a sequence in $N$ of the form

$$u_k = c_k \sin x$$

such that $|c_k| \to \infty$. Then

$$\rho_k^2 = \pi c_k^2, \quad u_k(x)^2 = c_k^2 \sin^2 x.$$

Hence,

$$2G_0(u_k)/\rho_k^2 = 1 - 2\int_\Omega F(x, u_k)/\rho_k^2 dx$$

$$= 1 - 2\int_\Omega [F(x, u_k)/u_k^2]\sin^2 x\,dx$$

$$\to 1 - \int_\Omega \beta(x)\sin^2 x\,dx$$

$$= \int_\Omega [2 - \beta(x)]\sin^2 x\,dx < 0$$

by hypothesis. Thus,

$$G_0(u_k) \to -\infty \quad \text{as} \quad k \to \infty.$$

Since $G_0$ is continuous, we see that $m_1 < \infty$.

We can now apply Theorem 2.5 to conclude that there is a sequence $\{u_k\} \subset H_0^1$ satisfying (2.27). By (1.37),

$$(G_0'(u_k), v)_H = (u_k, v)_H - (f(\cdot, u_k), v) = o(\|v\|_H), \quad \|v\|_H \to 0, \quad v \in H_0^1.$$

Assume first that

$$\rho_k = \|u_k\|_H \to \infty. \tag{3.42}$$

Set $\tilde{u}_k = u_k/\rho_k$. Then $\|\tilde{u}_k\|_H = 1$, and consequently, by Lemma 3.14, there is a renamed subsequence such that

$$\tilde{u}_k \rightharpoonup \tilde{u} \text{ in } H_0^1, \quad \tilde{u}_k \to \tilde{u} \text{ uniformly on } I. \tag{3.43}$$

Thus,

$$(\tilde{u}_k, v)_H - (f(\cdot, u_k)/\rho_k, v) \to 0, \quad v \in H_0^1.$$

As we saw before, this implies in the limit that

$$(\tilde{u}, v)_H = (\beta \tilde{u}, v), \quad v \in H_0^1.$$

Take

$$\tilde{u} = \tilde{w} + \gamma, \ v = \tilde{w} - \gamma, \quad \text{where } \tilde{w} \in M, \ \gamma \in N.$$

Then

$$([\tilde{w} + \gamma], [\tilde{w} - \gamma])_H = (\beta[\tilde{w} + \gamma], \tilde{w} - \gamma).$$

This gives

$$\|\tilde{w}\|_H^2 - \|\gamma\|_H^2 = (\beta \tilde{w}, \tilde{w}) - (\beta \gamma, \gamma).$$

We write this as

$$(\|\tilde{w}'\|^2 - 5\|\tilde{w}\|^2) + \int_0^\pi [5 - \beta(x)]\tilde{w}(x)^2 dx + \int_0^\pi [\beta(x) - 2]\gamma(x)^2 dx$$
$$= A + B + C = 0.$$

This follows from the fact that

$$\gamma(x) = c \sin x.$$

Consequently,

$$\|\gamma\|_H^2 = 2\|\gamma\|^2.$$

Note that $A, B, C$ are all nonnegative. Since their sum is 0, they must each vanish. If $A = 0$, then we must have, in view of Lemma 1.23,

$$\tilde{w} = b \sin 2x.$$

If $b \neq 0$, then $\tilde{w} \neq 0$ a.e. If $B = 0$, then $[5 - \beta(x)]\tilde{w}(x)^2 \equiv 0$ a.e., and since $\beta(x) \not\equiv 2$, we must have $\tilde{w}(x) \equiv 0$. If $C = 0$, we must have $\gamma \equiv 0$. Hence, $\tilde{u}(x) \equiv 0$.

On the other hand, we also have

$$2G_0(u_k)/\rho_k^2 = 1 - 2 \int_\Omega F(x, u_k)/\rho_k^2 \to 0.$$

Thus,

$$1 - \int_\Omega \beta(x)\tilde{u}^2 \, dx = 0.$$

This cannot happen if $\tilde{u} \equiv 0$. Thus, (3.42) cannot be true, and the $\rho_k$ are bounded. Consequently, there is a renamed subsequence such that

$$u_k \rightharpoonup u \text{ in } H_0^1, \quad u_k \to u \text{ uniformly in } I \qquad (3.44)$$

(Lemma 3.14). By (2.27),

$$(G_0'(u_k), v) = (u_k, v)_H - (f(\cdot, u_k), v) \to 0, \quad v \in H_0^1,$$

and we have in the limit

$$(u, v)_H - (f(\cdot, u), v) = 0, \quad v \in H_0^1.$$

Thus, (2.29) holds with $u_0 = u$. Since $u \in H_0^1$, it is continuous in $\Omega$. If $f(x, t)$ is continuous in both variables, then $f(x, u(x))$ is continuous in $\Omega$. Thus, $u'' = u - f(x, u)$ in the usual sense by Theorem 3.8. Hence, $u$ is a solution of (3.1),(3.2).                                    □

## 3.9    Other intervals

Suppose $f(x, t)$ satisfies (1.78), but $\beta(x)$ does not satisfy (3.37). Are there other intervals $(a, b)$ such that a solution of (3.1),(3.2) can be found when $a \leq \beta(x) \leq b$? (We have asked this question before.) We are going to show that this is indeed the case. In fact we have the following counterpart of Theorem 2.24.

**Theorem 3.23.** *Let $n$ be an integer $\geq 0$. Assume that (1.78) holds with $\beta(x)$ satisfying*

$$1 + n^2 \leq \beta(x) \leq 1 + (n+1)^2, \quad 1 + n^2 \not\equiv \beta(x) \not\equiv 1 + (n+1)^2 \text{ a.e.} \tag{3.45}$$

*If (2.37) holds, then (3.1),(3.2) has a solution.*

*Proof.* As you guessed, we follow the proof of Theorem 2.24. First, we note by (3.31) and (3.32), that

$$\|u\|_H^2 = \sum (1 + k^2)|\alpha_k|^2, \quad u \in H_0^1, \tag{3.46}$$

where the $\alpha_k$ are given by (1.43) and (1.44). Let

$$N = \{u \in H_0^1 : \alpha_k = 0 \text{ for } |k| > n\}.$$

Thus,

$$\|u\|_H^2 = \sum_{|k| \leq n} (1 + k^2)|\alpha_k|^2 \leq (1 + n^2)\|u\|^2, \quad u \in N. \tag{3.47}$$

Let

$$M = \{u \in H_0^1 : \alpha_k = 0 \text{ for } |k| \leq n\}.$$

In this case,

$$\|u\|_H^2 = \sum_{|k| \geq n+1} (1 + k^2)|\alpha_k|^2 \geq (1 + (n+1)^2)\|u\|^2, \quad u \in M. \tag{3.48}$$

Note that $M, N$ are closed subspaces of $H_0^1$ and that $M = N^\perp$. Note also that $N$ is finite dimensional. Next, we consider the functional (3.15) and show that

$$G_0(v) \to -\infty \text{ as } \|v\|_H \to \infty, \quad v \in N, \tag{3.49}$$

and

$$G_0(w) \to \infty \text{ as } \|w\|_H \to \infty, \quad w \in M. \tag{3.50}$$

Assuming these for the moment, we note that they imply

$$\inf_M G_0 > -\infty; \quad \sup_N G_0 < \infty. \tag{3.51}$$

This is easily seen from the fact that (3.50) implies that there is an $R > 0$ such that

$$G_0(w) > 0, \quad \|w\|_H > R, \ w \in M.$$

Consequently, if the first statement in (3.51) were false, there would be a sequence satisfying

$$G_0(w_k) \to -\infty, \quad \|w_k\|_H \leq R, \ w_k \in M.$$

But this would imply that there is a renamed subsequence converging uniformly to a limit $w_0$ in $\Omega$. Thus,

$$G_0(w_k) \geq -\int_\Omega F(x, w_k)dx \to -\int_\Omega F(x, w_0)dx > -\infty.$$

This contradiction verifies the first statement in (3.51). The second is verified similarly by (3.49).

We are now in a position to apply Theorem 2.5. This produces a sequence in $H_0^1$ satisfying

$$G_0(u_k) \to c, \quad G_0'(u_k) \to 0, \tag{3.52}$$

where $c$ is finite. In particular, this implies

$$(G_0'(u_k), v)_H = (u_k, v)_H - (f(\cdot, u_k), v) = o(\|v\|_H), \quad \|v\|_H \to 0, v \in H_0^1. \tag{3.53}$$

Assume first that

$$\rho_k = \|u_k\|_H \to \infty. \tag{3.54}$$

Set $\tilde{u}_k = u_k/\rho_k$. Then $\|u_k\|_H = 1$, and consequently, by Lemma 3.14, there is a renamed subsequence satisfying (3.20) and (3.21). Thus

$$(\tilde{u}_k, v)_H - (f(\cdot, u_k)/\rho_k, v) \to 0, \quad v \in H_0^1. \tag{3.55}$$

As we saw before, this implies in the limit that

$$(\tilde{u}, v)_H = (\beta\tilde{u}, v), \quad v \in H_0^1. \tag{3.56}$$

Let

$$\tilde{u} = \tilde{w} + \tilde{v}, \quad \hat{u} = \tilde{w} - \tilde{v}. \tag{3.57}$$

Then

$$(\tilde{u}, \hat{u})_H = (\beta\tilde{u}, \hat{u}).$$

This implies

$$\|\tilde{w}\|_H^2 - \|\tilde{v}\|_H^2 = (\beta[\tilde{w} + \tilde{v}], \tilde{w} - \tilde{v}) = (\beta\tilde{w}, \tilde{w}) - (\beta\tilde{v}, \tilde{v}),$$

since

$$(\beta \tilde{v}, \tilde{w}) = (\beta \tilde{w}, \tilde{v}) = \int_\Omega \beta(x)\tilde{v}(x)\tilde{w}(x)dx.$$

Thus,

$$\|\tilde{w}\|_H^2 - (\beta \tilde{w}, \tilde{w}) = \|\tilde{v}\|_H^2 - (\beta \tilde{v}, \tilde{v}).$$

Consequently,

$$\left(\|\tilde{w}\|_H^2 - (1 + (n+1)^2)\|\tilde{w}\|^2\right) + \int_\Omega [1 + (n+1)^2 - \beta(x)]\tilde{w}^2 \, dx$$
$$= \left(\|\tilde{v}\|_H^2 - (1 + n^2)\|\tilde{v}\|^2\right) + \int_\Omega [1 + n^2 - \beta(x)]\tilde{v}^2 \, dx.$$

We write this as $A + B = C + D$. In view of (3.45), (3.47), and (3.48), $A \geq 0$, $B \geq 0$, $C \leq 0$, $D \leq 0$. But this implies $A = B = C = D = 0$. If

$$\tilde{u} = \sum \tilde{\alpha}_k \varphi_k,$$

then in view of (3.48) the only way $A$ can vanish is if

$$\tilde{w} = b \, \sin{(n+1)}x.$$

If $b$ is not 0, then this function can vanish only at a finite number of points in $\Omega$. But then, $B$ cannot vanish in view of (3.45). Hence, $\tilde{w} \equiv 0$. Similar reasoning shows that $C = D = 0$ implies that $\tilde{v} \equiv 0$. On the other hand, (3.52) implies

$$2G_0(u_k)/\rho_k^2 = \|\tilde{u}\|_H^2 - 2\int_\Omega F(x, \tilde{u})dx/\rho^2 \to 1 - 2\int_\Omega \beta(x)\tilde{u}^2 dx = 0,$$

from which we conclude that $\tilde{u} \not\equiv 0$. This contradiction shows that the assumption (3.54) is incorrect. Once this is known, we can conclude that there is a renamed subsequence such that (3.20) and (3.21) hold (Lemma 3.14). It then follows from (3.53) that

$$(u, v)_H - (f(\cdot, u), v) = 0, \quad v \in H_0^1. \qquad (3.58)$$

It remains to prove (3.49) and (3.50). Let $\{w_k\} \subset M$ be any sequence such that $\rho_k = \|w_k\|_H \to \infty$. Let $\tilde{w}_k = w_k/\rho_k$. Then $\|\tilde{w}_k\|_H = 1$. Thus, there is a renamed subsequence such that (3.20) and (3.21) hold. This implies

$$2G_0(w_k)/\rho_k^2 = 1 - 2\int_\Omega \frac{f(x, w_k)}{w_k^2}\tilde{w}_k^2 dx \to 1 - \int_\Omega \beta(x)\tilde{w}^2(x)dx$$

$$\geq (1 - \|\tilde{w}\|_H^2) + (\|\tilde{w}\|_H^2 - (1 + (n + 1)^2)\|\tilde{w}\|^2)$$

$$+ \int_\Omega [1 + (n + 1)^2 - \beta(x)]\tilde{w}^2(x)dx$$

$$= A + B + C.$$

As before, we note that $A \geq 0$, $B \geq 0$, $C \geq 0$. The only way $G_0(w_k)$ can fail to become infinite is if $A = B = C = 0$. As before, $B = C = 0$ implies that $\tilde{w} \equiv 0$. But this contradicts the fact that $A = 0$. Thus, $G_0(w_k) \to \infty$ for each such sequence. This proves (3.50). The limit (3.49) is proved in a similar fashion. This completes the proof of Theorem 3.23.  □

**Remark 3.24.** *It is rather surprising that Theorems 2.24 and 3.23 are practically the same since the spaces of functions are so different. This is especially true if we compare Theorems 2.6 and 3.22. The crux of the matter is that the spectrum of the operator*

$$Lu = -u'' + u$$

*in the space $H$ is*

$$\lambda_n = 1 + n^2, \quad n = 0, 1, \ldots,$$

*while that of the same (formal) operator in $H_0^1$ is*

$$\lambda_n = 1 + n^2, \quad n = 1, 2, \ldots$$

## 3.10    Super-linear problems

Up until now in our study of the Dirichlet problem we have assumed that $f(x, t)$ satisfies (1.35). As we did before, in this case we refer to problem (3.1),(3.2) as sub-linear. If $f(x, t)$ does not satisfy (1.35), we call problem (3.1),(3.2) super-linear. We now want to consider this problem in such a case. However, if we want to use the functional $G_0(u)$ given by (3.15) and we want this functional to be continuous on $H_0^1$ and have a continuous derivative on this space, we will have to make some assumptions on $f(x, t)$. However, we have shown that the assumption (1.62) is sufficient for this purpose in the periodic case (Theorem 1.20).

As we saw before, we now come to a situation which causes a serious departure from the sub-linear case. In the sub-linear periodic case we assumed (1.78) with $\beta(x)$ having certain properties. From these properties we were able to infer that either $G$ is bounded from below (Theorem 1.27), or (2.40) holds with $m_0, m_1$ finite. If (1.78) does not hold, these configurations are not true. The same is essentially true for the Dirichlet problem.

We must look for another "geometry." The simplest one employs the ideas used previously in connection with uniqueness. As in the case of periodic functions, if we can show that 0 is in a "valley" surrounded by "mountains" and that there are villages beyond the mountains, then we can adapt to this situation the splitting subspaces method that we used before. For suppose we can show that

$$G_0(u) \geq \varepsilon, \quad \|u\|_H = \rho, \tag{3.59}$$

holds for some positive $\varepsilon, \rho$, and that $G_0$ is bounded from above on $V_1 = \{v \in V : v = c \sin x, \ c > 0\}$. Now, $G_0(0) = 0$. Hence, 0 would be in a valley surrounded by mountains. If we can draw a curve of bounded length from each $v \in V_1$ along which $G_0$ decreases and such that

(a) the endpoint of each curve depends continuously on the beginning point and

(b) at the endpoint of each curve $G$ is less than $\varepsilon$ (the height of the mountains),

then we will have the desired contradiction. The reason is simple. Since $G_0$ decreases along the curves and $G(0) = 0$, curves emanating from points in $V_1$ near 0 will be trapped inside the mountain sphere $\|u\|_H = \rho$. Moreover, there will be points in $V_1$ so far away from the origin that the curves emanating from them will remain outside the sphere $\|u\|_H = \rho$. As before, the continuity of the endpoint curve will imply that there is an endpoint on the sphere, providing the contradiction.

We now need three sets of hypotheses.

(a) those that will imply that (3.59) holds,

(b) those that imply that $G$ is bounded from above on $V_1$ and

(c) those that imply that for each $v \in V_1$ there is a curve of bounded length emanating from $v$ such that the endpoint depends continuously on $v$ and $G_0 < \varepsilon$ at the endpoint.

### 3.11    More mountains

As we did in Theorem 2.21, we now want to give sufficient conditions on $F(x,t)$ which will imply that the origin is surrounded by mountains for the functional $G_0$. This can be done as follows.

**Theorem 3.25.** *Assume that (1.62) holds and that there is a $\delta > 0$ such that*

$$F(x,t) \le t^2, \quad |t| \le \delta. \tag{3.60}$$

*Then for each positive $\rho \le \delta/2$, we have either*

(**a**) *there is an $\varepsilon > 0$ such that*

$$G_0(u) \ge \varepsilon, \quad \|u\|_H = \rho, \tag{3.61}$$

*or*

(**b**) *there is a constant $c \in \mathbb{R}$ such that $|c| = \rho/\pi^{\frac{1}{2}} \le \delta/2$, and*

$$f(x, c \sin x) \equiv 2c \sin x \text{ a.e.}, \quad x \in \Omega. \tag{3.62}$$

*Moreover, such a function is a solution of (3.1),(3.2).*

*Proof.* For each $u \in H$ write $u = v + w$, where $v \in V$, $w \in W$. Then

$$2G_0(u) = \|u\|_H^2 - 2 \int_\Omega F(x,u)\, dx$$

$$= \|u'\|^2 - \|u\|^2 - 2 \int_\Omega [F(x,u) - u^2]\, dx$$

$$\ge \|w'\|^2 - \|w\|^2 - 2 \int_{|u|>\delta} [F(x,u) - u^2]\, dx.$$

Now

$$v(x) = c \sin x$$

and

$$\|v\|_H^2 = \pi c^2, \quad |v(x)| \le |c| \le \|v\|_H/\pi^{\frac{1}{2}}.$$

Consequently,

$$\|u\|_H \le \rho \Rightarrow \|v\|_H^2 + \|w\|_H^2 \le \rho^2 \Rightarrow \pi^{\frac{1}{2}}|v(x)| \le \rho.$$

Thus if $\rho \le \delta/2$, then $|v(x)| \le \delta/2$. Hence, if

$$\|u\|_H \le \rho, \quad |u(x)| \ge \delta,$$

then
$$\delta \le |u(x)| \le |v(x)| + |w(x)| \le \delta/2 + |w(x)|.$$

Consequently,
$$|v(x)| \le \delta/2 \le |w(x)|$$

and
$$\delta \le |u(x)| \le 2|w(x)|$$

for all such $x$. Thus

$$
\begin{aligned}
2G_0(u) &\ge \frac{3}{5}\|w\|_H^2 - C \int_{|u|>\delta} (|u|^{q+1} + u^2 + |u|)\, dx \\
&\ge \frac{3}{5}\|w\|_H^2 - C(1 + \delta^{1-q} + \delta^{-q}) \int_{|u|>\delta} |u|^{q+1} dx \\
&\ge \frac{3}{5}\|w\|_H^2 - C' \int_{2|w|>\delta} |w|^{q+1}\, dx \\
&\ge \frac{3}{5}\|w\|_H^2 - C'' \int_{\Omega} \|w\|_H^{q+1}\, dx \\
&\ge \frac{3}{5}\|w\|_H^2 - C'''\|w\|_H^{q+1} \\
&= \left(\frac{3}{5} - C'''\|w\|_H^{q-1}\right) \|w\|_H^2
\end{aligned}
$$

by Lemma 1.11 and (1.71). Hence,

$$G_0(u) \ge \frac{2}{5}\|w\|_H^2, \quad \|u\|_H \le \rho, \tag{3.63}$$

for $\rho > 0$ sufficiently small. Now suppose alternative (b) of the theorem did not hold. Then there would be a sequence such that

$$G_0(u_k) \to 0, \quad \|u_k\|_H = \rho. \tag{3.64}$$

If $\rho$ is taken sufficiently small, (3.63) implies that $\|w_k\|_H \to 0$. Consequently, $\|v_k\|_H \to \rho$. Now,

$$v_k(x) = c_k \sin x$$

and
$$\|v_k\|_H = \pi c_k^2.$$

Thus, $|c_k| \to \rho/\pi^{\frac{1}{2}}$. Since the $c_k$ are bounded, there is a renamed subsequence such that

$$c_k \to c_0.$$

Let

$$v_0 = c_0 \sin x.$$

Then we have

$$\|v_0\|_H = \rho, \quad G_0(v_0) = 0, \quad |v_0(x)| \le \delta/2, \quad x \in \Omega.$$

Consequently, (3.60) implies

$$F(x, v_0(x)) \le v_0(x)^2, \quad x \in \Omega. \tag{3.65}$$

Since

$$\int_\Omega \{v_0(x)^2 - F(x, v_0(x))\} \, dx = G_0(v_0) = 0$$

and the integrand is $\ge 0$ a.e. by (3.65), we see that

$$F(x, v_0(x)) \equiv v_0(x)^2, \quad x \in \Omega.$$

Let $\varphi(x)$ be any function in $C_0^\infty(\Omega)$. Then for $t > 0$ sufficiently small

$$t^{-1}[F(x, v_0 + t\varphi) - (v_0 + t\varphi)^2 - F(x, v_0) + v_0^2] \le 0.$$

Taking the limit as $t \to 0$, we have

$$(f(x, v_0) - 2v_0)\varphi(x) \le 0, \quad x \in \Omega.$$

Since this is true for every $\varphi \in C_0^\infty(\Omega)$, we see that

$$f(x, v_0(x)) \equiv 2v_0(x), \quad x \in \Omega.$$

Since $v_0 \in V$, it follows that (3.62) holds. Since $\rho$ was any sufficiently small constant, we see that (b) holds. This completes the proof. $\square$

We note that (b) implies that every function $v \in V$ satisfying $v = t_k$ is a solution of $G_0'(v) = 0$. We therefore have

**Corollary 3.26.** *Under the hypotheses of Theorem 3.25, either (a) holds for all $\rho > 0$ sufficiently small, or (3.1),(3.2) has an infinite number of solutions.*

We are now able to prove

**Theorem 3.27.** *Assume that (1.78), (2.28), (2.37), and (3.60) hold. Then there is a nontrivial solution of (3.1),(3.2).*

*Proof.* By Corollary 3.26, either (3.61) holds for some positive constants $\varepsilon$, $\rho$, or (3.1),(3.2) has an infinite number of solutions. Thus, we may assume that (3.61) holds. Then by Lemma 2.19 there is a solution of (3.1),(3.2) satisfying (2.75). But such a solution cannot be trivial since $G_0(0) = 0$. □

## 3.12    Satisfying the Palais–Smale condition

In solving the problem (3.1),(3.2), as well as problem (1.1),(1.2) our approach has been to find a sequence $\{u_k\}$ such that (2.27) holds and then show that this implies that $\{u_k\}$ has a convergent subsequence. So far we have shown this only when $f(x,t)$ satisfies (1.78). In Chapter 2 we allowed $f(x,t)$ to satisfy (1.62) with $q < \infty$ and gave sufficient conditions which guarantee that the PS condition holds for $G(u)$ given by (1.63) (Theorem 1.37). The identical results hold in the case of the Dirichlet problem. In fact, we have

**Theorem 3.28.** *If there are constants $\mu > 2, C$ such that*

$$H_\mu(x,t) := \mu F(x,t) - tf(x,t) \leq C(t^2 + 1) \tag{3.66}$$

*and*

$$\limsup_{|t| \to \infty} H_\mu(x,t)/t^2 \leq 0, \tag{3.67}$$

*then (2.27) implies that $\{u_k\}$ has a convergent subsequence which converges to a solution of (3.1),(3.2).*

The proof of Theorem 3.28 is almost identical to that of Theorem 1.37, and is omitted.

We can now combine Theorems 3.25 and 3.28 to solve a super-linear problem. We have

**Theorem 3.29.** *Under the hypotheses of Theorems 3.25, 3.28, and Lemma 2.7, if either*

$$t^2 - 2F(x,t) \leq W(x) \in L^1(\Omega), \quad t > 0$$

*or*

$$t^2 - 2F(x,t) \leq W(x) \in L^1(\Omega), \quad t < 0,$$

*then problem (3.1),(3.2) has at least one nontrivial solution.*

*Proof.* Use Theorem 2.26. We have

$$G(t \sin x) = \int_\Omega [t^2 \sin^2 x - 2F(x, t \sin x)] \, dx \leq \int_\Omega W(x) \, dx$$

for either $t > 0$ or $t < 0$. $\qquad\qquad\qquad\qquad\qquad\qquad\qquad\qquad\qquad$ $\square$

## 3.13    The linear problem

As in the periodic case, you may be curious about the linear Dirichlet problem corresponding to (3.1),(3.2), namely

$$-u''(x) + u(x) = f(x), \quad x \in \Omega = (0, \pi), \tag{3.68}$$

under the conditions

$$u(0) = u(\pi) = 0, \tag{3.69}$$

where the function $f(x)$ is continuous in $\bar{\Omega}$ and we took $a = 0$ and $b = \pi$. Again after a substantial calculation one finds that there is a unique solution given by

$$u(x) = A \sinh x + \int_0^x \sinh(t - x) \, f(t) \, dt, \tag{3.70}$$

where

$$A = \frac{-1}{\sinh \pi} \int_0^\pi \sinh(t - \pi) f(t) \, dt. \tag{3.71}$$

Can this solution be used to solve (3.1), (3.2)? Again, the answer is yes if $f(x, t)$ is bounded for all $x$ and $t$. For then we can define

$$Tu(x) = A(u) \sinh x + \int_0^x \sinh(t - x) \, f(t, u(t)) \, dt, \tag{3.72}$$

where

$$A(u) = \frac{-1}{\sinh \pi} \int_0^\pi \sinh(t - \pi) f(t, u(t)) \, dt. \tag{3.73}$$

Then a solution of (3.1),(3.2) will exist if we can find a function $u(x)$ such that

$$Tu(x) = u(x), \quad x \in \bar{\Omega}. \tag{3.74}$$

As we mentioned, such a function is called a **fixed point** of the operator $T$, and in Chapter 6 we shall study techniques of obtaining fixed points of operators in various spaces. In the present case, one can show that there is indeed a fixed point for the operator $T$ when $f(x, t)$ is bounded.

It is also of interest to note that in this case as well, the linear problem (3.68),(3.69) can be solved easily by the Hilbert space techniques of this chapter. To see this note that

$$Fv = (v, f), \quad v \in H_0^1, \tag{3.75}$$

is a bounded linear functional on $H_0^1$ (see Appendix A). By the Riesz representation theorem (Theorem A.12), there is an element $u \in H_0^1$ such that

$$Fv = (v, u)_H, \quad v \in H_0^1.$$

Hence,

$$(u, v)_H = (f, v), \quad v \in H_0^1. \tag{3.76}$$

Since $u \in H_0^1$, it is continuous and satisfies (3.69) (Corollary 3.12). Since $f$ is continuous, Theorem 3.8 tells us that $u''$ is continuous in $\Omega$ and satisfies $u'' = u - f$ there. Thus, $u$ is a solution of (3.68),(3.69).

### 3.14    Exercises

1. Show that the function

$$j(x) = \begin{cases} ae^{-1/(1-|x|^2)}, & |x| < 1, \\ 0, & |x| \geq 1, \end{cases}$$

is in $C^\infty(\mathbb{R}^n)$.

2. Show that

$$\int j_\varepsilon(x)\, dx = 1,$$

where $j_\varepsilon(x)$ is given by

$$j_\varepsilon(x) = \varepsilon^{-n} j(x/\varepsilon).$$

3. If $u \in L^1(\mathbb{R}^n)$, show that

$$\int_{|x|>R} |u(x)|\, dx \to 0 \text{ as } R \to \infty.$$

4. Show that $C(\mathbb{R}^n) \cap L^p(\mathbb{R}^n)$ is dense in $L^p(\mathbb{R}^n)$ for $1 \leq p < \infty$.

5. Show that

$$J_\varepsilon u(x) = \int j_\varepsilon(x - y)u(y)\, dy = \int j(z)u(x - \varepsilon z)\, dz \tag{3.77}$$

is in $C^\infty(\mathbb{R}^n)$.

6. Prove: If $\tilde{u}$ is an odd function in $L^2(-\pi, \pi)$, then

$$\int_{-\pi}^{0} \tilde{u}(t)e^{-ikt}\,dt = -\int_{0}^{\pi} u(s)e^{iks}\,ds.$$

7. Show that the functions $\sin kx$ are orthonormal in $L^2(0, \pi)$.

8. Show that (3.12) implies (3.13).

9. Prove that

$$x_\varepsilon \to x \text{ as } \varepsilon \to 0, \text{ uniformly in } \bar{\Omega},$$

   where

$$x_\varepsilon = \begin{cases} 0, & 0 \le x \le \varepsilon, \\ b(x - \varepsilon)/(b - 2\varepsilon), & \varepsilon \le x \le b - \varepsilon, \\ b, & b - \varepsilon \le x \le b. \end{cases}$$

10. If $h \in C(\bar{\Omega})$, show that

$$h_\varepsilon(x) = h(x_\varepsilon) \to h(x) \text{ as } \varepsilon \to 0, \text{ uniformly in } \bar{\Omega}.$$

11. Show that for each $h \in L^2(\Omega)$ there is a $v \in H_0^1(\Omega)$ such that $v'$ has a weak derivative satisfying $v - v'' = h$.

12. Prove: If $u = w + v$, where $w$ satisfies (3.26) (and consequently (3.27)) and $v$ is of the form (3.25), then

$$\|u\|_H^2 \ge \frac{3}{5}\|w\|_H^2 + 2\|v\|^2.$$

13. Why, in Theorem 3.25, were we able to obtain mountains under the assumption

$$F(x, t) \le t^2, \quad |t| \le \delta, \tag{3.78}$$

   while in Theorem 2.21 we needed

$$2F(x, t) \le t^2, \quad |t| \le \delta? \tag{3.79}$$

14. Show that (3.70),(3.71) give a solution of (3.68),(3.69).

15. Show that the solution is unique.

# 4

# Saddle points

## 4.1 Game theory

An interesting problem arising in the theory of games involves two players, $P$ and $Q$. In this model, they are given a number $\lambda$ and a "machine" $G(v, w)$ defined on $N \times M$, where $M, N$ are given sets. When an element $v \in N$ and an element $w \in M$ are input into the machine, it produces a number $G(v, w)$. The player $P$ inputs an element $v \in N$ (called his **strategy**), and player $Q$ inputs an element $w \in M$. (Neither player knows what the other did.) If it turns out that $G(v, w) > \lambda$, then player $P$ wins. If $G(v, w) < \lambda$, then player $Q$ wins. (If $G(v, w) = \lambda$, then it is a draw.) An important question is whether a player can pick a strategy that will better his chances of winning?

The following discussions will provide no help whatsoever in picking a strategy. All that we shall do is describe a situation in which selecting an optimum strategy is possible.

## 4.2 Saddle points

Let $M, N$ be as above and let $G(v, w)$ be a map of $M \times N \to \mathbb{R}$. First, we note that

$$\sup_{v \in N} \inf_{w \in M} G(v, w) \le \inf_{w \in M} \sup_{v \in N} G(v, w). \tag{4.1}$$

To see this, note that

$$\inf_{z \in M} G(v, z) \le G(v, w), \quad v \in N, \ w \in M.$$

Hence,

$$\sup_{v \in N} \inf_{z \in M} G(v, z) \le \sup_{v \in N} G(v, w), \quad w \in M.$$

123

Since the left-hand side does not involve $w$, it is

$$\leq \inf_{w \in M} \sup_{v \in N} G(v, w).$$

This proves (4.1).

We say that $(v_0, w_0)$ is a **saddle point** of $G$ if

$$G(v, w_0) \leq G(v_0, w_0) \leq G(v_0, w), \quad v \in N, w \in M. \qquad (4.2)$$

We note that

**Lemma 4.1.** *If there exist* $v_0 \in N, w_0 \in M, \lambda \in \mathbb{R}$ *such that*

$$G(v, w_0) \leq \lambda, \quad v \in N,$$

*and*

$$G(v_0, w) \geq \lambda, \quad w \in M,$$

*then* $(v_0, w_0)$ *is a saddle point of* $G$ *and*

$$\lambda = \inf_{w \in M} \sup_{v \in N} G(v, w) = \sup_{v \in N} \inf_{w \in M} G(v, w). \qquad (4.3)$$

*Proof.* Clearly

$$G(v_0, w_0) \leq \lambda \leq G(v_0, w_0).$$

Thus, $\lambda = G(v_0, w_0)$ and (4.2) holds. Hence, $(v_0, w_0)$ is a saddle point by definition. To prove (4.3) note that (4.2) implies

$$\sup_{v \in N} G(v, w_0) \leq G(v_0, w_0) \leq \inf_{w \in M} G(v_0, w).$$

Thus,

$$\inf_{w \in M} \sup_{v \in N} G(v, w) \leq \sup_{v \in N} G(v, w_0)$$
$$\leq G(v_0, w_0)$$
$$\leq \inf_{w \in M} G(v_0, w)$$
$$\leq \sup_{v \in N} \inf_{w \in M} G(v, w). \qquad (4.4)$$

But this inequality is reversed by (4.1). Hence, we have equality throughout, and (4.3) holds. $\qquad \square$

We also have

**Lemma 4.2.** *The functional $G$ has a saddle point if, and only if,*

$$\max_{v \in N} \inf_{w \in M} G(v, w) = \min_{w \in M} \sup_{v \in N} G(v, w). \tag{4.5}$$

*Proof.* If $(v_0, w_0)$ is a saddle point of $G$, then equality in (4.4) gives

$$\inf_{w \in M} \sup_{v \in N} G(v, w) = \sup_{v \in N} G(v, w_0)$$
$$= G(v_0, w_0)$$
$$= \inf_{w \in M} G(v_0, w)$$
$$= \sup_{v \in N} \inf_{w \in M} G(v, w).$$

This implies that

$$\min_{w \in M} \sup_{v \in N} G(v, w) = \sup_{v \in N} G(v, w_0)$$
$$= G(v_0, w_0)$$
$$= \inf_{w \in M} G(v_0, w)$$
$$= \max_{v \in N} \inf_{w \in M} G(v, w).$$

Thus the minimum and maximum are attained. This implies (4.5). Conversely, if (4.5) holds, then

$$\inf_{w \in M} G(v, w) \le \inf_{w \in M} G(\bar{v}, w) = \lambda = \sup_{v \in N} G(v, \bar{w}) \le \sup_{v \in N} G(v, w)$$

for some $(\bar{v}, \bar{w})$, where $\lambda$ is the common value in (4.5). Consequently,

$$G(v, \bar{w}) \le \lambda \le G(\bar{v}, w), \quad v \in N, w \in M.$$

Thus, by Lemma 4.1, $(\bar{v}, \bar{w})$ is a saddle point of $G$ and $G(\bar{v}, \bar{w}) = \lambda$. □

## 4.3 Convexity and lower semi-continuity

Let $M$ be a convex subset of a Hilbert space $E$, and let $G$ be a functional (real valued function) defined on $M$. We call $G$ **convex** on $M$ if

$$G((1 - t)w_0 + tw_1) \le (1 - t)G(w_0) + tG(w_1), \quad w_0, w_1 \in M, \, 0 \le t \le 1.$$

We call it **strictly convex** if the inequality is strict when $t = \frac{1}{2}$, $w_0 \ne w_1$.

We call $G(v)$ **upper semi-continuous** (u.s.c.) at $w_0 \in M$ if $w_k \to$

$w_0 \in M$ implies

$$G(w_0) \geq \limsup G(w_k).$$

It is called **lower semi-continuous** (l.s.c.) if the inequality is reversed and lim sup is replaced by lim inf. We have

**Lemma 4.3.** *If $M$ is closed, convex, and bounded in $E$, and $G$ is convex and l.s.c., then there is a point $w_0 \in M$ such that*

$$G(w_0) = \min_M G. \tag{4.6}$$

*If $G$ is strictly convex, then $w_0$ is unique.*

In proving Lemma 4.3 we shall make use of

**Lemma 4.4.** *If $u_k \rightharpoonup u$ in $E$, then there is a renamed subsequence such that $\bar{u}_k \to u$, where*

$$\bar{u}_k = (u_1 + \cdots + u_k)/k. \tag{4.7}$$

*Proof.* We may assume that $u = 0$. Take $n_1 = 1$, and inductively pick $n_2, n_3, \ldots$, so that

$$|(u_{n_k}, u_{n_1})| \leq \frac{1}{k}, \ldots, |(u_{n_k}, u_{n_{k-1}})| \leq \frac{1}{k}.$$

This can be done since

$$(u_n, u_{n_j}) \to 0 \quad \text{as} \quad n \to \infty, \ 1 \leq j \leq k.$$

Since

$$\|u_k\| \leq C$$

for some $C$, we have

$$\|\bar{u}_k\|^2 = \left[ \sum_{j=1}^{k} \|u_j\|^2 + 2 \sum_{j=1}^{k} \sum_{i=1}^{j} (u_i, u_j) \right] / k^2$$

$$\leq \left[ kC^2 + 2 \sum_{j=1}^{k} \sum_{i=1}^{j} \frac{1}{j} \right] / k^2$$

$$\leq (C^2 + 2)/k \to 0.$$

$\square$

**Lemma 4.5.** *It $G(u)$ is convex and l.s.c. on $E$, and $u_k \rightharpoonup u$, then*

$$G(u) \leq \liminf G(u_k).$$

*Proof.* Let

$$L = \liminf G(u_k).$$

Then there is a renamed subsequence such that $G(u_k) \to L$. Let $\varepsilon > 0$ be given. Then

$$L - \varepsilon < G(u_k) < L + \varepsilon \qquad (4.8)$$

for all but a finite number of $k$. Remove a finite number and rename it so that (4.8) holds for all $k$. Moreover, there is a renamed subsequence such that $\bar{u}_k \to u$ by Lemma 4.4, where $\bar{u}_k$ is given by (4.7). Thus,

$$G(u) \leq \liminf G(\bar{u}_k) = \liminf G\left(\frac{1}{k}\sum_{j=1}^{k} u_j\right)$$

$$\leq \liminf \frac{1}{k}\sum_{j=1}^{k} G(u_j) \leq \liminf \frac{1}{k} \cdot k(L + \varepsilon) = L + \varepsilon.$$

Since $\varepsilon$ was arbitrary, we see that $G(u) \leq L$, and the proof is complete. $\square$

A subset $M \subset E$ is called **weakly closed** if $u \in M$ whenever there is a sequence $\{u_k\} \subset M$ converging weakly to $u$ in $E$. This terminology is very unfortunate and misleading. A weakly closed set is closed in a stronger sense than an ordinary closed set. It follows from Lemma 4.4 that

**Lemma 4.6.** *If $M$ is a closed convex subset of $E$, then it is weakly closed in $E$.*

*Proof.* Suppose $\{u_k\} \subset M$ and $u_k \rightharpoonup u$ in $E$. Then by Lemma 4.4, there is a renamed subsequence such that $\bar{u}_k \to u$, where $\bar{u}_k$ is given by (4.7). Since $M$ is convex, each $\bar{u}_k$ is in $M$. Since $M$ is closed, we see that $u \in M$. $\square$

We can now give the proof of Lemma 4.3.

*Proof.* Let

$$\alpha = \inf_{M} G.$$

(At this point we do not know if $\alpha \neq -\infty$.) Let $\{w_k\} \subset M$ be a sequence such that $G(w_k) \to \alpha$. Since $M$ is bounded, we see that there is a renamed subsequence such that $w_k \rightharpoonup w_0$. Since $M$ is closed and convex, it is weakly closed (Lemma 4.6). Hence, $w_0 \in M$. By Lemma 4.5,

$G(w_0) \leq \liminf G(w_k) = \alpha$. Since $G(w_0) \geq \alpha$, we see that (4.6) holds. So far, we have only used the convexity of $G$. We use the strict convexity to show that $w_0$ is unique. If there were another element $w_1 \in M$ such that $G(w_1) = \alpha$, then we have

$$G\left(\frac{1}{2}w_0 + \frac{1}{2}w_1\right) < \frac{1}{2}[G(w_0) + G(w_1)] = \alpha,$$

which is impossible from the definition of $\alpha$. This completes the proof. $\square$

We also have

**Lemma 4.7.** *If $M$ is closed and convex, $G$ is convex, l.s.c., and satisfies*

$$G(u) \to \infty \quad \text{as} \quad \|u\| \to \infty, \quad u \in M, \tag{4.9}$$

*if $M$ is unbounded, then $G$ is bounded from below on $M$ and has a minimum there.*

*Proof.* If $M$ is bounded, then Lemma 4.7 follows from Lemma 4.3. Otherwise, let $u_0$ be any element in $M$. By (4.9), there is an $R \geq \|u_0\|$ such that

$$G(u) \geq G(u_0), \quad u \in M, \ \|u\| \geq R.$$

By Lemma 4.3, $G$ is bounded from below on the set

$$M_R = \{w \in M : \|w\| \leq R\}$$

and has a minimum there. A minimum of $G$ on $M_R$ is a minimum of $G$ on $M$. Hence, $G$ is bounded from below on $M$ and has a minimum there. $\square$

## 4.4    Existence of saddle points

We now present some sufficient conditions for the existence of saddle points. Let $M, N$ be closed, convex subsets of a Hilbert space, and let $G(v, w) : M \times N \to \mathbb{R}$ be a functional such that $G(v, w)$ is convex and l.s.c. in $w$ for each $v \in N$, and concave and u.s.c. in $v$ for each $w \in M$. Assume also that there is a $v_0 \in N$ such that

$$G(v_0, w) \to \infty \quad \text{as} \quad \|w\| \to \infty, \quad w \in M, \tag{4.10}$$

and there is a $w_0 \in M$ such that

$$G(v, w_0) \to -\infty \quad \text{as} \quad \|v\| \to \infty, \quad v \in N. \tag{4.11}$$

(If $M$ is bounded, then (4.10) is automatically satisfied; the same is true for (4.11) when $N$ is bounded.) We have

**Theorem 4.8.** *Under the above hypotheses, $G$ has at least one saddle point.*

*Proof.* Assume first that $M, N$ are bounded and that $G(v, w)$ is strictly convex with respect to $w$. Then for each $v \in N$, there is a point $\sigma(v) \in M$ where $G(v, w)$ achieves its minimum (Lemma 4.3). Since $G$ is strictly convex in $w$, this minimum point is unique. Let

$$J(v) = G(v, \sigma(v)) = \min_{w \in M} G(v, w).$$

Since $J(v)$ is the minimum of a family of functionals which are concave and u.s.c., it is also concave and u.s.c. In fact, if

$$v_t = (1 - t)v_0 + tv_1, \quad t \in [0, 1],$$

then

$$G(v_t, w) \geq (1 - t) \min_{\hat{w} \in M} G(v_0, \hat{w}) + t \min_{\hat{w} \in M} G(v_1, \hat{w}), \quad w \in M.$$

Since this is true for each $w \in M$, we finally obtain

$$J(v_t) \geq (1 - t)J(v_0) + tJ(v_1). \tag{4.12}$$

Similarly, if $v_k \to v \in N$, then we have

$$J(v_k) \leq G(v_k, w), \quad w \in M.$$

Thus,

$$\limsup J(v_k) \leq \limsup G(v_k, w) \leq G(v, w), \quad w \in M.$$

Since this is true for each $w \in M$, we have

$$\limsup J(v_k) \leq \inf_{w \in M} G(v, w) = J(v). \tag{4.13}$$

Consequently, $J(v)$ has a maximum point $\bar{v}$ satisfying

$$J(v) \leq J(\bar{v}), \quad v \in N$$

(Lemma 4.3). In particular, we have

$$J(\bar{v}) = \min_{\hat{w} \in M} G(\bar{v}, \hat{w}) \leq G(\bar{v}, w), \quad w \in M. \tag{4.14}$$

Let $v$ be an arbitrary point in $N$, and let

$$v_\theta = (1 - \theta)\bar{v} + \theta v, \quad w_\theta = \sigma(v_\theta), \quad 0 \leq \theta \leq 1.$$

Since $G$ is concave in $v$, we have

$$G(v_\theta, w) \geq (1 - \theta)G(\bar{v}, w) + \theta G(v, w).$$

Consequently,

$$
\begin{aligned}
J(\bar{v}) &\geq J(v_\theta) \\
&= G(v_\theta, w_\theta) \\
&\geq (1 - \theta)G(\bar{v}, w_\theta) + \theta G(v, w_\theta) \\
&\geq (1 - \theta)J(\bar{v}) + \theta G(v, w_\theta).
\end{aligned}
$$

This gives

$$J(\bar{v}) \geq G(v, w_\theta), \quad v \in N, \ 0 < \theta \leq 1. \tag{4.15}$$

Let $\{\theta_k\}$ be a sequence converging to 0, and let $w_k = w_{\theta_k}$. Since $M$ is bounded, there is a renamed subsequence such that $w_k \rightharpoonup \bar{w}$. Since

$$(1 - \theta)G(\bar{v}, w_\theta) + \theta G(v, w_\theta) \leq G(v_\theta, w_\theta) \leq G(v_\theta, w), \quad w \in M,$$

we have

$$(1 - \theta_k)G(\bar{v}, w_k) + \theta_k J(v) \leq G(v_{\theta_k}, w), \quad w \in M.$$

In the limit this gives

$$G(\bar{v}, \bar{w}) \leq G(\bar{v}, w), \quad w \in M$$

(cf. Lemma 4.5). This tells us that $\bar{w} = \sigma(\bar{v})$ and does not depend on $v$. Since

$$J(\bar{v}) \geq G(v, w_k),$$

we have

$$G(v, \bar{w}) \leq J(\bar{v}) \leq G(\bar{v}, w), \quad v \in N, \ w \in M$$

in view of (4.14) and (4.15). The result now follows from Lemma 4.1.

Now we remove the assumption that $G$ is strictly convex in $w$. For $\varepsilon > 0$, let

$$G_\varepsilon(v, w) = G(v, w) + \varepsilon \|w\|^2.$$

Now $G_\varepsilon$ satisfies all of the hypotheses of the theorem and is also strictly convex with respect to $w$. This follows from the fact that

$$\|(1 - \theta)w_0 + \theta w_1\|^2 = (1 - \theta)\|w_0\|^2 + \theta\|w_1\|^2 - \theta(1 - \theta)\|w_0 - w_1\|^2.$$

We can now apply the theorem to $G_\varepsilon$ and obtain saddle points $(\bar{v}_\varepsilon, \bar{w}_\varepsilon)$ satisfying

$$G_\varepsilon(v, \bar{w}_\varepsilon) \le G_\varepsilon(\bar{v}_\varepsilon, \bar{w}_\varepsilon) \le G_\varepsilon(\bar{v}_\varepsilon, w), \quad v \in N, \ w \in M.$$

Let $\{\varepsilon_k\}$ be a sequence tending to 0. Then there are renamed subsequences $\bar{v}_k = \bar{v}_{\varepsilon_k}$, $\bar{w}_k = \bar{w}_{\varepsilon_k}$ such that $\bar{v}_k \rightharpoonup \bar{v}$, $\bar{w}_k \rightharpoonup \bar{w}$. By Lemma 4.5, we have

$$G(v, \bar{w}) \le G(\bar{v}, w), \quad v \in N, \ w \in M.$$

Thus $(\bar{v}, \bar{w})$ is a saddle point of $G$. Next, we remove the restriction that $M, N$ are bounded. Let $R$ be so large that $\|v_0\| < R$, $\|w_0\| < R$. The sets

$$M_R = \{w \in M : \|w\| \le R\}, \quad N_R = \{v \in N : \|v\| \le R\}$$

are closed, convex, and bounded. By what we have already proved, there is a saddle point $(\bar{v}_R, \bar{w}_R)$ such that

$$G(v, \bar{w}_R) \le G(\bar{v}_R, \bar{w}_R) \le G\bar{v}_R, w), \quad v \in N_R, \ w \in M_R. \tag{4.16}$$

In particular, we have

$$G(v_0, \bar{w}_R) \le G(\bar{v}_R, \bar{w}_R) \le G(\bar{v}_R, w_0).$$

Since $G(v_0, w)$ is convex, l.s.c., and satisfies (4.10), it is bounded from below on $M$ (Lemma 4.7). Thus,

$$G(v_0, \bar{w}_R) \ge A > -\infty.$$

Similarly, $G(v, w_0)$ is bounded from above. Hence,

$$G(\bar{v}_R, w_0) \le B < \infty.$$

Combining these with (4.16), we have

$$A \le G(v_0, \bar{w}_R) \le G(\bar{v}_R, w_0) \le B.$$

By (4.10) and (4.11), the sequences $\{\bar{v}_R\}$, $\{\bar{w}_R\}$ are bounded. Hence, there are renamed subsequences such that

$$\bar{v}_R \rightharpoonup \bar{v}, \ \bar{w}_R \rightharpoonup \bar{w}, \ \text{as} \ R \to \infty,$$

and

$$G(\bar{v}_R, \bar{w}_R) \to \lambda, \ \text{as} \ R \to \infty.$$

In view of (4.16) we have in the limit

$$G(v, \bar{w}) \le \lambda \le G(\bar{v}, w), \quad v \in N, \ w \in M.$$

This shows that $(\bar{v}, \bar{w})$ is a saddle point, and the theorem is completely proved. □

## 4.5     Criteria for convexity

Recall that a functional $G$ is called **convex** if

$$G([(1-t)u_0 + tu_1]) \le (1-t)G(u_0) + tG(u_1), \quad t \in [0,1].$$

It is **strictly convex** if the inequality is strict when $u_0 \ne u_1$ and $t = \frac{1}{2}$. This implies that the inequality is strict when $u_0 \ne u_1$ and $t \in (0,1)$.

If $G$ is a differentiable functional on a Hilbert space $E$, there are simple criteria which can be used to verify convexity of $G$. We gave one such criterion in Lemma 1.29. We also have

**Theorem 4.9.** *Let $G$ be a differentiable functional on a closed, convex subset $M$ of $E$. Then $G$ is convex on $E$ iff it satisfies any of the following inequalities for $u_0, u_1 \in M$.*

$$(G'(u_0), u_1 - u_0) \le G(u_1) - G(u_0) \tag{4.17}$$

$$(G'(u_1), u_1 - u_0) \ge G(u_1) - G(u_0) \tag{4.18}$$

$$(G'(u_1) - G'(u_0), u_1 - u_0) \ge 0. \tag{4.19}$$

*Moreover, it will be strictly convex iff there is strict inequality in any of them for $u_0 \ne u_1$.*

*Proof.* Let $u_t = (1-t)u_0 + tu_1$, $0 \le t \le 1$, and $\varphi(t) = G(u_t)$. If $G$ is convex, then

$$G(u_t) \le (1-t)G(u_0) + tG(u_1), \tag{4.20}$$

or

$$\varphi(t) \le (1-t)\varphi(0) + t\varphi(1), \quad 0 \le t \le 1. \tag{4.21}$$

In particular, the slope of $\varphi$ at $t = 0$ is $\le$ the slope of the straight line connecting $(0, \varphi(0))$ and $(1, \varphi(1))$. Thus $\varphi'(0) \le \varphi(1) - \varphi(0)$, and this is merely (4.17). Alternatively, we can use Lemma 1.29. Reversing the roles of $u_0, u_1$ produces (4.18). We obtain (4.19) by substracting (4.17) from (4.18). Conversely, (4.19) implies

$$\varphi'(t) - \varphi'(s) = (G'(u_t) - G'(u_s), u_1 - u_0)$$
$$= (G'(u_t) - G'(u_s), u_t - u_s)/(t - s) \geq 0, \quad 0 \leq s < t \leq 1.$$

Thus,

$$\varphi'(t) \geq \varphi'(s), \quad 0 \leq s \leq t \leq 1,$$

which implies (4.21). Since this is equivalent to (4.20), we see that $G$ is convex. If $G$ is strictly convex, we obtain strict inequalities in (4.17)–(4.19), and strict inequalities in any of them implies strict inequalities in (4.21) and (4.20). $\qquad\square$

**Corollary 4.10.** *Let $G$ be a differentiable functional on a closed, convex subset $M$ of $E$. Then $G$ is concave on $E$ iff it satisfies any of the following inequalities for $u_0, u_1 \in M$.*

$$(G'(u_0), u_1 - u_0) \geq G(u_1) - G(u_0) \tag{4.22}$$

$$(G'(u_1), u_1 - u_0) \leq G(u_1) - G(u_0) \tag{4.23}$$

$$(G'(u_1) - G'(u_0), u_1 - u_0) \leq 0. \tag{4.24}$$

*Moreover, it will be strictly concave iff there is strict inequality in any of them for $u_0 \neq u_1$.*

*Proof.* Note that $G(u)$ is concave iff $-G(u)$ is convex. $\qquad\square$

## 4.6     Partial derivatives

Let $M, N$ be closed subspaces of a Hilbert space $H$ satisfying $H = M \oplus N$. Let $G(u)$ be a functional on $H$. We can consider "partial" derivatives of $G$ in the same way we considered total derivatives. We keep $w = w_0 \in M$ fixed and consider $G(u)$ as a functional on $N$, where $u = v + w_0$, $v \in N$. If the derivative of this functional exists at $v = v_0 \in N$, we call it the **partial derivative** of $G$ at $u_0 = v_0 + w_0$ with respect to $v \in N$ and denote it by $G'_N(u_0)$. Similarly, we can define the partial derivative $G'_M(u_0)$. We have

**Lemma 4.11.** *If $G'$ exists at $u_0 = v_0 + w_0$, then $G'_M(u_0)$ and $G'_N(u_0)$ exist and satisfy*

$$(G'(u_0), u) = (G'_M(u_0), w) + (G'_N(u_0), v), \quad v \in N, \ w \in M. \tag{4.25}$$

*Proof.* By definition

$$G(u_0 + u) = G(u_0) + (G'(u_0), u) + o(\|u\|), \quad u \in H.$$

Therefore,

$$G(u_0 + v) = G(u_0) + (G'(u_0), v) + o(\|v\|), \quad v \in N,$$

and

$$G(u_0 + w) = G(u_0) + (G'(u_0), w) + o(\|w\|), \quad w \in M.$$

But

$$G(u_0 + v) = G(u_0) + (G'_N(u_0), v) + o(\|v\|), \quad v \in N,$$

and

$$G(u_0 + w) = G(u_0) + (G'_M(u_0), w) + o(\|w\|), \quad w \in M.$$

In particular, we have

$$(G'(u_0) - G'_N(u_0), v) = o(\|v\|) \text{ as } \|v\| \to 0, \quad v \in N.$$

Thus,

$$(G'(u_0) - G'_N(u_0), tv) = o(|t|) \text{ as } |t| \to 0$$

for each fixed $v \in N$. This means that

$$(G'(u_0) - G'_N(u_0), v) = \frac{o(|t|)}{t} \to 0 \text{ as } t \to 0.$$

Hence,

$$(G'(u_0), v) = (G'_N(u_0), v), \quad v \in N.$$

Similarly,

$$(G'(u_0), w) = (G'_M(u_0), w), \quad w \in M.$$

These two identities combine to give (4.25). $\qquad\square$

**Lemma 4.12.** *Under the hypotheses of Lemma 4.11, assume that $G$ is differentiable on $H$, convex on $M$ and concave on $N$. Then,*

$$G(u) - G(u_0) \leq (G'_N(u_0), v - v_0) + (G'_M(u), w - w_0),$$
$$u = v + w, \ u_0 = v_0 + w_0, \ v, v_0 \in N, \ w, w_0 \in M. \qquad (4.26)$$

*Proof.* This follows from Theorem 4.9 and its corollary. In fact, we have

$$G(u) - G(u_0) = G(u) - G(v + w_0) + G(v + w_0) - G(u_0)$$
$$\leq (G'(u), w - w_0) + (G'(u_0), v - v_0).$$

Apply Lemma 4.11. $\square$

We also have

**Lemma 4.13.** *Under the hypotheses of Lemma 4.11, if $G'(u_0)$ exists and $u_0 = v_0 + w_0$ is a saddle point, then*

$$G'(u_0) = G'_M(u_0) = G'_N(u_0) = 0.$$

*Proof.* By definition

$$G(v + w_0) \leq G(u_0) \leq G(v_0 + w), \quad v \in N, \ w \in M.$$

Since $v_0$ is a maximum point on $N$, we see that $G'_N(u_0) = 0$ (Lemma 1.1). Since $w_0$ is a minimum point on $M$, we have $G'_M(u_0) = 0$ for the same reason. We then apply Lemma 4.11. $\square$

We also have

**Theorem 4.14.** *Under the hypotheses of Lemma 4.11, let $a(u)$ be a differentiable functional on $H$ which is convex on $M$ and concave on $N$. Assume that there are a Hilbert space $H_1$ and linear operators $S$, $T$, with $S$ mapping $N$ into $H_1$ and $T$ mapping $M$ into $H_1$ such that*

$$Sv \in D(T^*), \ Tw \in D(S^*), \quad v \in N, \ w \in M,$$

*and*

$$a(v_1 + w) - (Sv_1, Tw)_1 \to \infty \text{ as } \|w\| \to \infty, \quad w \in M,$$
$$a(v + w_1) - (Sv, Tw_1)_1 \to -\infty \text{ as } \|v\| \to \infty, \quad v \in N,$$

*for some $v_1 \in N$ and $w_1 \in M$. Then there is a solution $u_0 = v_0 + w_0$ of*

$$T^*Sv_0 = a'_M(u_0), \quad S^*Tw_0 = a'_N(u_0). \tag{4.27}$$

*Proof.* Let

$$G(u) = a(u) - (Sv, Tw)_1, \quad u = v + w, \ v \in N, \ w \in M.$$

Note that $v_k \to v$ in $N$ implies

$$(Sv_k, Tw)_1 = (v_k, S^*Tw) \to (v, S^*Tw) = (Sv, Tw)_1, \quad w \in M,$$

and $w_k \to w$ in $M$ implies

$$(Sv, Tw_k)_1 = (T^*Sv, w_k) \to (T^*Sv, w) = (Sv, Tw)_1, \quad v \in N.$$

Consequently, $G(u)$ is continuous and convex on $M$ for each $v \in N$, and continuous and concave on $N$ for each $w \in M$. Thus, all of the hypotheses of Theorem 4.8 are satisfied. It therefore follows that $G(u)$ has a saddle point $u_0$ satisfying

$$G(v + w_0) \leq G(u_0) \leq G(v_0 + w), \quad v \in N, \ w \in M.$$

Consequently,

$$a(v + w_0) - (Sv, Tw_0)_1 \leq a(v_0 + w_0) - (Sv_0, Tw_0)_1$$
$$\leq a(v_0 + w) - (Sv_0, Tw)_1, \quad v \in N, \ w \in M,$$

or

$$a(v + w_0) - a(v_0 + w_0) \leq (S(v - w_0), Tw_0)_1, \quad v \in N,$$

and

$$a(v + w_0) - a(v_0 + w_0) \geq (Sv_0, T(w - w_0))_1, \quad w \in M.$$

Let $\tilde{v} \in N$, $\tilde{w} \in M$, $t > 0$ be arbitrary, and take $v = v_0 + t\tilde{v}$, $w = w_0 + t\tilde{w}$. Then

$$a(u_0 + t\tilde{v}) - a(u_0) \leq t(S\tilde{v}, Tw_0)_1, \quad \tilde{v} \in N, \ t > 0,$$

and

$$a(u_0 + t\tilde{w}) - a(u_0) \geq t(Sv_0, T\tilde{w})_1, \quad \tilde{w} \in M, \ t > 0,$$

Letting $t \to 0$, we obtain

$$(a'_N(u_0) - S^*Tw_0, \tilde{v}) \leq 0, \quad \tilde{v} \in N,$$

and

$$(a'_M(u_0) - T^*Sv_0, \tilde{w}) \geq 0, \quad \tilde{w} \in M.$$

By picking $\tilde{v}$, $\tilde{w}$ judiciously, we see that these imply (4.27). The proof is complete.                                                                    $\square$

## 4.7    Nonexpansive operators

If $M$ is a closed subset of a Banach space, and $f(x)$ maps $M$ into itself and satisfies

$$\|f(x) - f(y)\| \le \theta \|x - y\|, \quad x, y \in M$$

for some $\theta < 1$, then we know that there is a unique $x_0 \in M$ such that $f(x_0) = x_0$ (Theorem 2.12). The mapping $f$ is called a **contraction** because $\theta < 1$, and $f(x), f(y)$ are closer together than $x, y$ (unless $x = y$). A natural question is whether or not the theorem is true if one relaxes the hypotheses to

$$\|f(x) - f(y)\| \le \|x - y\|, \quad x, y \in M. \tag{4.28}$$

In this case we call $f$ **nonexpansive**. Unfortunately, the theorem is false in this case as the following example shows.

Let $X$ be the set of bounded sequences $x = \{x_n\}$ such that $x_n \to 0$ as $n \to \infty$. With the norm

$$\|x\| = \max_n |x_n|,$$

it becomes a Banach space. Let $M = \{x \in X : \|x\| \le 1\}$, and define $f$ by

$$f(x) = \{1, x_1, x_2, \dots\}.$$

If $y = \{y_n\} \in M$, then

$$\|f(x) - f(y)\| = \|\{0, x_1 - y_1, x_2 - y_2, \dots\}\| = \|x - y\|.$$

Thus $f$ satisfies (4.28). However, if $f(x) = x$, then

$$\{x_1, x_2, \dots\} = \{1, x_1, x_2, \dots\},$$

and this implies $x_1 = 1, x_2 = 1, \dots, x_n = 1, \dots$ But then $x \notin X$ since $x_n = 1 \not\to 0$. However, the following is true.

**Theorem 4.15.** *Let $M$ be a bounded, closed, convex subset of a Hilbert space $X$, and let $f$ be a map of $M$ into itself that satisfies (4.28). Then there is at least one point $x_0 \in M$ such that*

$$f(x_0) = x_0. \tag{4.29}$$

*The set of such points is convex.*

**Remark 4.16.** *Recall that a set $Q$ is convex if $tu + (1-t)v \in Q$ whenever $u, v \in Q$ and $0 \le t \le 1$. Note that we added the hypotheses that $M$ is bounded and convex and that $X$ is a Hilbert space. However, we cannot claim that a solution of (4.29) is unique.*

*Proof.* By shifting everything, we may assume that $0 \in M$. Now, for each $t$ such that $0 \le t < 1$, the mapping $tf(x)$ is a contraction. Hence, there is a unique $x_t \in M$ such that

$$tf(x_t) = x_t.$$

Since $M$ is a closed, bounded, and convex subset of a Hilbert space, there is a sequence $t_n \to 1$ such that $x_n = x_{t_n}$ converges weakly to a limit $x_0 \in M$ (Theorem A.61). Let

$$A_n(x) = x - t_n f(x), \quad A(x) = x - f(x).$$

Then,

$$(A_n(x) - A_n(y), x - y) = \|x - y\|^2 - t_n(f(x) - f(y), x - y) \ge 0 \tag{4.30}$$

by (4.28). If we take $y = x_n$, we have

$$(A_n(x) - A_n(x_n), x - x_n) \ge 0.$$

Since $A_n(x_n) = 0$, this gives

$$(A_n(x), x - x_n) \ge 0, \quad x \in M.$$

Letting $n \to \infty$, we obtain

$$(A(x), x - x_0) \ge 0, \quad x \in M. \tag{4.31}$$

This implies

$$(A(x_0), x - x_0) \ge 0, \quad x \in M. \tag{4.32}$$

Assuming this for the moment, we see that

$$(x_0 - f(x_0), x - x_0) \ge 0, \quad x \in M.$$

If we take $x = f(x_0)$, we obtain

$$-\|A(x_0)\|^2 \ge 0,$$

which implies that $x_0$ satisfies (4.29).

It therefore remains only to show that (4.32) holds and that the set of solutions of (4.29) is convex. To see that (4.32) holds, let $y$ be any element of $M$, and take $x = \theta y + (1 - \theta)x_0$, $0 < \theta < 1$. Then by (4.31),

$$(A(\theta y + (1 - \theta)x_0), y - x_0) \geq 0.$$

If we let $\theta \to 0$, we obtain (4.32) with $x$ replaced by $y$. In particular, we see that $x$ is a solution of $A(x) = 0$ iff it satisfies (4.31).

To show that that the set of solutions of (4.29) is convex, note that (4.28) implies

$$(A(x) - A(y), x - y) \geq 0, \quad x, y \in M. \tag{4.33}$$

If $x_0$ is a solution of (4.29), it satisfies

$$(A(x), x - x_0) \geq (A(x_0), x - x_0) = 0, \quad x \in M.$$

If $x_1$ is also a solution, then

$$(A(x), x - [\theta x_0 + (1 - \theta)x_1]) = \theta(A(x), x - x_0)$$
$$+ (1 - \theta)(A(x), x - x_1) \geq 0,$$

showing that $x = \theta x_0 + (1 - \theta)x_1$ satisfies (4.31) and consequently, it is also a solution for $0 \leq \theta \leq 1$. This completes the proof. $\square$

## 4.8     The implicit function theorem

Before we prove a general theorem which is useful in many applications, we shall prove a simple, well known version.

**Theorem 4.17.** *Suppose $f(x,y)$ is a continuous function mapping an open set $\Omega \subset \mathbb{R}^2 \to \mathbb{R}$. Assume that $f_y(x,y)$ is continuous on $\Omega$ and that there is a point $(x_0, y_0) \in \Omega$ such that*

$$f(x_0, y_0) = 0, \quad A \equiv f_y(x_0, y_0) \neq 0.$$

*Then there is an $r > 0$ and a continuous function $g(x)$ on*

$$I_r(x_0) = \{x \in \mathbb{R} : |x - x_0| < r\}$$

*such that $g(x_0) = y_0$,*

$$(x, g(x)) \subset \Omega, \quad x \in I_r(x_0), \tag{4.34}$$

*and*

$$f(x, g(x)) \equiv 0, \quad x \in I_r(x_0). \tag{4.35}$$

*Proof.* We may assume that $x_0 = y_0 = 0$. Define

$$R(x, y) = Ay - f(x, y).$$

For $x, y_1, y_2$ close to 0, we have

$$f(x, y_1) - f(x, y_2) = \int_0^1 \frac{d}{d\theta} f(x, \theta y_1 + (1-\theta)y_2)\, d\theta$$

$$= (y_1 - y_2) \int_0^1 f_y(x, \theta y_1 + (1-\theta)y_2)\, d\theta.$$

Consequently,

$$R(x, y_1) - R(x, y_2) = (y_1 - y_2) \int_0^1 [f_y(0,0) - f_y(x, \theta y_1 + (1-\theta)y_2)]\, d\theta,$$

since $A = f_y(0,0)$. By continuity, there is a $\delta > 0$ such that

$$|R(x, y_1) - R(x, y_2)| \le |y_1 - y_2|/2|A|, \quad |x|, |y_j| \le \delta.$$

Let

$$h(x, y) = R(x, y)/A.$$

Then,

$$|h(x, y_1) - h(x, y_2)| \le |y_1 - y_2|/2, \quad |x|, |y_j| \le \delta. \qquad (4.36)$$

Moreover,

$$|h(x, y)| \le |h(x, y) - h(x, 0)| + |h(x, 0) - h(0,0)|$$
$$\le \delta/2 + \delta/2 = \delta, \quad |x| \le r, \ |y| \le \delta$$

for some $r \le \delta$. Consequently, for each $x$ satisfying $|x| \le r$, $h(x, y)$ is a continuous map of $I_\delta(0)$ into itself satisfying (4.36). We can now apply the contraction mapping theorem (Theorem 2.12) to conclude that for each $x \in I_r(0)$ there is a unique $y \in I_\delta(0)$ such that $h(x, y) = y$. This means $f(x, y) = 0$. Define $g(x) = y$. Then $g(x)$ is a unique function satisfying $g(0) = 0$ and (4.34), (4.35).

To show that $g(x)$ is continuous, let $x_1, x_2$ be any points in $I_r(0)$. Then

$$|g(x_1) - g(x_2)| = |h(x_1, g(x_1)) - h(x_2, g(x_2))|$$
$$\le |h(x_1, g(x_1)) - h(x_1, g(x_2))|$$
$$\quad + |h(x_1, g(x_2)) - h(x_2, g(x_2))|$$
$$\le \frac{1}{2}|g(x_1) - g(x_2)| + |h(x_1, g(x_2)) - h(x_2, g(x_2))|.$$

From this we conclude that

$$|g(x_1) - g(x_2)| \le 2|h(x_1, g(x_2)) - h(x_2, g(x_2))|.$$

Let $x_1 \to x_2$. Since $h(x, y)$ is continuous,

$$h(x_1, g(x_2)) \to h(x_2, g(x_2)).$$

This implies that $g(x_1) \to g(x_2)$, and the proof is complete. $\qquad\square$

We also have

**Theorem 4.18.** *If, in addition, $f \in C^1(\Omega, \mathbb{R})$, then $g \in C^1(I_r(x_0), \mathbb{R})$ and*

$$g'(x) = -f_y(x, g(x))^{-1} \cdot f_x(x, g(x)). \tag{4.37}$$

*Proof.* Again we assume that $x_0 = y_0 = 0$. Let $x$, $x + \xi \in I_r(0)$ and $\eta = g(x + \xi) - g(x)$. Then $\eta \to 0$ as $\xi \to 0$. Let $\varepsilon < |A|/4$ be given. Since $f \in C^1$, we have

$$|f(x+\xi, g(x + \xi)) - f(x, g(x)) - f_x(x, g(x))\xi - f_y(x, g(x))\eta| < \varepsilon(|\xi| + |\eta|)$$

for $|\xi|$ sufficiently small. Since

$$f_y(x, g(x)) \to A \ne 0 \quad as \ x \to 0,$$

we have

$$|f_y(x, g(x))| \ge \frac{|A|}{2}$$

for $x$ close to 0. Since $f(x + \xi, g(x + \xi)) = f(x, g(x)) = 0$, we have

$$|f_x(x, g(x))\xi + f_y(x, g(x))\xi| \le \varepsilon(|\xi| + |\eta|),$$

and consequently,

$$|f_y(x, g(x))^{-1} f_x(x, g(x))\xi + \eta| \le \frac{2\varepsilon}{|A|}(|\xi| + |\eta|).$$

Let $v = f_y(x, g(x))^{-1} f_x(x, g(x))\xi$. Then there is a constant $C$ such that

$$|v| \le C|\xi|.$$

Thus,

$$|\eta + v| \le \frac{2\varepsilon}{|A|}(|\xi| + |\eta + v| + |v|) \le \frac{1}{2}|\eta + v| + \frac{2\varepsilon}{|A|}(C + 1)|\xi|.$$

Consequently,

$$|\eta + v| \le \frac{4\varepsilon}{|A|}(C+1)|\xi|.$$

Since $\varepsilon$ was arbitrary, we see that

$$|\frac{\eta}{\xi} + \frac{v}{\xi}| \to 0 \ \text{ as } \ |\xi| \to 0.$$

But this is

$$\frac{g(x+\xi) - g(x)}{\xi} + f_y(x, g(x))^{-1}f_x(x, g(x)) \to 0 \ \text{ as } \ \xi \to 0.$$

Thus, $g'(x)$ exists and equals

$$g'(x) = f_y(x, g(x))^{-1}f_x(x, g(x)).$$

But the right-hand side is continuous in $x$. Thus, the same is true of $g'(x)$. This completes the proof.                                                □

We also have

**Corollary 4.19.** *If, in addition, $f \in C^k(\Omega, \mathbb{R})$, then $g \in C^k(I_r(x_0), \mathbb{R})$.*

*Proof.* We merely differentiate (4.37).                                □

Now we generalize Theorem 4.17 to a Hilbert space setting.

**Theorem 4.20.** *Let $X, Y, Z$ be Hilbert spaces and let $\Omega$ be an open set in $X \times Y$. Let $f$ be a continuous map from $\Omega$ to $Z$ such that there is a point $(x_0, y_0) \subset \Omega$ for which $f(x_0, y_0) = 0$. Assume that $f'$ exists and is continuous in $\Omega$ and $A = f_y(x_0, y_0)$ is invertible from $X \times Y$ to $Z$. Then there exists an $r > 0$ and a unique mapping $g(x)$ from $B_r(x_0) = \{x \in X : \|x - x_0\| < r\}$ to $Y$ such that*

*(a) $y_0 = g(x_0)$*
*(b) $f(x, g(x)) \equiv 0, \quad x \in B_r(x_0)$*
*(c) If $f \in C^1(\Omega, Z)$, then*

$$g'(x) = -f_y(x, g(x))^{-1}f_x(x, g(x)), \quad x \in B_r(x_0) \quad (4.38)$$

*(d) If $f \in C^k(\Omega, Z)$, then $g \in C^k(B_r(x_0), Y)$.*

*Proof.* We follow the proof of Theorem 4.17 replacing $\mathbb{R}^2$ with $X \times Y$ and absolute value signs with norms. The same is true with the proof of Theorem 4.18 and Corollary 4.19.                                □

## 4.9    Exercises

1. Why does

$$\inf_{w \in M} \sup_{v \in N} G(v, w) = \sup_{v \in N} G(v, w_0)$$

$$= G(v_0, w_0)$$

$$= \inf_{w \in M} G(v_0, w)$$

$$= \sup_{v \in N} \inf_{w \in M} G(v, w)$$

imply that

$$\min_{w \in M} \sup_{v \in N} G(v, w) = \sup_{v \in N} G(v, w_0)$$

$$= G(v_0, w_0)$$

$$= \inf_{w \in M} G(v_0, w)$$

$$= \max_{v \in N} \inf_{w \in M} G(v, w)?$$

2. Verify

$$\|\bar{u}_k\|^2 = \left[ \sum_{j=1}^{k} \|u_j\|^2 + 2 \sum_{j=1}^{k} \sum_{i=1}^{j} (u_i, u_j) \right] / k^2$$

$$\leq \left[ kC^2 + 2 \sum_{j=1}^{k} \sum_{i=1}^{j} \frac{1}{j} \right] / k^2$$

$$\leq (C^2 + 2)/k \to 0$$

when

$$|(u_{n_k}, u_{n_1})| \leq \frac{1}{k}, \dots, |(u_{n_k}, u_{n_{k-1}})| \leq \frac{1}{k}.$$

3. Verify (4.15).

4. If $G(v, w)$ satisfies the hypotheses of Theorem 4.8, show that

$$G_\varepsilon(v, w) = G(v, w) + \varepsilon \|w\|^2$$

satisfies the hypotheses of Theorem 4.8 and is also strictly convex with respect to $w$.

5. Verify that

$$\|(1-\theta)w_0 + \theta w_1\|^2 = (1-\theta)\|w_0\|^2 + \theta\|w_1\|^2 - \theta(1-\theta)\|w_0 - w_1\|^2.$$

6. Show that
$$G_\varepsilon(v, \bar{w}_\varepsilon) \leq G_\varepsilon(\bar{v}_\varepsilon, \bar{w}_\varepsilon) \leq G_\varepsilon(\bar{v}_\varepsilon, w), \quad v \in N, \ w \in M$$
implies
$$G(v, \bar{w}) \leq G(\bar{v}, w), \quad v \in N, \ w \in M$$
when $\varepsilon \to 0$, $\bar{v}_\varepsilon \to \bar{v}$, $\bar{w}_\varepsilon \to \bar{w}$.

7. Why is there a subsequence such that
$$G(\bar{v}_R, \bar{w}_R) \to \lambda \text{ as } R \to \infty$$
in the proof of Theorem 4.8?

8. Verify
$$G(v, \bar{w}) \leq \lambda \leq G(\bar{v}, w), \quad v \in N, \ w \in M.$$

9. Show that a functional $G(u)$ is strictly convex iff
$$G([(1-t)u_0 + tu_1]) < (1-t)G(u_0) + tG(u_1)$$
when $u_0 \neq u_1$ and $t \in (0, 1)$.

10. Show that
$$(a'_N(u_0) - S^*Tw_0, \tilde{v}) \leq 0, \quad \tilde{v} \in N,$$
and
$$(a'_M(u_0) - T^*Sv_0, \tilde{w}) \geq 0, \quad \tilde{w} \in M$$
imply (4.27).

11. Show that the set of bounded sequences $x = \{x_n\}$ such that $x_n \to 0$ as $n \to \infty$ with the norm
$$\|x\| = \max_n |x_n|,$$
is a Banach space.

12. Prove Corollary 4.19.

13. Carry out the proof of Theorem 4.20.

# 5

# Calculus of variations

## 5.1    Introduction

In the first three chapters we wanted to solve a specific problem. Our attack was to find a functional such that the vanishing of its derivative at a point is equivalent to providing a solution of the original problem. Many problems arise in which one is given a functional and one is searching for an extremum. Before we present the general theory we give some examples.

## 5.2    The force of gravity

We assume that a particle is travelling in the plane starting at the point $(a_0, b_0)$, $b_0 > 0$, and ending at the point $(0, 0)$. The only force that is to act on the particle is that of gravity in the negative $y$ direction. Assume that the particle is moving along a wire connecting the two points, but there is no friction. We would like to know if there is a fixed path that will minimize the time that it will take the particle to make its descent. If it exists, the path is known as the "brachistochrone."

If $m$ is the mass of the particle and $g$ is the acceleration due to gravity, then the particle satisfies

$$\frac{1}{2}mv^2 + mgy = mgb_0, \tag{5.1}$$

where $v$ is its velocity. If $t$ represents time, then

$$v(t)^2 = \left(\frac{dx(t)}{dt}\right)^2 + \left(\frac{dy(t)}{dt}\right)^2 = \left(\frac{ds(t)}{dt}\right)^2,$$

where $s(t)$ is the arc length starting from $t = 0$. Thus $ds/dt = v$. If we assume that the wire is such that the curve it makes can be expressed

145

as $x = f(y)$, then we have

$$T = \int_0^{b_0} \frac{(1 + f'(y)^2)^{\frac{1}{2}}}{(2g(b_0 - y))^{\frac{1}{2}}}\, dy = \int_0^{b_0} F(y, f'(y))\, dy, \qquad (5.2)$$

where $F(y, z)$ is given by

$$F(y, z) = \frac{(1 + z^2)^{\frac{1}{2}}}{(2g(b_0 - y))^{\frac{1}{2}}} \qquad (5.3)$$

and $T$ represents the total time taken for the particle to reach $(0, 0)$. We want $T$ for the given path to be $\leq$ the time taken if another path is chosen. In particular if $\varepsilon > 0$ and $\eta(y)$ is a smooth function in $[0, b_0]$ such that $\eta(0) = \eta(b_0) = 0$, then

$$x = f_\varepsilon(y) = f(y) + \varepsilon\eta(y)$$

is a competing path. We want $T \leq T_\varepsilon$, where

$$T_\varepsilon = \int_0^{b_0} F(y, f'_\varepsilon(y))dy.$$

In particular, we should have

$$\frac{dT_\varepsilon}{d\varepsilon} = 0$$

when $\varepsilon = 0$. But

$$\frac{dT_\varepsilon}{d\varepsilon} = \int_0^{b_0} \frac{\partial F}{\partial z}(y, f'_\varepsilon(y))\eta'(y)\, dt = -\int_0^{b_0} \frac{d}{dy}\frac{\partial F}{\partial z}(y, f'_\varepsilon(y))\eta(y)\, dy.$$

The reason we integrated by parts is because of

**Lemma 5.1.** *If $h(t)$ is a continuous function in the interval $[a, b]$ and*

$$\int_a^b h(t)\eta(t)dt = 0$$

*for all smooth functions $\eta(t)$ vanishing at $a$ and $b$, then $h(t) \equiv 0$ in $[a, b]$.*

We shall prove Lemma 5.1 later at the end of this section. Applying it to our case, we see that

$$\int_0^{b_0} \frac{d}{dy}\frac{\partial F}{\partial z}(y, f'(y))\eta(y)\, dy = 0$$

for every such $\eta(y)$. Hence, we have

$$\frac{d}{dy}\frac{\partial F}{\partial z}(y, f'(y)) \equiv 0, \quad y \in [0, b_0].$$

This means that

$$\frac{\partial F}{\partial z}(y, f'(y)) \equiv c$$

for some constant $c$. Now

$$\frac{\partial F}{\partial z}(y, z) = \frac{z}{[2g(b_0 - y)(1 + z^2)]^{1/2}}.$$

Hence,

$$\frac{f'(y)}{[2g(b_0 - y)(1 + f'(y)^2)]^{1/2}} \equiv c.$$

Thus,

$$f'(y)^2 = \frac{2gc^2(b_0 - y)}{1 - 2gc^2(b_0 - y)}.$$

Let

$$4gc^2(b_0 - y) = 1 - \cos\theta.$$

We can do this because

$$2gc^2(b_0 - y) \le 1$$

(in fact, it equals $f'(y)^2/[1 + f'(y)^2]$). Then

$$f'(y)^2 = \frac{\frac{1}{2}(1 - \cos\theta)}{1 - \frac{1}{2}(1 - \cos\theta)]} = \frac{1 - \cos\theta}{1 + \cos\theta}.$$

Consequently,

$$f(y) = \frac{-1}{4gc^2} \int \left[\frac{1 - \cos\theta}{1 + \cos\theta}\right]^{1/2} \sin\theta \, d\theta + \text{const.}$$

$$= \frac{1}{4gc^2} \int (\cos\theta - 1)d\theta + \text{const.}$$

$$= \frac{1}{4gc^2}(\sin\theta - \theta) + \text{const.}$$

Since $\theta = 0$ when $y = b_0$ and $f(b_0) = a_0$, we have

$$f(y) = \frac{1}{2gc^2}(\sin\theta - \theta) + a_0.$$

Since

$$\frac{dy}{d\theta} = -\frac{1}{4gc^2}\sin\theta,$$

we have

$$y = \frac{1}{4gc^2}\cos\theta + \text{const.}$$

Since $\theta = 0$ when $y = b_0$, this constant must be $b_0 - (1/4gc^2)$. Consequently, we have

$$y = \frac{1}{4gc^2}(\cos\theta - 1) + b_0.$$

From these equations, we see that there should be a number $\theta_0$ such that

$$4gc^2 b_0 = 1 - \cos\theta_0, \quad 4gc^2 a_0 = \theta_0 - \sin\theta_0.$$

We leave it as a simple exercise to show that we can always find such a $\theta_0$. The equations we have found describe a **cycloid**. Note that

$$c^2 = \frac{1 - \cos\theta_0}{4gb_0} = \frac{\theta_0 - \sin\theta_0}{4ga_0}.$$

It should be stressed that we have not provided a minimum. All we have shown is that **if a minimum function exists**, it is a cycloid.

The proof of Lemma 5.1 is simple.

*Proof.* Functions in $C_0^\infty(a,b)$ meet the requirements of $\eta$. Hence, Lemma 3.3 tells us that $h(t) = 0$ a.e. Since $h(t)$ is continuous, it must vanish everywhere in $[a,b]$. □

## 5.3    Hamilton's principle

This principle states that a dynamical system will move along a path which minimizes the integral with respect to time of the difference between the kinetic and potential energies of the system.

If $T, V$ represent the kinetic and potential energies of the system, respectively, then the principle states that the path followed by the system from $t = t_1$ to $t = t_2$ minimizes

$$\int_{t_1}^{t_2} (T - V)dt.$$

Consider the motion of a particle of mass $m$ along the $x$-axis. If the particle is subject to a force $f(x)$, then

$$T = \frac{1}{2}m\dot{x}^2, \quad V = -\int_{x_0}^{x} f(x)dx,$$

where

$$\dot{x} = \frac{dx}{dt}.$$

We want to minimize the integral

$$\int_{t_0}^{t_1} L(x, \dot{x}) \, dt,$$

where

$$L = T - V = \frac{1}{2}m\dot{x}^2 + \int_{x_0}^{x} f(x)dx$$

is called the **Lagrangian**. We shall use

**Lemma 5.2.** *If the integral*

$$J = \int_{a}^{b} F(x, \dot{x}, t)dt$$

*has a minimum at the function*

$$x = x(t), \quad a \leq t \leq b,$$

*then the function $x(t)$ is a solution of*

$$\frac{\partial F}{\partial x} - \frac{d}{dt}\left(\frac{\partial F}{\partial \dot{x}}\right) = 0. \tag{5.4}$$

This equation is known as **Euler's equation**. We shall prove Lemma 5.2 later (Corollary 5.8). Applying it to our situation, we look for solutions of

$$\frac{\partial L}{\partial x} - \frac{d}{dt}\left(\frac{\partial L}{\partial \dot{x}}\right) = 0.$$

This gives

$$f(x) - \frac{d}{dt}(m\dot{x}) = 0.$$

Hence, we are looking for solutions of

$$m\ddot{x} = f(x).$$

This is the usual Newtonian equation of motion.

As another example, consider the problem of a pendulum under the force of gravity. Let $\phi$ denote the angle between the pendulum and the downward direction, and let $\ell$ denote the length of the pendulum. In this case

$$T = \frac{1}{2}m(\ell\dot{\phi})^2, \quad V = mg\ell(1 - \cos\phi),$$

so that

$$L = \frac{1}{2}m\ell^2\dot{\phi}^2 - mg\ell(1 - \cos\phi).$$

In this case the Euler equation becomes

$$\frac{\partial L}{\partial \phi} - \frac{d}{dt}\left(\frac{\partial L}{\partial \dot{\phi}}\right) = 0.$$

From this we obtain

$$-mg\ell\sin\phi - \frac{d}{dt}(m\ell^2\dot{\phi}) = 0,$$

or

$$\ddot{\phi} + \frac{g}{\ell}\sin\phi = 0.$$

If we assume that the vertical position of the pendulum is always small, we can approximate this equation by

$$\ddot{\phi} = -\left(\frac{g}{\ell}\right)\phi.$$

This represents **simple harmonic motion** about $\phi = 0$. The general solution is

$$\phi = A\cos(\omega t) + B\sin(\omega t) = C\cos(\omega t + \alpha),$$

where $\omega^2 = (g/\ell)$ and $A$, $B$ (or $C$, $\alpha$) are arbitrary constants. Again, we must emphasize that we have not shown the existence of minima.

As another example, consider

$$J = \int_0^1 (1 + \dot{x}^2)^2\,dt.$$

Any curve $x = x(t)$ which minimizes $J$ must satisfy

$$4\dot{x}(1 + \dot{x}^2) = \text{const.}$$

The only solution of this is

$$\dot{x} = \text{const.},$$

or

$$x(t) = At + B,$$

where the constants $A$, $B$ are to be determined by the given conditions of the problem.

## 5.4    The Euler equations

Suppose $H(x, y, z)$ is a function on $\bar{\Omega} \times \mathbb{R} \times \mathbb{R}$, where $\Omega = (a, b)$, and we want to find a function $y(x) \in C^1(\bar{\Omega})$ satisfying

$$y(a) = a_1, \quad y(b) = b_1 \tag{5.5}$$

and minimizing the expression

$$J(y) = \int_\Omega H(x, y(x), y'(x))dx. \tag{5.6}$$

This is a typical situation dealt with in the calculus of variations. Here, one not only wants to know that a minimum exists, but one wants to find one of them (and hopefully it should be unique). The "brachistochrone" problem was discussed at the beginning of this chapter. It was posed by John Bernoulli in 1696. In it, one is given two points $(a, a_1)$, $(b, b_1)$ in a vertical plane such that $b_1 < a_1$. One wants to find a curve joining them so that a particle starting from rest at $(a, a_1)$ will traverse the curve (without friction) from $(a, a_1)$ to $(b, b_1)$ in the shortest possible time. One can reduce the problem to that of minimizing (5.6) under the conditions (5.5) when

$$H(x, y, z)^2 = \frac{c(1 + z^2)}{(a_1 - y)}. \tag{5.7}$$

In dealing with such problems one is again tempted to follow the logic used in elementary calculus which notes that the derivative vanishes at an extreme point. If we want to follow this line of reasoning, we are again confronted with the need to define a derivative for the expression (5.6). If $y(x) \in C^1(\bar{\Omega})$ satisfying (5.5) makes $J$ a minimum, then

$$J(y) \le J(z)$$

for every other $z(x) \in C^1(\bar{\Omega})$ satisfying (5.5). The difference $\eta(x) = z(x) - y(x)$ is in $C_0^1(\Omega)$. Thus, we want

$$J(y) \le J(y + \eta), \quad \eta \in C_0^1(\Omega). \tag{5.8}$$

For this purpose, we note that if

$$A(y, \eta) = \lim_{t \to 0} \frac{J(y + t\eta) - J(y)}{t} \tag{5.9}$$

exists for each $\eta \in C_0^1(\Omega)$ and $y \in C^1(\bar{\Omega})$ satisfying (5.5), then we must have

$$A(y, \eta) = 0, \quad \eta \in C_0^1(\Omega). \tag{5.10}$$

To see this, note that if $A(y, \eta) \neq 0$ for some $\eta \in C_0^1(\Omega)$, then

$$J(y + t\eta) - J(y) = tA(y, \eta) + o(t), \tag{5.11}$$

and the right-hand side can be made $< 0$ for $|t|$ sufficiently small by taking $tA(y, \eta) < 0$. Thus, if we want to follow this line of reasoning, we are looking for a $y(x) \in C^1(\bar{\Omega})$ satisfying (5.5) such that the limit (5.9) exists and vanishes for each $\eta \in C_0^1(\Omega)$.

You may be wondering why we did not suggest using the Fréchet derivative that we used earlier. There are several reason for this. (1) There, we were trying to solve an equation with boundary conditions. We wanted the derivative of the functional to coincide with the equation. (2) Here, we have no equation. We are trying to find a function which minimizes the functional under the given boundary conditions. (3) The function $y(x)$ is required to satisfy (5.5). It need not be in $C_0^1(\Omega)$. The derivative as defined earlier would require the limits to exist for all $\eta \in C^1(\bar{\Omega})$. (4) It will be shown later that the expression (5.9) is related to the Fréchet derivative we used earlier (Theorem 5.9).

Let us consider the limit (5.9) for the functional (5.6). We have

$$J(y + t\eta) - J(y) = \int_\Omega [H(x, y + t\eta, \ y' + t\eta') - H(x, y, y')]\, dx$$

$$= \int_\Omega \int_0^1 \frac{d}{d\theta} H(x, y + t\theta\eta, \ y' + t\theta\eta')\, d\theta\, dx$$

$$= t \int_\Omega \int_0^1 [H_y(x, y + t\theta\eta, \ y' + t\theta\eta')\eta$$

$$+ H_z(x, y + t\theta\eta, \ y' + t\theta\eta')\eta']\, d\theta\, dx,$$

provided the derivatives $H_y$, $H_z$ exist and are continuous. If we divide by $t$, then the limit will exist as $t \to 0$. What we need is

**Theorem 5.3.** *Assume that $H(x, y, z)$, $H_y(x, y, z)$, $H_z(x, y, z)$ are continuous in $\bar{\Omega} \times \mathbb{R} \times \mathbb{R}$ and satisfy*

$$|H(x, y, z)| + |H_y(x, y, z)| + |H_z(x, y, z)| \tag{5.12}$$

$$\leq C(|y|^q + |z|^p + 1), \quad x \in \bar{\Omega}, \ y, z \in \mathbb{R}$$

*for some $p, q < \infty$. Then the limit (5.9) exists and equals*

$$A(y, \eta) = \int_\Omega [H_y(x, y, y')\eta + H_z(x, y, y')\eta']\, dx, \quad y \in C^1(\bar{\Omega}), \ \eta \in C_0^1(\Omega). \tag{5.13}$$

*Proof.* We note that the right-hand side of

$$\frac{[J(y + t\eta) - J(y)]}{t}$$

$$= \int_\Omega \int_0^1 [H_y(x, y + t\theta\eta, y' + t\theta\eta')\eta + H_z(x, y + t\theta\eta, y' + t\theta\eta')\eta'] d\theta\, dx$$

converges to the right-hand side of (5.13) as $t \to 0$. Clearly, the integrand converges to the integrand of (5.13) a.e. (Since $\theta$ disappeared from the integrand in (5.13), the integral with respect to $\theta$ can be suppressed.) Moreover, it is majorized by

$$C(|y|^q + |\eta|^q + |y'|^p + |\eta'|^p + 1)$$

when $|t| \leq 1$. Since $y, \eta \in C^1(\bar\Omega)$, this is in $L^1(\Omega)$. Hence, the integral converges (Theorem B.18), and the proof is complete. □

Thus we have

**Corollary 5.4.** *Under the hypotheses of Theorem 5.3, if $y(x) \in C^1(\bar\Omega)$ satisfies (5.5) and minimizes $J(y)$ given by (5.6), then it must satisfy (5.10).*

Now our path leads us to search for functions $y(x) \in C^1(\bar\Omega)$ which satisfy (5.10). This is not an easy task. It would be much easier if $H_z(x, y, z) \equiv 0$. For then (5.10) would reduce to

$$\int_\Omega H_y(x, y, y')\eta\, dx = 0, \quad \eta \in C_0^1(\Omega).$$

But we know that $C_0^\infty(\Omega)$ is dense in $L^2(\Omega)$, and this would lead to

$$H_y(x, y, y') = 0.$$

This is an equation which may be solvable for $y$ in terms of $x$. However, the restriction $H_z(x, y, z) \equiv 0$ is very severe. It says that $H(x, y, z)$ does not really depend on $z$, a situation which rarely comes up in practice. We then need only solve $H_y(x, y) = 0$ for $y$ in terms of $x$. This can be done, for instance, if $H_{yy}(x, y) \neq 0$.

However, in the general case we have a more complicated situation. One way of attacking the problem is to try to get (5.10) to reduce to something of the form

$$\int_\Omega h(x)\eta'dx = 0, \quad \eta \in C_0^1(\Omega),$$

then we could conclude that $h(x)$ is a constant by means of the following lemma.

**Lemma 5.5.** *If* $h(x) \in L^p(\Omega)$, $1 \le p < \infty$, *and*

$$\int_\Omega h(x)\eta' dx = 0, \quad \eta \in C_0^1(\Omega), \tag{5.14}$$

*then* $h(x) \equiv$ *constant a.e.*

*Proof.* Let

$$\psi_\varepsilon(x) = J_\varepsilon h(x) = \int j_\varepsilon(x-y) h(y)\, dy.$$

If $\varepsilon < d(x, \partial\Omega) = \min\{|x-a|, |x-b|\}$, then $j_\varepsilon(x-y) \in C_0^\infty(\Omega)$, and consequently

$$\psi'_\varepsilon(x) = \int \frac{\partial}{\partial x} j_\varepsilon(x-y) h(y)\, dy = -\int \frac{\partial}{\partial y} j_\varepsilon(x-y) h(y)\, dy = 0.$$

Thus, $\psi_\varepsilon(x)$ is a constant $c_\varepsilon$ in $[a+\varepsilon, b-\varepsilon]$ for each $\varepsilon > 0$. On the other hand, $\psi_\varepsilon(x)$ converges to $h(x)$ in $L^p(\Omega)$ as $\varepsilon \to 0$ (c.f. (3.6)). This implies that there is a sequence $\{\varepsilon_k\}$ such that $\varepsilon_k \to 0$ and $\psi_{\varepsilon_k}(x) \to h(x)$ a.e. (Theorem B.25). Thus, $c_k = \psi_{\varepsilon_k}(x) \to h(x)$ a.e. as $k \to \infty$. This shows that $h(x) \equiv$ constant a.e. The proof is complete. $\qquad\square$

This leads to

**Theorem 5.6.** *Under the hypotheses of Theorem 5.3, if* $y \in C^1(\bar\Omega)$ *satisfies (5.10), then*

$$H_z(x, y(x), y'(x)) - \int_a^x H_y(t, y(t), y'(t))\, dt \equiv \text{constant}, \quad x \in \bar\Omega. \tag{5.15}$$

*Proof.* Let $h(x)$ denote the left-hand side of (5.15). Then

$$\int_\Omega h(x)\eta' dx = \int_\Omega H_z(x, y, y')\eta' dx - \int_\Omega \eta' \int_a^x H_y(t, y(t), y'(t))\, dt\, dx$$

$$= \int_\Omega [H_z(x, y, y')\eta' + H_y(x, y, y')\eta]\, dx$$

$$= A(y, \eta), \quad \eta \in C_0^1(\Omega)$$

by integration by parts. Hence, $h(x)$ satisfies (5.14), and we can conclude via Lemma 5.5 that $h(x) \equiv$ constant. $\qquad\square$

We also have

**Corollary 5.7.** *Under the same hypotheses,* $H_z(x, y(x), y'(x))$ *is differentiable on* $\bar{\Omega}$ *and satisfies*

$$H_y(x, y(x), y'(x)) - \frac{d}{dx} H_z(x, y(x), y'(x)) = 0, \quad x \in \bar{\Omega}. \tag{5.16}$$

*Proof.* Apply Theorem B.36. □

Combining these results, we have

**Corollary 5.8.** *Under the hypotheses of Theorem 5.3, if* $y(x) \in C^1(\bar{\Omega})$ *minimizes the functional (5.6) under the conditions (5.5), then it must satisfy (5.15) and (5.16).*

## 5.5 The Gâteaux derivative

Let $G(u)$ be a functional on a Hilbert space $H$. According to our definition, the Fréchet derivative of $G$ at an element $u \in H$ exists if there is an element $g \in H$ such that

$$\frac{[G(u + v) - G(u) - (v, g)_H]}{\|v\|_H} \to 0 \text{ as } \|v\|_H \to 0, \quad v \in H. \tag{5.17}$$

We defined $G'(u)$ to be $g$. If there is a symmetric bilinear from $\langle v, w \rangle$ satisfying (1.18), then

$$(v, g)_H = \langle v, w \rangle, \quad v \in H.$$

This means that

$$(v, G'(u))_H = \langle v, w \rangle, \quad v \in H. \tag{5.18}$$

There is another definition of derivative which is sometimes more convenient to use. Recall that we had great difficulty defining the derivative because of multiplication and division in a vector space. Another approach is to consider the limit

$$\lim_{t \to 0} \frac{G(u + tv) - G(u)}{t}.$$

We say that $G$ is Gâteaux differentiable at a point $u \in H$ if this limit exists for each $v \in H$ and satisfies

$$\lim_{t \to 0} \frac{G(u + tv) - G(u)}{t} = (v, h)_H, \quad v \in H \tag{5.19}$$

for some element $h \in H$. We call $h$ the Gâteaux derivative of $G$ at $u$ and denote it (temporarily) by $G'_1(u)$. There is an important relationship between the two derivatives. We have

**Theorem 5.9.** *If the Gâteaux derivative of $G$ exists in the neighborhood of a point $u \in H$ and is continuous at $u$, then the Fréchet derivative of $G$ exists at $u$ and equals the Gâteau derivative at $u$.*

*Proof.* For $v \in H$, let

$$\varphi(t) = G(u + tv), \quad t \in [0, 1].$$

Let $\varepsilon > 0$ be given. If the Gâteaux derivative of $G$ exists in the neighborhood of a point $u \in H$ and is continuous at $u$, then there is a $\delta > 0$ such that $\varphi'(t)$ exists and satisfies

$$|\varphi'(t) - \varphi'(0)| < \varepsilon$$

for $t \in I = [0, 1]$ when $\|v\| < \delta$. Thus, $\varphi'(t)$ is bounded on $I$. Hence, $\varphi'(t)$ is Lebesgue integrable on $I$ and satisfies

$$\varphi(1) - \varphi(0) = \int_0^1 \varphi'(s)ds$$

(Theorem B.20). Consequently,

$$|G(u + v) - G(u) - (v, G_1'(u))_H| = \left| \int_0^1 (v, G_1'(u + tv) - G_1'(u))dt \right|$$

$$\leq \|v\|_H \int_0^1 \|G_1'(u + tv) - G_1'(u)\| dt.$$

Thus,

$$\frac{[G(u + v) - G(u) - (v, G_1'(u))_H]}{\|v\|_H} \to 0, \quad v \in H.$$

This shows that $G'(u)$ exists and equals $G_1'(u)$. $\qquad\square$

**Remark 5.10.** *When $G_1'(u)$ equals $G'(u)$, we shall designate both of them by $G'(u)$.*

### 5.6    Independent variables

In some situations we are required to minimize expressions such as

$$J = \int_a^b F(x_1, x_2, \ldots, x_n, \dot{x}_1, \dot{x}_2, \ldots, \dot{x}_n, t)dt, \qquad (5.20)$$

where

$$x_i = x_i(t), \quad i = 1, 2, \ldots, n$$

are independent functions. Corresponding to Lemma 5.2 we have

**Lemma 5.11.** *If the functions* $x_1(t), x_2(t), \ldots, x_n(t)$ *minimize the functional (5.20), then they satisfy*

$$\frac{\partial F}{\partial x_i} - \frac{d}{dt}\left(\frac{\partial F}{\partial \dot{x}_i}\right) = 0, \quad i = 1, 2, \ldots, n. \tag{5.21}$$

The proof of Lemma 5.11 is similar to that of Lemma 5.2 and is omitted. As an example, consider a particle of mass $m$ moving in the plane under an attractive force $mk/r^2$ towards the origin, where $(r, \phi)$ are the polar coordinates. We apply Hamilton's principle. In this case the potential energy is

$$V = -\frac{mk}{r},$$

and the kinetic energy is

$$T = \frac{1}{2}m(\dot{r}^2 + r^2\dot{\phi}).$$

Hence,

$$L = T - V = \frac{1}{2}m(\dot{r}^2 + r^2\dot{\phi}) + mk/r.$$

By Lemma 5.11, curves $r(t), \phi(t)$ which minimize

$$\int_{t_1}^{t_2} L dt$$

must satisfy

$$\frac{\partial L}{\partial r} - \frac{d}{dt}\left(\frac{\partial L}{\partial \dot{r}}\right) = 0$$

and

$$\frac{\partial L}{\partial \phi} - \frac{d}{dt}\left(\frac{\partial L}{\partial \dot{\phi}}\right) = 0.$$

Thus we must have

$$-\frac{mk}{r^2} + mr\dot{\phi}^2 - \frac{d}{dt}(m\dot{r}) = 0$$

and

$$mr^2\dot{\phi} = \frac{\partial L}{\partial \dot{\phi}} = \text{const.} = M.$$

Eliminating $\dot{\phi}$ from these equations, we obtain

$$\ddot{r} - M^2/m^2r^3 = -k/r^2,$$

which is the equation of motion of such a particle. In particular, this is the equation of motion of a planet as it travels around the Sun.

## 5.7 A useful lemma

If the integrand in (5.6) does not contain $x$ explicitly (i.e., it is only to be found in $y(x)$ and $y'(x)$), then it is often useful to utilize the following

**Lemma 5.12.** *If $y = y(x)$ is a critical point of*

$$J = \int_a^b F(y, y')dx, \tag{5.22}$$

*then it is a solution of*

$$y'\frac{\partial F}{\partial y'} - F = \text{constant}. \tag{5.23}$$

*Proof.* We know that it must satisfy the Euler equation

$$\frac{\partial F}{\partial y} - \frac{d}{dx}\left(\frac{\partial F}{\partial y'}\right) = 0$$

(Corollary 5.8). Hence, we must have

$$\frac{d}{dx}\left(y'\frac{\partial F}{\partial y'}\right) = y'\frac{d}{dx}\left(\frac{\partial F}{\partial y'}\right) + y''\frac{\partial F}{\partial y'} = y'\frac{\partial F}{\partial y} + y''\frac{\partial F}{\partial y'}$$

and

$$\frac{d}{dx}F = \frac{\partial F}{\partial y}y' + \frac{\partial F}{\partial y'}y''.$$

Subtracting, we obtain

$$\frac{d}{dx}\left(F - y'\frac{\partial F}{\partial y'}\right) = 0,$$

which is precisely the conclusion of the lemma. $\qquad\square$

As an example, let us find a curve that is a critical point of

$$J = \int_a^b y(1 + y'^2)^{1/2}dx.$$

In this case $F = y(1 + y'^2)^{1/2}$, and we can apply Lemma 5.12. Thus, we are looking for a solution of (5.23). Hence,

$$yy'^2(1 + y'^2)^{-1/2} - y(1 + y'^2)^{1/2} = \text{constant} = K.$$

This implies

$$y^2 = K^2(1 + y'^2),$$

or

$$y' = (y^2 - K^2)^{1/2}/K.$$

Thus,

$$\int \frac{dy}{(y^2 - K^2)^{1/2}} = \frac{1}{K}\left(\int dx + C\right).$$

Integrating, we find

$$\cosh^{-1}(y/K) = (x + C)/K,$$

or

$$y = K \cosh\left(\frac{x + C}{K}\right).$$

The constants $C, K$ are to be determined from the boundary conditions.

## 5.8    Sufficient conditions

Our next step is to try to find a minimum for $J(y)$ under the conditions (5.5). Of course, we must make sure that $J(y)$ is bounded from below on this set. A fairly simple condition which will accomplish this is

$$H(x, y, z) \geq \rho|z|^p - W(x), \quad x \in \bar{\Omega}, \ y, z \in \mathbb{R}, \qquad (5.24)$$

where $\rho > 0$, $p > 1$, and $W(x) \in L^1(\Omega)$. In this case, $J(y)$ is clearly bounded from below. So now we can try to find a minimum of $J(y)$ on

$$D = \{y(x) \in C^1(\bar{\Omega}) : y(a) = a_1, \ y(b) = b_1\}.$$

Let

$$\alpha = \inf_D J.$$

It is clear that $\alpha > -\infty$. Let $\{y_n(x)\} \subset D$ be a minimizing sequence:

$$J(y_n) \searrow \alpha.$$

Since

$$\rho \int_\Omega |y_n'|^p dx - \int_\Omega W(x)dx \leq J(y_n) \leq C,$$

we see that there is a constant $B$ such that

$$|y_n'|_p \leq B.$$

Since

$$|y_n(x) - y_n(x')| = \left|\int_{x'}^x y_n'(t)dt\right| \leq \int_{x'}^x |y_n'(t)|dt, \quad a \leq x' \leq x \leq b,$$

we have

$$|y_n(x) - y_n(x')| \le |x - x'|^{1/p'} \left( \int_{x'}^{x} |y_n'(t)|^p dt \right)^{1/p}$$

$$\le |x - x'|^{1/p'} B, \quad a \le x' < x \le b,$$

where $p' = p/(p-1)$. In particular, we have

$$|y_n(x) - a_1| \le |x - a|^{1/p'} B, \quad |y_n(x) - b_1| \le |x - b|^{1/p'} B.$$

These last inequalities show that the sequence $\{y_n(x)\}$ is uniformly bounded and equicontinuous on $\bar{\Omega}$. By the Arzelà–Ascoli theorem (Theorem C.6), there is a renamed subsequence that converges uniformly on $\bar{\Omega}$ to a continuous function $\bar{y}(x)$. In the limit we have

$$|\bar{y}(x) - \bar{y}(x')| \le |x - x'|^{1/p'} B, \quad a \le x' < x \le b,$$

and

$$|\bar{y}(x) - a_1| \le |x - a|^{1/p'} B, \quad |\bar{y}(x) - b_1| \le |x - b|^{1/p'} B, \quad x \in \bar{\Omega}.$$

In particular, we have

$$\bar{y}(a) = a_1, \quad \bar{y}(b) = b_1.$$

Since the $L^p(\Omega)$ norms of the $y_n'$ are uniformly bounded, there is a renamed subsequence which converges weakly in $L^p(\Omega)$ to a function $h(x) \in L^p(\Omega)$ (Theorems B.23 and A.61). This means that

$$\int_{\Omega} [y_n'(x) - h(x)]v(x)\, dx \to 0 \quad \text{as } n \to \infty, \quad v \in L^{p'}(\Omega).$$

Therefore, if $\eta \in C_0^1(\Omega)$, then

$$(\bar{y}, \eta') \leftarrow (y_n, \eta') = -(y_n', \eta) \to -(h, \eta).$$

Thus, $h$ is the weak $L^p(\Omega)$ derivative of $\bar{y}$. We shall denote it by $\bar{y}'(x)$.

Now we come to our final assumption. We assume that $H_{zz}(x, y, z)$ exists and is continuous and satisfies

$$H_{zz}(x, y, z) > 0, \quad x \in \Omega, \ y, z \in \mathbb{R}. \tag{5.25}$$

In particular, $H_z(x, y, z)$ is an increasing function of $z$ for each $x$ and $y$. We now proceed to show that the minimum exists.

For fixed $M$, let

$$\Omega_M = \{x \in \bar{\Omega} : |\bar{y}'(x)| \le M\}.$$

Let $\varepsilon > 0$ be given. Since $y_n(x) \to \bar{y}(x)$ uniformly, there is an integer $N$ such that

$$|H(x, y_n, \bar{y}') - H(x, \bar{y}, \bar{y}')| < \varepsilon, \quad |H_z(x, y_n, \bar{y}') - H_z(x, \bar{y}, \bar{y}')| < \varepsilon,$$
$$x \in \Omega_M, \ n > N.$$

Thus, for $x \in \Omega_M$, we have

$$
\begin{aligned}
H(x, y_n, y_n') - H(x, \bar{y}, \bar{y}') =& H(x, y_n, y_n') - H(x, y_n, \bar{y}') \\
&+ H(x, y_n, \bar{y}') - H(x, \bar{y}, \bar{y}') \\
\geq& (y_n' - \bar{y}')H_z(x, y_n, \bar{y}') - \varepsilon \\
\geq& (y_n' - \bar{y}')[H_z(x, y_n, \bar{y}') - H_z(x, \bar{y}, \bar{y}')] \\
&+ (y_n' - \bar{y}')H_z(x, \bar{y}, \bar{y}') - \varepsilon \\
\geq& (y_n' - \bar{y}')H_z(x, \bar{y}, \bar{y}') - \varepsilon(|y_n'| + |\bar{y}'| + 1)
\end{aligned}
$$

for $n > N$. Here we used the fact that

$$H(x, y_n, y_n') - H(x, y_n, \bar{y}') = (y_n' - \bar{y}')H_z(x, y_n, \xi),$$

where $\xi(x)$ lies between $y_n'(x)$ and $\bar{y}'(x)$. Since $H_z(x, y, z)$ is an increasing function of $z$ for each $x$ and $y$, we have

$$H_z(x, y_n, \xi) \geq H_z(x, y_n, \bar{y}')$$

when $y_n'(x) \geq \bar{y}'(x)$ and

$$H_z(x, y_n, \xi) \leq H_z(x, y_n, \bar{y}')$$

when $y_n'(x) \leq \bar{y}'(x)$. Hence,

$$H(x, y_n, y_n') - H(x, y_n, \bar{y}') \geq (y_n' - \bar{y}')H_z(x, y_n, \bar{y}').$$

The function

$$
g(x) = \begin{cases} H_z(x, \bar{y}, \bar{y}'), & x \in \Omega_M, \\ 0, & x \in \Omega \backslash \Omega_M, \end{cases}
$$

is bounded in $\Omega$. Consequently,

$$\int_{\Omega_M} (y_n'(x) - \bar{y}'(x))H_z(x, \bar{y}, \bar{y}')dx = \int_{\Omega} (y_n' - \bar{y}')g(x)dx \to 0$$

as $n \to \infty$. On the other hand,

$$\int_{\Omega_M} (|y_n'| + |\bar{y}'| + 1)dx \leq \left(\int_{\Omega_M} |y_n'|^p dx\right)^{1/p}(b-a)^{1/p'} + M(b-a) + (b-a)$$
$$\leq B(b-a)^{1/p'} + (M+1)(b-a) = B_M.$$

Thus

$$\int_{\Omega_M} [H(x, y_n, y_n') - H(x, \bar{y}, \bar{y}')] \, dx \geq \int_{\Omega_M} (y_n' - \bar{y}') H_z(x, \bar{y}, \bar{y}') \, dx - \varepsilon B_M,$$

for $n > N$. Thus for $n$ sufficiently large

$$J(y_n) \geq \int_{\Omega_M} H(x, \bar{y}, \bar{y}') \, dx - \varepsilon(B_M + 1).$$

Letting $n \to \infty$, we find

$$\alpha \geq \int_{\Omega_M} H(x, \bar{y}, \bar{y}') \, dx - \varepsilon(B_M + 1),$$

and since $\varepsilon$ was arbitrary, we see that

$$\int_{\Omega_M} H(x, \bar{y}, \bar{y}') \, dx \leq \alpha.$$

This is true for each $M$. Consequently,

$$J(\bar{y}) \leq \alpha.$$

This does not yet produce a minimum. We do not yet know if $\bar{y} \in D$. We know that it is continuous on $\bar{\Omega}$ and satisfies (5.5). It also has a weak derivative in $L^p(\Omega)$. But in order for $\bar{y}$ to be in $D$, it must have a continuous derivative.

In this regard, let us try to determine if $\bar{y}$ has any further regularity properties. Our first step in this direction is to show that $\bar{y}(x)$ is absolutely continuous on $\bar{\Omega}$ (see Appendix B). Recall that we have shown that for the minimizing sequence

$$|y_n(x) - y_n(x')| \leq |x - x'|^{1/p'} \left( \int_{x'}^{x} |y_n'(t)|^p dt \right)^{1/p}, \quad a \leq x' < x \leq b.$$

Let $[x_k', x_k] \subset \bar{\Omega}$, $1 \leq k \leq m$, be non-overlapping intervals. Then by Hölder's inequality (Theorem B.23),

$$\sum_{k=1}^{m} |y_n(x_k) - y_n(x_k')| \leq \sum_{k=1}^{m} |x_k - x_k'|^{1/p'} \left( \int_{x_k'}^{x_k} |y_n'(t)|^p dt \right)^{1/p}$$

$$\leq \left( \sum_{k=1}^{m} |x_k - x_k'| \right)^{1/p'} \left( \sum_{k=1}^{m} \int_{x_k'}^{x_k} |y_n'(t)|^p dt \right)^{1/p}$$

$$\leq \left( \sum_{k=1}^{m} |x_k - x_k'| \right)^{1/p'} B.$$

Taking the limit as $n \to \infty$, we find

$$\sum_{k=1}^{m} |\bar{y}(x_k) - \bar{y}(x_k')| \le \left( \sum_{k=1}^{m} |x_k - x_k'| \right)^{1/p'} B,$$

which shows that $\bar{y}(x)$ is absolutely continuous on $\Omega$. In particular, the true derivative $\bar{y}'(x)$ of $\bar{y}(x)$ exists a.e., is in $L^1(\Omega)$, and satisfies

$$\bar{y}(x) - a_1 = \int_a^x \bar{y}'(t)\, dt, \quad x \in \bar{\Omega} \tag{5.26}$$

(Theorem B.36). (Until now, $\bar{y}'(x)$ was only known to be the weak $L^p$ derivative of $\bar{y}(x)$.) Moreover,

$$|y_n(x+h) - y_n(x)| \le h^{1/p'} \left( \int_x^{x+h} |y_n'(t)|^p dt \right)^{1/p}, \quad a \le x < x + h \le b.$$

Consequently,

$$\left| \frac{y_n(x+h) - y_n(x)}{h} \right|^p \le \frac{1}{h} \int_x^{x+h} |y_n'(t)|^p dt = \frac{1}{h} \int_0^h |y_n'(x+s)|^p\, ds.$$

This implies

$$\int_a^{b-h} \left| \frac{y_n(x+h) - y_n(x)}{h} \right|^p dx \le \frac{1}{h} \int_a^{b-h} dx \int_0^h |y_n'(x+s)|^p ds$$

$$\le \frac{1}{h} \int_0^h ds \int_a^{b-h} |y_n'(x+s)|^p\, dx$$

$$\le B^p.$$

Since $y_n(x) \to \bar{y}(x)$ uniformly, we obtain

$$\int_a^{b-h} \left| \frac{\bar{y}(x+h) - \bar{y}(x)}{h} \right|^p dx \le B^p.$$

The integrand converges to $|\bar{y}'(x)|^p$ a.e. as $h \to 0$, showing that

$$\int_a^b |\bar{y}'(x)|^p dx \le B^p.$$

Hence, $\bar{y}$ is differentiable a.e. and its true derivative $\bar{y}'$ is in $L^p(\Omega)$.

To recapitulate, we have shown that $\bar{y}(x)$ is absolutely continuous in $\bar{\Omega}$ and that its derivative is in $L^p(\Omega)$. This is not enough to declare that we have a minimum. So we try again. We replace $D$ with the set

$$D' = \{y \in C(\bar{\Omega}) : y(a) = a_1, \ y(b) = b_1,$$

$$y(x) \text{ is absolutely continuous in } \bar{\Omega} \text{ and } y' \in L^p(\Omega).\}$$

Note that $\bar{y} \in D'$. We now seek to find a minimum for $J$ on $D'$. "But then we shall not have solved our problem," you object. "We were looking for a minimum in $C^1(\bar{\Omega})$." You have a point, but be patient. We let

$$\alpha' = \inf_{D'} J.$$

If we now retrace our steps, we will discover that all of the proofs we have given go through even if we replace $y \in C^1(\bar{\Omega})$ with $y \in D'$ (we require the constant $p$ appearing in (5.12) to be the same as the $p$ appearing in (5.24)). In particular, there is a $\bar{y} \in D'$ such that $J(\bar{y}) = \alpha'$, and a minimum is attained on $D'$. Moreover, the limit (5.9) exists and equals (5.13) for $y \in D'$. Thus, if $J(\bar{y})$ is a minimum on $D'$, then $A(\bar{y}, \eta) = 0$ for all $\eta \in C_0^1(\Omega)$. Consequently, $\bar{y}$ satisfies

$$H_z(x, \bar{y}(x), \bar{y}'(x)) - \int_a^x H_y(t, \bar{y}(t), \bar{y}'(t)) \, dt \equiv C_0, \qquad x \in \bar{\Omega} \text{ a.e.}$$

for some constant $C_0$ (Theorem 5.6). However, in view of hypothesis (5.25), for each $x \in \bar{\Omega}$ there is exactly one value $z(x)$ satisfying

$$H_z(x, \bar{y}(x), z(x)) - \int_a^x H_y(t, \bar{y}(t), \bar{y}'(t)) dt = C_0.$$

Moreover, Theorem 4.17 tells us that the function $z(x)$ is continuous in $\bar{\Omega}$. Thus, $\bar{y}'(x)$ is a.e. equal to the continuous function $z(x)$. Consequently,

$$\bar{y}(x) = \int_a^x \bar{y}'(t) \, dx = \int_a^x z(t) \, dt,$$

showing that $\bar{y}(x)$ has a continuous derivative in $\bar{\Omega}$. This means that $\bar{y} \in C^1(\bar{\Omega})$. Hence

$$J(\bar{y}) = \min_{D'} J = \min_D J.$$

To summarize, we have

**Theorem 5.13.** *Assume that* $H, H_y, H_z, H_{zz}$ *are continuous on* $\bar{\Omega} \times \mathbb{R} \times \mathbb{R}$ *and satisfy (5.12) for* $1 < p, q < \infty$*. Assume also that (5.24) and (5.25) hold (with the same* $p$*). Then the functional* $J(y)$ *given by (5.6)* *has a minimum on the set D consisting of functions in $C^1(\bar{\Omega})$ satisfying (5.5). Any minimizing function is a solution of (5.15) and (5.16).*

We also note

**Remark 5.14.** *All of the examples mentioned so far in this chapter meet the requirements of Theorem 5.13. Hence, in each case a minimum was attained.*

## 5.9    Examples

Since all of the applications described earlier meet the requirements of Theorem 5.13, all of the solutions of the Euler equations which we obtained actually produced minima for their corresponding functionals. We now discuss some other examples, some of which do not qualify and do not produce extrema.

**Example 5.15.** *Consider the problem to minimize*

$$J(y) = \int_0^1 \dot{y}^2 dx,$$

*under the conditions*

$$y(0) = y(1) = 0.$$

*The solution of the Euler equation satisfying the boundary conditions exists, is unique, and yields an absolute minimum.*

*Proof.* The Euler equation is $\ddot{y}(x) = 0$. The solution that satisfies the boundary conditions is $y(x) \equiv 0$. It provides an absolute minimum.    □

**Example 5.16.** *Consider the problem to minimize*

$$J(y) = \int_0^1 x^{2/3} \dot{y}^2 dx,$$

*under the conditions*

$$y(0) = y(1) = 0.$$

*The solution of the Euler equation exists and is unique, but it is not in $C^1(\bar{\Omega})$. Although the functional is bounded from below, there is no minimum.*

*Proof.* The Euler equation is

$$2\frac{d}{dx}(x^{2/3}\dot{y}(x)) = 0.$$

The unique solution that satisfies the boundary conditions is

$$y_0(x) = x^{1/3}.$$

This is not in $C^1(\bar{\Omega})$. However,

$$J(y_0) = \inf J(y),$$

where the infimum is taken over the class of $y \in C^1(\bar{\Omega})$ satisfying the boundary conditions. $\square$

**Example 5.17.** *Consider the problem to minimize*

$$J(y) = \int_0^T [\dot{y}^2 - y^2]\, dx,$$

*under the conditions*

$$y(0) = y(T) = 0.$$

We consider two cases.

Case 1. $T \leq \pi$. First we note that

$$\int_0^T [\dot{y} - y \cot x]^2\, dx = \int_0^T [\dot{y}^2 + y^2 \cot^2 x - 2y\dot{y} \cot x]\,dx$$

$$= \int_0^T [\dot{y}^2 + y^2(\cot^2 x - \csc^2 x)]\, dx$$

$$= \int_0^T [\dot{y}^2 - y^2]\,dx.$$

Note that $y(x) \cot x \in C^1(\bar{\Omega})$ if $y$ satisfies the boundary conditions. For, we have

$$\lim_{x \to 0} \frac{y(x)}{\sin x} = \lim_{x \to 0} \frac{y'(x)}{\cos x} = y'(0).$$

Thus $J(y) \geq 0$ for $y \in C^1(\bar{\Omega})$ satisfying the boundary conditions. If $T < \pi$, the only solution of the Euler equation satisfying the boundary conditions is $y(x) \equiv 0$. If $T = \pi$, all such solutions are of the form $y(x) = c \sin x$, for which $J(y) = 0$.

Case 2. $T > \pi$. Let

$$y_k(x) = k \sin(\pi x/T), \quad k = 1, 2, \ldots$$

Then

$$J(y_k) = Tk^2(\pi^2/T^2 - 1)/2 \to -\infty \quad \text{as } k \to \infty.$$

Thus, $J(y)$ is not bounded from below in this case. Hence, we have

1. If $T < \pi$, then there is a unique solution of the Euler equation satisfying the boundary conditions and producing a minimum.
2. If $T = \pi$, then there is an infinite number of such solutions producing a minimum.
3. If $T > \pi$, there is no minimum.

**Example 5.18.** *Consider the problem to minimize*

$$J(y) = \int_0^1 x^2 \dot{y}^2 \, dx,$$

*under the conditions*

$$y(0) = 0, \ y(1) = 1.$$

Clearly, $J(y) \geq 0$. If we take $y_k(x) = x^{1/k}$, then we have

$$J(y_k) \to 0 \quad \text{as } k \to \infty.$$

Since $J(y) > 0$ when $y(x) \not\equiv 0$, there is no minimum.

## 5.10    Exercises

1. In the brachistochrone problem, show that if we assume that the wire is such that the curve it makes can be expressed as $x = f(y)$, then we have

$$T = \int_0^{b_0} \frac{(1 + f'(y)^2)^{1/2}}{(2g(b_0 - y))^{1/2}} \, dy, \qquad (5.27)$$

where $T$ represents the total time taken for the particle to reach $(0,0)$.

2. If

$$f'(y)^2 = \frac{\frac{1}{2}(1 - \cos\theta)}{1 - \frac{1}{2}(1 - \cos\theta)]} = \frac{1 - \cos\theta}{1 + \cos\theta}$$

and

$$4gc^2(b_0 - y) = 1 - \cos\theta,$$

show that

$$f(y) = \frac{-1}{4gc^2} \int \left[ \frac{1 - \cos\theta}{1 + \cos\theta} \right]^{1/2} \sin\theta \, d\theta + \text{const.}$$

$$= \frac{1}{4gc^2} \int (\cos\theta - 1) d\theta + \text{const.}$$

$$= \frac{1}{4gc^2} (\sin\theta - \theta) + \text{const.}$$

3. Show that there is a number $\theta_0$ such that

$$4gc^2 b_0 = 1 - \cos\theta_0, \quad 4gc^2 a_0 = \theta_0 - \sin\theta_0.$$

4. Show that if

$$L = \frac{1}{2} m\ell^2 \dot\phi^2 - mg\ell(1 - \cos\phi)$$

and $L$ satisfies

$$\frac{\partial L}{\partial \phi} - \frac{d}{dt} \left( \frac{\partial L}{\partial \dot\phi} \right) = 0,$$

then

$$\ddot\phi + \frac{g}{\ell} \sin\phi = 0.$$

5. Show that the only solution of

$$4\dot x(1 + \dot x^2) = \text{const}$$

is

$$x(t) = At + B.$$

6. Prove Lemma 5.11.

7. Derive

$$\ddot r - M^2/m^2 r^3 = -k/r^2$$

for a particle moving in the plane.

8. Show that the only solution of

$$yy'^2(1 + y'^2)^{-1/2} - y(1 + y'^2)^{1/2} = \text{constant} = K$$

is

$$y = K \cosh\left( \frac{x + C}{K} \right).$$

9. Define a weak $L^p(\mathbb{R}^n)$ derivative for $1 < p < \infty$. Show that it is unique and satisfies

$$|J_\varepsilon u|_p \leq |u|_p, \quad u \in L^p$$

and

$$|J_\varepsilon u - u|_p \to 0 \text{ as } \varepsilon \to 0, \quad u \in L^p.$$

10. Verify

$$
\begin{aligned}
H(x, y_n, y_n') - H(x, \bar{y}, \bar{y}') &= H(x, y_n, y_n') - H(x, y_n, \bar{y}') \\
&\quad + H(x, y_n, \bar{y}') - H(x, \bar{y}, \bar{y}') \\
&\geq (y_n' - \bar{y}')H_z(x, y_n, \bar{y}') - \varepsilon \\
&\geq (y_n' - \bar{y}')[H_z(x, y_n, \bar{y}') - H_z(x, \bar{y}, \bar{y}')] \\
&\quad + (y_n' - \bar{y}')H_z(x, \bar{y}, \bar{y}') - \varepsilon \\
&\geq (y_n' - \bar{y}')H_z(x, \bar{y}, \bar{y}') - \varepsilon(|y_n'| + |\bar{y}'| + 1)
\end{aligned}
$$

for $n > N$.

11. Verify

$$\int_{\Omega_M} (y_n'(x) - \bar{y}'(x))H_z(x, \bar{y}, \bar{y}')dx = \int_\Omega (y_n' - \bar{y}')g(x)dx \to 0$$

as $n \to \infty$.

12. Verify

$$\sum_{k=1}^m |\bar{y}(x_k) - \bar{y}(x_k')| \leq \left( \sum_{k=1}^m |x_k - x_k'| \right)^{1/p'} B$$

13. Verify

$$\left| \frac{y_n(x+h) - y_n(x)}{h} \right|^p \leq \frac{1}{h} \int_x^{x+h} |y_n'(t)|^p dt = \frac{1}{h} \int_0^h |y_n'(x+s)|^p ds$$

and

$$
\begin{aligned}
\int_a^{b-h} \left| \frac{y_n(x+h) - y_n(x)}{h} \right|^p dx &\leq \frac{1}{h} \int_a^{b-h} dx \int_0^h |y_n'(x+s)|^p ds \\
&\leq \frac{1}{h} \int_0^h ds \int_a^{b-h} |y_n'(x+s)|^p dx \\
&\leq B^p.
\end{aligned}
$$

14. Use Theorem 4.17 to show that there is exactly one continuous function $z(x)$ satisfying

$$H_z(x, \bar{y}(x), z(x)) - \int_a^x H_y(t, \bar{y}(t), \bar{y}'(t))dt = C_0.$$

15. Show that

$$J(y_0) = \inf J(y),$$

where

$$J(y) = \int_0^1 x^{2/3} \dot{y}^2 dx,$$

$$y_0(x) = x^{1/3}$$

and the infimum is taken over the class of $y \in C^1(\bar{\Omega})$ satisfying the boundary conditions

$$y(0) = y(1) = 0.$$

16. Prove:

$$\int_0^T [\dot{y} - y \cot x]^2 dx = \int_0^T [\dot{y}^2 + y^2 \cot^2 x - 2y\dot{y} \cot x]dx$$

$$= \int_0^T [\dot{y}^2 + y^2(\cot^2 x - \csc^2 x)]dx$$

$$= \int_0^T [\dot{y}^2 - y^2]dx$$

when

$$y(0) = y(T) = 0.$$

# 6

# Degree theory

## 6.1    The Brouwer degree

Let $\Omega$ be an open, bounded subset of $\mathbb{R}^n$, and let $p$ be any point in $\mathbb{R}^n$. Assume that for each continuous map $\varphi : \bar{\Omega} \to \mathbb{R}^n$ such that $p \notin \varphi(\partial\Omega)$ there is an integer $d(\varphi, \Omega, p)$ with the following properties:

  **(a)** $d(I, \Omega, p) = 1$. If both $p$ and $-p$ are in $\Omega$, then $d(\pm I, \Omega, p) = (\pm 1)^n$.
  **(b)** If $d(\varphi, \Omega, p) \neq 0$, then there is an $x \in \Omega$ such that $\varphi(x) = p$.
  **(c)** If $h_t(x) = H(t, x)$ is a continuous map of $[0, 1] \times \bar{\Omega}$ into $\mathbb{R}^n$ and $p \notin h_t(\partial\Omega)$ for $0 \leq t \leq 1$, then $d(h_t, \Omega, p)$ is independent of $t$.

We shall prove later that such an integer exists satisfying (a)–(c) for each set $\varphi, \Omega, p$ described above. Unfortunately, the same is not true in infinite dimensional spaces without restricting the map $\varphi$. We shall have more to say about this later. We call $d(\varphi, \Omega, p)$ the **Brouwer degree** of $\varphi$ at $p$ relative to $\Omega$.

As an application of the degree, we have

**Theorem 6.1.** *Let $B$ be the unit ball*

$$B = \{x \in \mathbb{R}^n : \|x\| < 1\}.$$

*If $\varphi$ is a continuous map of $\bar{B}$ into itself, then there is a point $x \in \bar{B}$ such that*

$$\varphi(x) = x. \tag{6.1}$$

*Proof.* If there is a point $x \in \partial B$ such that (6.1) holds, then the theorem is true. Otherwise, let

$$h_t(x) = x - t\varphi(x), \quad x \in \bar{B}, \ t \in [0, 1]. \tag{6.2}$$

171

Then $h_t(x)$ is a continuous map of $[0,1] \times \bar{B}$ into $\mathbb{R}^n$. We note that

$$h_t(x) \neq 0, \quad x \in \partial B, \; t \in [0,1]. \tag{6.3}$$

For $t = 1$, this is true by assumption. For $0 \leq t < 1$, we have $t\varphi(x) \in B$, so that (6.3) holds in this case as well. Thus, $d(h_t, B, 0)$ exists for each $t \in [0,1]$. By properties (c) and (a),

$$d(h_1, B, 0) = d(h_0, B, 0) = d(I, B, 0) = 1. \tag{6.4}$$

By property (b), there is a point $x \in B$ such that $h_1(x) = 0$. This means that $x$ satisfies (6.1). The proof is complete.  $\square$

**Corollary 6.2.** *If $B$ is the unit ball*

$$B = \{x \in \mathbb{R}^n : \|x\| < 1\},$$

*then there does not exist a continuous map $\varphi$ of $\bar{B}$ into $\partial B$ such that*

$$\varphi(x) = x, \quad x \in \partial B. \tag{6.5}$$

*Proof.* If such a map existed, let

$$\psi(x) = -\varphi(x), \quad x \in \bar{B}.$$

Then $\psi$ is a continuous map of $\bar{B}$ into itself. By Theorem 6.1, there is an $x \in \bar{B}$ such that $\psi(x) = x$. This means that $x = -\varphi(x) \in \partial B$. But on $\partial B$, we have $\varphi(x) = x$. Hence, $x = -x$. This means that $x = 0$ and cannot be on $\partial B$. This contradiction proves the corollary.  $\square$

If $\varphi_0, \varphi_1$ are two continuous maps of $\bar{\Omega}$ to $\mathbb{R}^n$ and there is a continuous mapping $h_t$ of $[0,1] \times \bar{\Omega}$ to $\mathbb{R}^n$ such that

$$h_0(x) = \varphi_0(x), \quad h_1(x) = \varphi_1(x), \tag{6.6}$$

then we say that $\varphi_0, \varphi_1$ are **homotopic** and $h_t$ is a **homotopy**. Property (c) states that if $\varphi_0, \varphi_1$ are homotopic, then

$$d(\varphi_0, \Omega, p) = d(\varphi_1, \Omega, p).$$

Two subsets of $\mathbb{R}^n$ are called **homeomorphic** if there is a continuous map $h$ having a continuous inverse $h^{-1}$ which maps one onto the other. The mapping $h$ is called a **homeomorphism**. As a consequence of Theorem 6.1, we have

**Corollary 6.3.** *If $\Omega$ is an open subset of $\mathbb{R}^n$ such that $\bar{\Omega}$ is homeomorphic to $\bar{B}$ and $\varphi$ is a continuous map of $\bar{\Omega}$ into itself, then there is a point $x \in \bar{\Omega}$ such that (6.1) holds.*

*Proof.* Let $h : \bar{\Omega} \to \bar{B}$ be the homeomorphism, and let $\psi(x) = h(\varphi(h^{-1}x))$. Then $\psi(x)$ is a continuous map of $\bar{B}$ into itself. By Theorem 6.1, there is a $z \in \bar{B}$ such that

$$\psi(z) = z.$$

Since $h$ is a homeomorphism of $\bar{\Omega}$ onto $\bar{B}$, there is an $x \in \bar{\Omega}$ such that $h(x) = z$. Then

$$\varphi(x) = \varphi(h^{-1}(z)) = h^{-1}(\psi(z)) = h^{-1}(z) = x,$$

and $x \in \bar{\Omega}$ is a solution of (6.1). This completes the proof.     □

A solution of (6.1) is called a **fixed point** of the map $\varphi$. Theorem 6.1 and Corollary 6.3 state that any continuous map of a closed ball in $\mathbb{R}^n$ (or any set homeomorphic to it) into itself has a fixed point. These sets are said to have the **fixed point property**. These results are due to L. Brouwer.

We also have

**Theorem 6.4.** *Let $K$ be a closed, bounded, convex set in $\mathbb{R}^n$. Then $K$ has the fixed point property.*

*Proof.* Since $K$ is bounded, there is an $R > 0$ such that

$$K \subset B_R = \{u \in \mathbb{R}^n : \|u\| \leq R\}.$$

For each $x \in B_R$, let $N(x)$ be the closest point in $K$ to $x$. (There is one because $K$ is closed and bounded; cf. Lemma A.56.) Because $K$ is convex, there is only one such point. For, if

$$d(x, K) = \|x - y_1\| = \|x - y_2\|,$$

then $\frac{1}{2}(y_1 + y_2) \in K$, and because we are in $\mathbb{R}^n$,

$$\left\| x - \frac{1}{2}(y_1 + y_2) \right\| < \|x - y_1\| = d(x, K),$$

contrary to the definition of $d(x, K)$. Note also that $N(x)$ is a continuous map of $B_R$ into $K$. For if $x_k \in B_R$, $x_k \to x$, and $N(x_k)$ does not converge to $N(x)$, then there is a renamed subsequence such that $N(x_k) \to y \neq N(x)$. Since $N(x_k)$ is the nearest point to $x_k$ in $K$, we have

$$\|x_k - N(x_k)\| \leq \|x_k - N(x)\|.$$

Taking the limit, we find

$$\|x - y\| \leq \|x - N(x)\|.$$

This means that $y = N(x)$, showing that $N(x)$ is indeed continuous.

Let $f(x)$ be a continuous map of $K$ into itself, and let $f_1 = fN$. Then $f_1$ is a continuous map of $B_R$ into itself. By Theorem 6.1 there is an $x \in B_R$ such that $f_1(x) = x$. Since $f_1$ maps $B_R$ into $K$, we see that $x \in K$. But then $f_1(x) = f(N(x)) = f(x)$. Hence, $f(x) = x$, and $x$ is a fixed point of $f$.                                                       □

**Corollary 6.5.** *If $K$ is homeomorphic to a closed, bounded, convex set in $\mathbb{R}^n$, then $K$ has the fixed point property.*

**Theorem 6.6.** *Let $\Omega$ be a bounded open set in $\mathbb{R}^n$, and let $\varphi$ be a continuous map of $\bar{\Omega} \to \mathbb{R}^n$ such that for some fixed $z \in \Omega$*

$$\varphi(x) - z \neq \lambda(x - z), \qquad x \in \partial\Omega, \ \lambda > 1. \tag{6.7}$$

*Then $\varphi$ has a fixed point in $\bar{\Omega}$.*

*Proof.* We may assume that $\varphi$ has no fixed point on $\partial\Omega$. Let

$$h_t(x) = x - z - t(\varphi(x) - z), \qquad x \in \bar{\Omega}, \ t \in [0, 1].$$

We know that $0 \notin h_0(\partial\Omega)$ since $z \in \Omega$. Also, $0 \notin h_1(\partial\Omega)$ by assumption. The same is true for $h_t$, $t \in (0, 1)$. For if $x \in \partial\Omega$ and $h_t(x) = 0$, then (6.7) would be violated for $\lambda = 1/t$. Thus, $0 \notin h_t(\partial\Omega)$ for $t \in [0, 1]$. Hence, $d(h_t, \Omega, 0)$ exists and is independent of $t$ by property (c). Thus,

$$d(h_1, \Omega, 0) = d(h_0, \Omega, 0).$$

This means that

$$d(I - \varphi, \Omega, 0) = d(I - z, \Omega, 0) = 1,$$

since $z \in \Omega$. Hence, there must be an $x \in \Omega$ such that $x - \varphi(x) = 0$. This completes the proof.                                                       □

One can interpret (6.7) as saying that $\varphi(x)$ does not lie on the continuation of the straight line $[z, x]$ beyond the point $x$ for any $x \in \partial\Omega$.

**Theorem 6.7.** *No matter how well you comb a hedgehog, at least one hair will stand up.*

Another way of stating this theorem is

**Theorem 6.8.** *If $n$ is odd, $\Omega$ is a bounded, open subset of $\mathbb{R}^n$ containing the origin, and $\varphi$ is a continuous map of $\bar{\Omega} \to \mathbb{R}^n$ such that $0 \notin \varphi(\partial\Omega)$, then there is an $x \in \partial\Omega$ such that $\varphi(x) = \lambda x$ with $\lambda \neq 0$.*

*Proof.* Let

$$h_t(x) = (1 - t)\varphi(x) + tx, \quad k_t(x) = (1 - t)\varphi(x) - tx, \quad x \in \bar{\Omega}, \, t \in [0, 1].$$

If one cannot find $x \in \bar{\Omega}$, $\lambda \neq 0$ such that $\varphi(x) = \lambda x$, then $h_t(x) \neq 0$, $k_t(x) \neq 0$ for $x \in \partial\Omega$ and $0 < t \leq 1$. Since $0 \notin \varphi(\partial\Omega)$, we have $h_0(x) \neq 0$, $k_0(x) \neq 0$ for $x \in \partial\Omega$ as well. Thus by property (c),

$$d(h_0, \Omega, 0) = d(h_1, \Omega, 0), \quad d(k_0, \Omega, 0) = d(k_1, \Omega, 0).$$

Hence,

$$d(\varphi, \Omega, 0) = d(I, \Omega, 0) = 1, \quad d(\varphi, \Omega, 0) = d(-I, \Omega, 0) = (-1)^n.$$

These can be equal only if $n$ is even, contrary to hypothesis. This completes the proof. $\qquad\square$

Note that Theorem 6.8 need not be true when $n$ is even. For instance, let $\varphi$ be the rotation of the unit disk in $\mathbb{R}^2$ into itself given by

$$\varphi(r, \theta) = (r, \theta + r), \quad 0 \leq r \leq 1, \, 0 \leq \theta < 2\pi.$$

Thus, you can comb a two dimensional hedgehog so that no hair stands up.

## 6.2 The Hilbert cube

We let $\ell^2$ be the set of sequences

$$x = (x_1, x_2, \dots)$$

satisfying

$$\|x\|^2 = \sum_{k=1}^{\infty} |x_k|^2.$$

It becomes a Hilbert space with the appropriate scalar product. The Hilbert cube $H_0$ is the subset of $\ell^2$ consisting of those sequences satisfying

$$|x_k| \leq \frac{1}{k}.$$

Clearly, it is closed, bounded, and convex. First we note

**Lemma 6.9.** *If $\{x^{(j)}\}$ is a sequence of points in $H_0$, then there is a subsequence that converges to a point in $H_0$.*

*Proof.* Let

$$P_n x = (x_1, \ldots, x_n, 0, 0, \ldots), \quad n = 1, 2, \ldots, \tag{6.8}$$

and let $\varepsilon > 0$ be given. Then there is an $n$ such that

$$\|P_n x - x\|^2 = \sum_{k=n+1}^{\infty} k^{-2} < \varepsilon^2, \quad x \in H_0. \tag{6.9}$$

Moreover, since the range of $P_n$ is finite dimensional, there is a renamed subsequence such that

$$\|P_n x^{(i)} - P_n x^{(j)}\| < \varepsilon, \quad i, j > N$$

(Corollary A.54). Hence,

$$\|x^{(i)} - x^{(j)}\| \leq \|x^{(i)} - P_n x^{(i)}\| + \|P_n(x^{(i)} - x^{(j)})\|$$
$$+ \|P_n x^{(j)} - x^{(j)}\| \leq \varepsilon + \varepsilon + \varepsilon.$$

Thus, there is subsequence that is Cauchy in $H_0$. Since $H_0$ is closed in $\ell^2$, the result follows (Lemma A.3). $\qquad \square$

This leads to

**Theorem 6.10.** *The Hilbert cube $H_0$ has the fixed point property.*

*Proof.* Let $\varphi$ be a continuous map of $H_0$ into itself. For each $n$, $P_n H_0$ is homeomorphic to a closed, bounded, convex subset of $\mathbb{R}^n$. Hence, by Theorem 6.4, there is a point $x^{(n)} \in P_n H_0$ such that

$$P_n \varphi(x^{(n)}) = x^{(n)}.$$

By Lemma 6.9 the sequence $\{x^{(n)}\}$ has a renamed subsequence which converges in $H_0$ to a limit $x$. Consequently,

$$\|\varphi(x) - x\| \leq \|\varphi(x) - P_n\varphi(x)\| + \|P_n\varphi(x) - P_n\varphi(x^{(n)})\| + \|x^{(n)} - x\|$$
$$\leq \|\varphi(x) - P_n\varphi(x)\| + \|\varphi(x) - \varphi(x^{(n)})\| + \|x^{(n)} - x\|$$
$$\to 0 \text{ as } n \to \infty.$$

This completes the proof. $\qquad \square$

**Corollary 6.11.** *Any closed, convex subset of $H_0$ has the fixed point property.*

*Proof.* Let $K$ be a closed, convex subset of $H_0$, and let $\varphi$ be a continuous map from $K \to K$. For each $n$ the set $K \cap P_n H_0$ is a closed, convex, bounded subset of $\mathbb{R}^n$, and $P_n\varphi$ is a continuous map of $K \cap P_n H_0$ to itself. By Theorem 6.4, there is a sequence $x^{(n)} \in K \cap P_n H_0$ such that $P_n \varphi(x^{(n)}) = x^{(n)}$. By Lemma 6.9 there is a renamed subsequence such that $x^{(n)} \to x \in H_0$. Since $K$ is closed, $x \in K$, and $P_n \varphi(x^{(n)}) \to \varphi(x)$. This implies that $\varphi(x) = x$, and $\varphi$ has a fixed point. Since $\varphi$ was arbitrary, we see that $K$ has the fixed point property. $\qquad\square$

A subset $K$ of a Hilbert space is called **compact** if every sequence in $K$ has a subsequence that converges to a limit in $K$. This is equivalent to the statement that every open cover of $K$ contains a finite subcover (Theorem C.3). A subset $V \subset W$ is called **dense** in W if for every $\varepsilon > 0$ and every $w \in W$ there is a $v \in V$ such that $\|v - w\| < \varepsilon$. A subset $W$ of a normed vector space is called **separable** if it has a dense subset that is denumerable. In other words, $W$ is separable if there is a sequence $\{x_k\}$ of elements of $W$ such that for each $x \in W$ and each $\varepsilon > 0$, there is an $x_k$ satisfying $\|x - x_k\| < \varepsilon$. A compact set is separable (Lemma A.46). For a subset $W$ of a normed vector space, the **linear span** of $W$ is the set of linear combinations of elements of $W$. It is a subspace. The **closed linear span** of a set is the closure of its linear span. The closed linear span of a separable set is separable (Lemma A.48).

**Lemma 6.12.** *Every convex, compact subset $K$ of a Banach space $X$ is homeomorphic to a closed, convex subset of $H_0$.*

*Proof.* Assume $K$ is a subset of the unit ball. Since $K$ is compact, it is separable (Lemma A.46). The linear span of $K$ is also separable (Lemma A.48). Hence, there is a sequence $\{x_n\}$ which is dense in the linear span of $K$. For each $n$ let $x'_n$ be a bounded linear functional on $X$ such that

$$x'_n(x_n) = \|x_n\|/n, \quad \|x'_n\| = 1/n$$

(Theorem A.15). Let

$$F(x) = \{x'_1(x), x'_2(x), \dots\} \tag{6.10}$$

be a map from $K$ to $H_0$. It is bounded and linear. It is also one-to-one. For let $x \neq y$ be two points in $K$. Then $z = x - y \neq 0$, and there is an $x_n$ such that

$$\|x_n - z\| < \frac{1}{2}\|z\|.$$

Then

$$\|x_n\| \geq \|z\| - \|z - x_n\| > \|z\| - \frac{1}{2}\|z\| = \frac{1}{2}\|z\| > \|x_n - z\|,$$

and consequently,

$$|x'_n(x) - x'_n(y)| \geq |x'_n(x_n)| - |x'_n(z - x_n)| \geq (\|x_n\| - \|z - x_n\|)/n > 0.$$

This shows that $F(x) \neq F(y)$. Finally, note that $R(F)$ is a closed, convex subset of $H_0$. Hence, $F$ is a homeomorphism of $K$ into $H_0$. $\qquad\square$

**Theorem 6.13.** *Any compact, convex subset $K$ of a Banach space has the fixed point property.*

*Proof.* By Lemma 6.12, every such set is homeomorphic to a convex, closed subset of $H_0$. But any such subset of $H_0$ has the fixed point property (Corollary 6.11). Thus $K$ has the fixed point property. $\qquad\square$

In contrast to the above we have

**Theorem 6.14.** *The unit ball in $\ell^2$ does not have the fixed point property.*

*Proof.* Elements in the unit ball are of the form

$$x = (x_1, x_2, \dots), \quad \|x\|^2 = \sum_{k=1}^{\infty} x_k^2 \leq 1.$$

Let

$$Tx = \left( \sqrt{1 - \|x\|^2}, x_1, x_2, \dots \right).$$

Then $T$ is a continuous map of the ball into itself. But $T$ does not have any fixed points. For, if $Tx = x$, then

$$x_1 = \sqrt{1 - \|x\|^2}, \quad x_2 = x_1, \quad x_3 = x_2, \dots, x_{k+1} = x_k, \dots$$

Thus all of the $x_k$ are equal. The norm $\|x\|$ can be finite only if they are all 0. But then $\|x\| = 0$, and consequently, $x_1 = 1$. This contradiction proves the theorem. $\qquad\square$

Contrast this with

**Theorem 6.15.** *Let $M$ be a convex subset of a normed vector space $X$. Let $T$ be a continuous map of $M$ into a compact subset $K$ of $M$. Then $T$ has a fixed point.*

This is known as **Schauder's fixed point theorem.** Before proving the theorem, we introduce several concepts. A map from a set to another is called **compact** if its range (image) is contained in a compact set. For any set $M$, we let $co(M)$ denote the smallest convex set containing $M$

(it is called the **convex hull** of $M$). The closure of $co(M)$ is denoted by $\overline{co}(M)$. We have

**Lemma 6.16.** *If $X$ is a normed vector space and $K$ is a compact subset, then for each $\varepsilon > 0$ there is a finite subset $Q$ of $K$ and a continuous map $P$ from $K$ to $co(Q)$ such that*

$$\|Px - x\| < \varepsilon, \quad x \in K. \tag{6.11}$$

*Proof.* Since $K$ is compact, we may cover it with a finite number of balls of radii $\varepsilon$. We may assume that the centers $x_1, \ldots, x_m$ are in $K$. Let $Q = \{x_1, \ldots, x_m\}$. Put

$$f_j(x) = \max[0, \varepsilon - \|x - x_j\|], \quad x \in X.$$

Then $f_j(x) \neq 0$ if and only if $\|x - x_j\| < \varepsilon$. Thus at each $x \in K$ there is a $j$ such that $f_j(x) \neq 0$. Put

$$Px = \frac{\displaystyle\sum_{j=1}^{m} f_j(x)x_j}{\displaystyle\sum_{j=1}^{m} f_j(x)}, \quad x \in K. \tag{6.12}$$

It is a continuous map of $K$ into $co(Q)$. Moreover, for each $x \in K$, $Px$ consists of a convex combination of those $x_k$ which are located in the ball of radius $\varepsilon$ and center $x$. Since this ball is convex, $Px$ must be located in the ball. Thus (6.11) holds. $\qquad\square$

**Corollary 6.17.** *Let $X$ be a normed vector space and $D$ a bounded subset of $X$. Let $T$ be a compact map from $D$ to $X$. For each $\varepsilon > 0$ there is a mapping $T_\varepsilon \in C(D, X)$ with finite dimensional range such that*

$$\|T(x) - T_\varepsilon(x)\| < \varepsilon, \quad x \in D.$$

*Proof.* Let $K = \overline{T(D)}$. Then $K$ is a compact subset of $X$. By Lemma 6.16 for each $\varepsilon > 0$ there is a finite subset $Q$ of $K$ and a continuous map $P$ from $K$ to $co(Q)$ such that (6.11) holds. Take

$$T_\varepsilon(x) = PT(x), \quad x \in D.$$

$\qquad\square$

We can now give the proof of Theorem 6.15.

*Proof.* For each $n = 1, 2, \ldots$, let $Q_n$ be the finite subset of $K$ and $P_n$ the mapping given by Lemma 6.16 corresponding to $\varepsilon = 1/n$. Since $Q_n \subset K \subset M$, we have $co(Q_n) \subset M$. Thus, for each $n$, $P_n T$ is a continuous map of the finite dimensional, convex, closed set $co(Q_n)$ into itself. By the Brouwer fixed point theorem (Theorem 6.4), there is a point $x_n \in co(Q_n)$ such that $P_n T x_n = x_n$. By (6.11)

$$\|x_n - Tx_n\| = \|P_n T x_n - Tx_n\| < 1/n$$

for each $n$. Since $T$ maps $M$ into the compact subset $K$, there is a renamed subsequence such that $Tx_n \to y \in K$. Thus, $x_n \to y$. In particular, $Ty$ is defined, and $Tx_n \to Ty$. Since $Tx_n \to y$, we see that $y$ is a fixed point of $T$.                                                              □

**Theorem 6.18.** *Let $B$ be the closed unit ball of a normed vector space $X$. Let $T$ be a continuous, compact map of $B$ into $X$ such that $T$ maps $\partial B$ into $B$. Then $T$ has a fixed point.*

In order to prove Theorem 6.18 we shall make use of the following lemma. Our notation is: $\overset{\circ}{M}$ is the interior of a set $M$, and $\partial M$ is its boundary.

**Lemma 6.19.** *Let $B$ be the closed ball of radius $k$ in a normed vector space $X$. Define*

$$r(x) = \begin{cases} x, & x \in B, \\ kx/\|x\|, & x \notin B. \end{cases} \tag{6.13}$$

*Then*

   (i) *$r$ is a continuous **retraction** of $X$ into $B$, that is, a mapping of $X$ into $B$ which is the identity on $B$,*
   (ii) *if $r(x) \in \overset{\circ}{B}$, then $r(x) = x$,*
   (iii) *if $x \notin B$, then $r(x) \in \partial B$.*

The proof of Lemma 6.19 is left as an exercise. We can now give the proof of Theorem 6.18.

*Proof.* Let $r$ be defined by (6.13) with $k = 1$, and let $T$ satisfy the hypotheses of Theorem 6.18. The mapping, $rT$ is a compact mapping of $B$ into itself. By Theorem 6.15, there is a $y \in B$ such that

$$rTy = y.$$

If $y \in \partial B$, then $Ty \in B$ by hypothesis, so that

$$y = rTy = Ty.$$

On the other hand, if $y \in \overset{\circ}{B}$, then $y = rTy \in \overset{\circ}{B}$, showing that $y = rTy = Ty$. Since $y \in B$, we have $Ty = y$ in any case.                                    $\square$

We can strengthen Theorem 6.18 by letting $B$ be any closed, convex subset of $X$. We have

**Theorem 6.20.** *Let $B$ be a closed, convex subset of $X$, and let $T$ be a continuous, compact map of $B$ into $X$ such that $T$ maps $\partial B$ into $B$. Then $T$ has a fixed point.*

In order to prove Theorem 6.20, we shall need the following.

**Lemma 6.21.** *Assume that $B$ is a closed, convex subset of $X$ such that $0 \in \overset{\circ}{B}$. Define*

$$g(x) = \inf_{\{c>0,\, cx \in B\}} c^{-1}. \tag{6.14}$$

*Then $g(x)$ is a continuous real valued function on $X$ and satisfies*

 (i) $g(cx) = cg(x), \quad c \geq 0,$
 (ii) $g(x + y) \leq g(x) + g(y),$
 (iii) $0 \leq g(x) < 1, \quad x \in \overset{\circ}{B},$
 (iv) $g(x) > 1, \quad x \notin B,$
 (v) $g(x) = 1, \quad x \in \partial B.$

We leave the proof of this as an exercise. We now show how it can be used to give the proof of Theorem 6.20.

*Proof.* If $B$ has no interior points, then $B = \partial B$. Thus, $T$ maps $B$ into $B$. The result now follows from Theorem 6.15. If $\overset{\circ}{B}$ is not empty, we may assume that $0 \in \overset{\circ}{B}$. Define $r(x)$ by

$$r(x) = x/\max[1, g(x)], \tag{6.15}$$

where $g(x)$ is given by (6.14). Then $r(x)$ has the properties described in Lemma 6.19. We can now follow the proof of Theorem 6.18.                  $\square$

One can sometimes deal with the question of fixed points by considering the more general problem of finding solutions of

$$x = \lambda Tx, \quad \lambda \in \mathbb{R}. \tag{6.16}$$

Along these lines we have

**Theorem 6.22.** *Let $X$ be a normed vector space, and let $T$ be a con-tinuous map on $X$ which is compact on bounded subsets of $X$. Then either*

    **(a)** *$T$ has a fixed point, or*
    **(b)** *the set of solutions of (6.16) for $0 < \lambda < 1$ is unbounded.*

*Proof.* Let $B$ be the closed ball of radius $k$ (center $0$), and defined $r(x)$ by (6.13). Then $rT$ is a compact map of $B$ into itself. By Theorem 6.15 it has a fixed point $x \in B$. Either

    **(i)** $\|Tx\| \le k$, or
    **(ii)** $\|Tx\| > k$.

In the first case,

$$Tx = rx = x.$$

In the second,

$$\|x\| = \|rTx\| = k. \tag{6.17}$$

Consequently,

$$x = rTx = (k/\|Tx\|)Tx = \lambda Tx, \ 0 < \lambda < 1. \tag{6.18}$$

Thus, either for some integer $k$ we obtain a fixed point of $T$ or we have a sequence $\{x_k\} \subset X$ such that (6.17) and (6.18) hold for $x = x_k$. Since, $\|x_k\| = k$, this gives alternative (b).     □

Another type of fixed point theorem is given by

**Theorem 6.23.** *Let $B$ be a closed, convex subset of a Banach space $X$, and suppose that $S, T$ map $B$ into $X$ and satisfy*

    **(a)** $Sx + Ty \in B, \quad x, y \in B,$
    **(b)** *$S$ is continuous and compact,*
    **(c)** *$T$ is a contraction.*

*Then there exists an $x$ in $B$ such that*

$$Sx + Tx = x.$$

*Proof.* For each $y \in B$, there is a unique $z \in B$ such that

$$z = Tz + Sy$$

(Theorem 2.12). The operator $(I - T)^{-1}S$ is continuous and compact on $B$. By Theorem 6.15, this operator has a fixed point $x \in B$. This point satisfies the requirements of the theorem.     □

## 6.3    The sandwich theorem

Once we have developed the Brouwer degree, we can now complete the proof of the sandwich theorem (Theorem 2.5). We follow the proof of Theorem 2.8. We take $E = H$, $M = V$, $N = W$ and assume that there is no sequence satisfying (2.27), where $m_0$, $m_1$ are given by (2.40). Then there must be a $\delta > 0$ such that (2.41) holds whenever $u \in H$ satisfies (2.42). By solving (2.48) we obtain a curve $\sigma(t)v$ emanating from each $v \in V$ such that (2.43) holds for each $v \in V$. Hence, if we take $T$ to satisfy (2.44), we see that (2.45) holds.

From (2.48) we see that

$$\|\sigma'(t)v\| \leq 1, \quad v \in H.$$

Consequently,

$$\|\sigma(t)v - v\|_H \leq \int_0^t \|\sigma'(s)v\|_H \, ds \leq t, \quad v \in H.$$

Let $P$ be the (orthogonal) projection of $H$ onto $V$. If $v \in V$ and $\|v\| = R$, then

$$\|P\sigma(t)v\| \geq \|v\| - \|v - P\sigma(t)v\| \geq R - t.$$

Thus, if

$$\varphi(t)v := P\sigma(t)v, \quad v \in V,$$

then $\varphi(t)$ is a continuous map of $V \times \mathbb{R}$ into $V$. Pick $R > T$. Then

$$\varphi(t)v \neq 0, \quad v \in V \cap \partial B_R, \ 0 \leq t \leq T.$$

Consequently, the Brouwer degree

$$d(\varphi(t), V \cap B_R, 0), \quad 0 \leq t \leq T$$

in defined. By property (c) of that degree

$$d(\varphi(t), V \cap B_R, 0) = d(\varphi(0), V \cap B_R, 0) = d(I, V \cap B_R, 0) = 1.$$

Hence, there is a $v_0 \in V \cap B_R$ such that

$$\varphi(T)v_0 = P\sigma(T)v_0 = 0.$$

This means that $\sigma(T)v_0 \in W$ and

$$G(\sigma(T)v_0) \leq m_0$$

by (2.45). But this contradicts (2.40). The proof of Theorem 2.5 is now complete.

**Remark 6.24.** *The requirement that $G'$ be locally Lipschitz continuous will be removed in Appendix D.*

## 6.4    Sard's theorem

We now begin our proof of the existence of the Brouwer degree. As we shall see later, it is extremely important for us to be able to move points slightly to avoid potholes without changing the degree. Our ability to do this is guaranteed by a theorem due to Sard which we describe next.

Let $\Omega \subset \mathbb{R}^n$ be an open set, and let $f \in C^1(\Omega, \mathbb{R}^n)$. Set

$$S_f = \{x \in \Omega : J_f(x) = 0\},$$

where $J_f$ denotes the Jacobian

$$J_f(x) = \det \begin{pmatrix} \dfrac{\partial f_1}{\partial x_1} & \cdots & \dfrac{\partial f_1}{\partial x_n} \\ \cdots & \cdots & \cdots \\ \dfrac{\partial f_n}{\partial x_1} & \cdots & \dfrac{\partial f_n}{\partial x_n} \end{pmatrix}.$$

We let $f(S_f)$ denote the images of the points of $S_f$ under the mapping $f$, that is,

$$f(S_f) = \{y \in \mathbb{R}^n : y = f(x),\ x \in S_f\}.$$

The aim of Sard's theorem is to show that this set is small in the sense that its measure is 0 (cf. Appendix B). Specifically, we have

**Theorem 6.25.** *Under the above hypotheses, for every $\varepsilon > 0$ there is a sequence of cuboids $R_k \subset \Omega$ such that*

$$f(S_f) \subset \bigcup_{k=1}^{\infty} R_k \quad \text{and} \quad \sum_{k=1}^{\infty} \mathrm{vol}\{R_k\} < \varepsilon. \tag{6.19}$$

*Proof.* Let $Q$ be any closed cube in $\Omega$, and let $\rho$ be the length of its side. Its diagonal has length $\rho\sqrt{n}$. Let $\varepsilon > 0$ be given. Since $\nabla f(x)$ is uniformly continuous in $Q$, there is an integer $m > 0$ such that

$$|\nabla f(x) - \nabla f(x')| < \varepsilon, \quad x, x' \in Q,\ |x - x'| \leq \delta = \frac{\sqrt{n}\,\rho}{m}.$$

Also, there is a constant $C$ such that

$$|\nabla f(x)| \leq C, \quad x \in Q.$$

Now,

$$f(x) - f(\bar{x}) = \int_0^1 \frac{d}{d\theta} f(\bar{x} + \theta(x - \bar{x})) \, d\theta$$

$$= \int_0^1 \sum_{k=1}^n \frac{\partial f}{\partial x_k} (\bar{x} + \theta(x - \bar{x})) \cdot (x_k - \bar{x}_k) \, d\theta.$$

Hence,

$$f(x) - f(\bar{x}) - \nabla f(\bar{x}) \cdot (x - \bar{x})$$

$$= \int_0^1 \{\nabla f(\bar{x} + \theta(x - \bar{x})) - \nabla f(\bar{x})\} \cdot (x - \bar{x}) \, d\theta$$

$$\equiv R(x, \bar{x}).$$

Thus,

$$|R(x, \bar{x})| \leq \int_0^1 |\nabla f(\bar{x} + \theta(x - \bar{x})) - \nabla f(\bar{x})| \cdot |x - \bar{x}| \, d\theta$$

$$\leq \varepsilon |x - \bar{x}|, \quad |x - \bar{x}| < \delta.$$

Divide $Q$ into $r = m^n$ cubes $Q_1, \ldots, Q_r$, each $Q_k$ having side length equal to $\rho/m$ and diameter $\delta = \sqrt{n}\rho/m$. Consequently,

$$|R(x, \bar{x})| \leq \varepsilon\delta, \quad x, \bar{x} \in Q_k.$$

Suppose $Q_k \cap S_f \neq \phi$. Pick $\bar{x} \in S_f$, and let $A = \nabla f(\bar{x})$. Take

$$g(x) = f(x) - f(\bar{x}) = A(x - \bar{x}) + R(x, \bar{x}).$$

Now, $\det A = J_f(\bar{x}) = 0$. This implies that the range of $A$ is contained in an $n-1$ dimensional subspace of $\mathbb{R}^n$, that is, there is an $n-1$ dimensional subspace $N$ of $\mathbb{R}^n$ such that

$$Az \in N, \quad z \in \mathbb{R}^n.$$

Thus, there is a unit vector $b \in \mathbb{R}^n$ such that

$$Az \perp b, \quad z \in \mathbb{R}^n.$$

Let $(b, b_2, \ldots, b_n)$ be an orthonormal basis for $\mathbb{R}^n$. Then

$$(g(x), b) = (A(x - \bar{x}), b) + (R(x, \bar{x}), b) = (R(x, \bar{x}), b).$$

Consequently,

$$|(g(x), b)| \leq |R(x, \bar{x})| \cdot |b| \leq \varepsilon\delta, \quad x \in Q_k,$$

while

$$|(g(x), b_j)| = |(A(x - \bar{x}), b_j)| + |(R(x, \bar{x}), b_j)|$$
$$\leq |A| \cdot |x - \bar{x}| + \varepsilon\delta$$
$$\leq (C + \varepsilon)\delta, \quad j = 2, \ldots, n.$$

This means that $g(x)$ is contained in a cuboid $R_k$ with one side of length $\varepsilon\delta$ and all other sides of length $(C + \varepsilon)\delta$. Thus, the volume of $R_k$ is $(C + \varepsilon)^{n-1}\delta^n\varepsilon$. Since

$$f(x) = f(\bar{x}) + g(x),$$

we have

$$f(Q_k) = f(\bar{x}) + g(Q_k) \subset R'_k,$$

where

$$vol\ R'_k = (C + \varepsilon)^{n-1}\delta^n\varepsilon.$$

Hence,

$$f(S_f(Q)) \subset \bigcup_{k=1}^{r} R'_k, \quad \sum_{k=1}^{r} vol\ R'_k = r(C + \varepsilon)^{n-1}\delta^n\varepsilon = n^{n/2}(C + \varepsilon)^{n-1}\rho^n\varepsilon.$$

Since $\varepsilon$ was arbitrary, we can take this to be as small as we like.

There is a sequence of cubes $\{Q_j\} \subset \Omega$ such that

$$\Omega = \bigcup_{j=1}^{\infty} Q_j.$$

Let $\eta > 0$ be given. Applying the argument above to each $Q_j$, we obtain

$$f(S_f(Q_j)) \subset \bigcup_{j=1}^{r_j} R_{jk}, \quad \sum_{k=1}^{r_j} vol\ R_{jk} < \eta/j^2.$$

Thus,

$$f(S_f(\Omega)) \subset \bigcup_{j=1}^{\infty} \bigcup_{k=1}^{r_k} R_{jk}, \quad \sum_{j=1}^{\infty} \sum_{k=1}^{r_j} vol\ R_{jk} < \sum_{j=1}^{\infty} \eta/j^2 = \eta.$$

This completes the proof.                                                    $\square$

**Corollary 6.26.** *Under the above hypotheses, every open set in $\mathbb{R}^n$ contains points not in $f(S_f)$.*

## 6.5    The degree for differentiable functions

Let $\Omega$ be a bounded, open set in $\mathbb{R}^n$, and let $p$ be a point in $\mathbb{R}^n$. Let $\varphi \in C^1(\bar{\Omega}, \mathbb{R}^n)$, and denote the set of critical points of $\varphi$ by $\gamma_\varphi$. First we note

**Lemma 6.27.** *If no critical points satisfy*

$$\varphi(x) = p, \tag{6.20}$$

*then there is only a finite number of solutions of (6.20).*

*Proof.* If $x_1$, $x_2$ are solutions of (6.20) then

$$0 = \varphi(x_2) - \varphi(x_1) = \det \varphi'(x_1)(x_2 - x_1) + o(|x_2 - x_1|),$$

where

$$\det \varphi'(x) = J_\varphi(x).$$

Since $\det \varphi'(x_1) \neq 0$, there is an $r > 0$ such that

$$|\varphi'(x_1)v| \geq r|v|, \quad v \in \mathbb{R}^n.$$

Thus

$$r|x_2 - x_1| \leq |\varphi'(x_1)(x_2 - x_1)| = o(|x_2 - x_1|).$$

Thus, if $x_2 \neq x_1$, then it cannot approach $x_1$. Hence, the solutions of (6.20) are isolated. This means that there is only a finite number of them. $\qquad\square$

If no critical points of $\varphi$ satisfy (6.20), we define the degree $d(\varphi, \Omega, p)$ of $\varphi$ at $p$ relative to $\Omega$ by

$$d(\varphi, \Omega, p) = \sum_{\varphi(x)=p} sign \det \varphi'(x), \tag{6.21}$$

where

$$sign \, \alpha = \begin{cases} \alpha/|\alpha|, & \alpha \neq 0, \\ 0, & \alpha = 0. \end{cases}$$

If there is no $x \in \Omega$ such that

$$\varphi(x) = p,$$

then we define $d(\varphi, \Omega, p) = 0$. Note the following

**Theorem 6.28.** *If* $p \in \Omega$, *then* $d(I, \Omega, p) = 1$; *if* $-p \in \Omega$, *then* $d(-I, \Omega, p) = (-1)^n$ ; *if* $p \notin \bar{\Omega}$, *then* $d(I, \Omega, p) = 0$.

*Proof.* This is immediate from the definition. $\qquad\square$

We also have

**Theorem 6.29.** *Assume that* $\varphi \in C^1(\bar{\Omega}, \mathbb{R}^n)$ *and that no critical or boundary points satisfy (6.20). Then there is a* $\delta > 0$ *such that no critical or boundary points satisfy*

$$\psi(x) = p$$

*whenever* $\psi \in C^1(\bar{\Omega}, \mathbb{R}^n)$ *satisfies* $\|\psi - \varphi\|_1 < \delta$, *where* $\| \cdot \|_1$ *is the norm of* $C^1(\bar{\Omega}, \mathbb{R}^n)$. *Moreover,*

$$d(\psi, \Omega, p) = d(\varphi, \Omega, p).$$

*Proof.* If there are no $x \in \bar{\Omega}$ satisfying (6.20), we can take $\delta > 0$ so small that

$$|\psi(x) - p| > \delta, \quad x \in \bar{\Omega}.$$

Then

$$d(\psi, \Omega, p) = d(\varphi, \Omega, p) = 0.$$

If there are such points, we note that by Lemma 6.27 there is only a finite number of points $a_1, \ldots, a_n \in \Omega$ satisfying $\varphi(a_i) = p$. Let $a$ be any one of them. Let

$$c = |\det \varphi'(a)| > 0,$$

and take $r > 0$ so small that

$$|\det \varphi'(x)| \geq \frac{2}{3}c, \quad |x - a| < r.$$

We can take $\delta > 0$ so small that

$$|\det \varphi'(x) - \det \psi'(x)| < \frac{1}{3}c, \quad x \in \bar{\Omega}.$$

Hence,

$$|\det \psi'(x)| > \frac{1}{3}c, \quad |x - a| < r.$$

We take $r > 0$ so small that the ball

$$B(a, r) = \{x \in \mathbb{R}^n : |x - a| < r\}$$

is contained in $\Omega$. Let

$$T(z) = \psi(z + a) - \psi'(a)z,$$

where $z = x - a$. Then

$$\begin{aligned}
T(z) - T(y) &= \int_0^1 \frac{d}{d\theta} T(\theta z + (1 - \theta)y) \, d\theta \\
&= \int_0^1 [\psi'(a + \theta z + (1 - \theta)y) - \psi'(a)](z - y) \, d\theta.
\end{aligned}$$

Consequently, there is a constant $K$ such that

$$|T(z) - T(y)| \le K(\delta + r)|z - y|, \quad |z|, |y| \le r.$$

Note that $\psi'(a)$ is an invertible operator on $\mathbb{R}^n$. Let

$$S(z) = \psi'(a)^{-1}[T(z) - \varphi(a)].$$

Then there are constants $K_1, K_2$ such that

$$|S(z)| \le K_1[(\delta + r)r + \delta]$$

and

$$|S(z) - S(y)| \le K_2(\delta + r)|z - y|$$

for

$$|z|, |y| \le r.$$

Shrink $\delta, r$ so that $K_1[(\delta + r)r + \delta] \le r$ and

$$K_2(\delta + r) < 1.$$

Then $-S(t)$ is a contraction mapping of $|z| \le r$ into itself. Consequently it has a unique fixed point. Hence, there is precisely one $z$ satisfying

$$-S(z) = z, \quad |z| \le r.$$

This is equivalent to the statement that there is precisely one $x$ satisfying

$$\psi(x) = \varphi(a) = p, \quad |x - a| \le r.$$

We follow the same procedure for each of the points $a_i$, shrinking $\delta$ if necessary. Let

$$B(r) = \bigcup_{j=1}^n B(a_j, r).$$

We can take $r > 0$ so small that none of the balls $B(a_j, r)$ overlap and

$\psi(x) = p$ has exactly one solution in each of them. We can also shrink $r$ so that

$$|\varphi(x) - p| \geq \eta > 0, \quad x \in \bar{\Omega} \setminus B(r).$$

We can then decrease $\delta$ to make

$$|\varphi(x) - \psi(x)| \leq \frac{1}{2}\eta, \quad x \in \bar{\Omega} \setminus B(r).$$

Consequently,

$$|\psi(x) - p| \geq \frac{1}{2}\eta, \quad x \in \bar{\Omega} \setminus B(r),$$

showing that $\psi(x) = p$ has no solutions in $\bar{\Omega}$ other than those in $B(r)$. But those in $B(r)$ satisfy

$$sign \ det \ \psi'(x) = sign \ det \ \varphi'(x).$$

Thus,

$$d(\psi, \Omega, p) = \sum_{\psi(x)=p} sign \ det \ \psi'(x) = \sum_{j=1}^{n} sign \ det \ \varphi'(a_j) = d(\varphi, \Omega, p).$$

$\square$

**Theorem 6.30.** *If $d(\varphi, \Omega, p) \neq 0$, then (6.20) has a solution $x \in \Omega$.*

*Proof.* If there exists no solution, then (6.21) shows that $d(\varphi, \Omega, p) = 0$.
$\square$

**Theorem 6.31.** *If $\varphi \in C^1(\bar{\Omega}, \mathbb{R}^n)$ and there are no critical or boundary points satisfying (6.20), then*

$$d(\varphi, \Omega, p) = \int_{\Omega} j_\varepsilon(\varphi(x) - p) \det \varphi'(x) \, dx \qquad (6.22)$$

*for $\varepsilon > 0$ sufficiently small.*

*Proof.* If (6.20) has no solutions, then both sides of (6.22) vanish. Otherwise, there is a finite number of points $a_1, \ldots, a_m \in \Omega$ satisfying $\varphi(a_j) = p$ (Lemma 6.27). Each $a_j$ is the center of a ball $B_j$ such that $\varphi$ is one-to-one on $B_j$, each $B_j$ is disjoint from $\varphi(\partial\Omega)$, and

$$\text{supp } j_\varepsilon(\varphi(x) - p) \subset \bigcup_{j=1}^{m} B_j.$$

(The **support** of a function is the closure of the set on which it does not vanish.) Hence,

$$\int_\Omega j_\varepsilon(\varphi(x) - p) \det \varphi'(x)\, dx$$

$$= \sum_{j=1}^m \int_{B_j} j_\varepsilon(\varphi(x) - p) \det \varphi'(x)\, dx$$

$$= \sum_{j=1}^m sign \det \varphi'(a_j) \int_{B_j} j_\varepsilon(\varphi(x) - p)|\det \varphi'(x)|\, dx$$

$$= \sum_{j=1}^m sign \det \varphi'(a_j) \int j_\varepsilon(y)\, dy$$

$$= d(\varphi, \Omega, p),$$

where we made the transformation $y = \varphi(x) - p$ and used the fact that

$$\int j_\varepsilon(y)\, dy = 1.$$

$\square$

**Lemma 6.32.** *Assume that $\varphi \in C^1(\bar{\Omega}, \mathbb{R}^n)$ and that $p_0, p_1$ are such that (6.20) has no solutions which are critical or boundary points when $p = p_0, p_1$. Assume also that there is a path $\gamma(s) \in \mathbb{R}^n$, $0 \le s \le 1$, such that $\gamma(0) = p_0$, $\gamma(1) = p_1$, and no critical or boundary points satisfy (6.20) with $p = \gamma(s)$, $0 \le s \le 1$. Then,*

$$d(\varphi, \Omega, p_0) = d(\varphi, \Omega, p_1).$$

*Proof.* By Theorem 6.29 for each $s \in [0,1]$ there is a $\delta_s > 0$ such that no critical or boundary points satisfy

$$\varphi(x) = p$$

whenever $p$ satisfies $\|p - \gamma(s)\| < \delta_s$ and

$$d(\varphi, \Omega, p) = d(\varphi, \Omega, \gamma(s)).$$

By compactness, there is one $\delta > 0$ which will serve for all $s \in I$. Moreover, we can cover $I = [0,1]$ with a finite number of intervals of length $> \eta > 0$ in which this holds in each interval. Let $0 = s_0 < s_1 < \cdots < s_m = 1$ be a partition of $I$ such that

$$\|\gamma(s_{k+1}) - \gamma(s_k)\| < \delta, \quad k = 0, 1, \ldots, m-1.$$

Hence,

$$d(\varphi, \Omega, p_0) = d(\varphi, \Omega, \gamma(s_0)) = d(\varphi, \Omega, \gamma(s_1))$$
$$= \cdots = d(\varphi, \Omega, \gamma(s_m)) = d(\varphi, \Omega, p_1).$$

This completes the proof. $\qquad\square$

We now define $d(\varphi, \Omega, p)$ for those points $p$ such that (6.20) is satisfied by critical points but not by boundary points. Let $p_0$ be such a point. Let $B_0$ be a neighborhood of $p_0$ such that (6.20) has no solutions in $\partial\Omega$ for $p \in B_0$. By Sard's theorem (Theorem 6.25), $B_0$ contains points $p$ such that no critical points satisfy (6.20). Let $p_1, p_2$ be any such points. Then by Lemma 6.32,

$$d(\varphi, \Omega, p_1) = d(\varphi, \Omega, p_2).$$

Thus, the degree is the same for all such points. We take $d(\varphi, \Omega, p_0)$ to be this constant value.

**Theorem 6.33.** *The degree $d(\varphi, \Omega, p)$ is constant on any component of $\mathbb{R}^n \backslash \varphi(\partial\Omega)$.*

*Proof.* If $\Sigma$ is any component of $\mathbb{R}^n \backslash \varphi(\partial\Omega)$, let $p_1, p_2$ be any two points in $\Sigma$. By Sard's theorem (Theorem 6.25) and the definition, there are points $q_1, q_2$ such that no critical points satisfy (6.20) when $p = q_j$, $j = 1, 2$ and

$$d(\varphi, \Omega, q_j) = d(\varphi, \Omega, p_j), \quad j = 1, 2.$$

But

$$d(\varphi, \Omega, q_1) = d(\varphi, \Omega, q_2)$$

by Lemma 6.32. Hence,

$$d(\varphi, \Omega, p_1) = d(\varphi, \Omega, p_2).$$

$\qquad\square$

**Theorem 6.34.** *If no boundary points satisfy (6.20), then there is a $\delta > 0$ such that $d(\psi, \Omega, p) = d(\varphi, \Omega, p)$ when $\psi \in C^1(\bar{\Omega}, \mathbb{R}^n)$ and $\|\psi - \varphi\|_1 < \delta$.*

*Proof.* There is a neighborhood of $p$ such that no boundary points satisfy $\varphi(x) = p'$ for $p'$ in this neighborhood. Let $q$ be a point in the neighborhood for which $\varphi(x) = q$ is not satisfied by any critical point. Take $\delta > 0$ so small that $\|\psi - \varphi\|_1 < \delta$ implies $d(\psi, \Omega, q) = d(\varphi, \Omega, q)$ (Theorem 6.29). If $x \in \partial\Omega$, then

$$|\psi(x) - p| \geq |\varphi(x) - p| - \delta.$$

Hence, we can take $\delta$ so small that $p, q$ are in the same component of $\mathbb{R}^n \setminus \psi(\partial\Omega)$. Consequently,

$$d(\psi, \Omega, p) = d(\psi, \Omega, q) = d(\varphi, \Omega, q) = d(\varphi, \Omega, p)$$

by Theorems 6.33 and 6.29. $\qquad\qquad\Box$

**Theorem 6.35.** *If $h_t(x) = H(t, x)$ is a continuous map of $[0, 1] \times \bar{\Omega}$ into $\mathbb{R}^n$ and $p \notin h_t(\partial\Omega)$, then $d(h_t, \Omega, p)$ is independent of $t$.*

*Proof.* By Theorem 6.34, $d(h_t, \Omega, p)$ is a continuous function of $t$. Since it only takes on integer values, it must be a constant. $\qquad\qquad\Box$

## 6.6 The degree for continuous functions

We now show that the degree can be defined for mappings that are only continuous. As before, we assume that $\Omega$ is a bounded open set in $\mathbb{R}^n$. Suppose $\varphi \in C(\bar{\Omega}, \mathbb{R}^n)$ and $p \notin \varphi(\partial\Omega)$, that is, no $x \in \partial\Omega$ satisfies (6.20). Let $\rho > 0$ be the distance from $p$ to $\varphi(\partial\Omega)$, and let $\psi_1, \psi_2 \in C^1(\bar{\Omega}, \mathbb{R}^n)$ be such that

$$\|\psi_j - \varphi\|_\infty < \rho, \quad j = 1, 2.$$

Let

$$h_t(x) = t\psi_1(x) + (1 - t)\psi_2(x), \quad x \in \bar{\Omega}, \ t \in [0, 1].$$

Then

$$\begin{aligned} |h_t(x) - \varphi(x)| &\leq t|\psi_1(x) - \varphi(x)| \\ &+ (1 - t)|\psi_2(x) - \varphi(x)| < t\rho + (1 - t)\rho = \rho. \end{aligned}$$

Thus,

$$|h_t(x) - p| \geq |\varphi(x) - p| - |h_t(x) - \varphi(x)| > \rho - \rho = 0.$$

Consequently,

$$p \notin h_t(\partial\Omega), \quad t \in [0, 1].$$

By Theorem 6.35,

$$d(\psi_1, \Omega, p) = d(\psi_2, \Omega, p).$$

This mean that $d(\psi, \Omega, p)$ is the same for all $\psi \in C^1(\bar{\Omega}, \mathbb{R}^n)$ satisfying $\|\psi - \varphi\|_\infty < \rho$. We can now define $d(\varphi, \Omega, p)$ to be this common value.

We note also that there exist $\psi \in C^1(\bar{\Omega}, \mathbb{R}^n)$ satisfying $\|\psi - \varphi\|_\infty < \rho$ such that no critical point satisfies $\psi(x) = p$. For let $\sigma(x) \in C^1(\bar{\Omega}, \mathbb{R}^n)$

be such that $\|\sigma - \varphi\|_\infty < \frac{1}{2}\rho$, and let $q \in \Omega$ be such that $|q - p| < \frac{1}{2}\rho$ and no critical point of $\sigma$ satisfies $\sigma(x) = q$. Let $\psi(x) = \sigma(x) + p - q$. Then, $\psi(x) = p$ if and only if $\sigma(x) = q$. Moreover,

$$\|\psi - \varphi\|_\infty \le \|\sigma - \varphi\|_\infty + |p - q| < \rho.$$

Since $x$ is a critical point of $\psi$ if and only if it is a critical point of $\sigma$, we see that our claim is correct.

We now have

**Theorem 6.36.** *If $\varphi \in C(\bar\Omega, \mathbb{R}^n)$ and $p \notin \varphi(\partial\Omega)$, then $d(\varphi, \Omega, p) \ne 0$ implies that there is an $x \in \Omega$ satisfying (6.20).*

*Proof.* Suppose $p \notin \varphi(\bar\Omega)$. If $\psi \in C^1(\bar\Omega, \mathbb{R}^n)$ is such that $\|\psi - \varphi\|_\infty$ is less than the distance from $p$ to $\varphi(\bar\Omega)$, then $p \notin \psi(\bar\Omega)$. By definition $d(\psi, \Omega, p) = 0$. Consequently, $d(\varphi, \Omega, p) = 0$. From this we see that $d(\varphi, \Omega, p) \ne 0$ implies that $p \in \varphi(\bar\Omega)$. Since $p \notin \varphi(\partial\Omega)$, we must have $p \in \varphi(\Omega)$. $\qquad\square$

We also have

**Theorem 6.37.** *Suppose $\varphi, \psi \in C(\bar\Omega, \mathbb{R}^n)$ and $p \notin \varphi(\partial\Omega)$. If $\|\psi - \varphi\|_\infty$ is less than the distance from $p$ to $\varphi(\partial\Omega)$, then $p \notin \psi(\partial\Omega)$ and $d(\psi, \Omega, p) = d(\varphi, \Omega, p)$.*

*Proof.* The first statement is obvious. Pick $\sigma \in C^1(\bar\Omega, \mathbb{R}^n)$ such that $\|\sigma - \psi\|_\infty + \|\psi - \varphi\|_\infty$ is less than the distance from $p$ to $\varphi(\partial\Omega)$. Then $d(\varphi, \Omega, p) = d(\sigma, \Omega, p)$. Moreover, $\|\sigma - \psi\|_\infty$ is less than the distance from $p$ to $\psi(\partial\Omega)$. Hence, $d(\psi, \Omega, p) = d(\sigma, \Omega, p)$. This proves the second statement. $\qquad\square$

**Theorem 6.38.** *If $h_t(x) = H(x, t)$ is a continuous map of $[0, 1] \times \bar\Omega$ into $\mathbb{R}^n$ and $p \notin h_t(x)$, $t \in [0, 1]$, then $d(h_t, \Omega, p)$ is independent of $t$.*

*Proof.* By Theorem 6.37, $d(h_t, \Omega, p)$ is a continuous function of $t$. Since it is integer valued, it must be a constant. $\qquad\square$

**Theorem 6.39.** *If $p_1, p_2 \in \Omega$ are in the same component of $\mathbb{R}^n \setminus \varphi(\partial\Omega)$, then $d(\varphi, \Omega, p_1) = d(\varphi, \Omega, p_2)$.*

*Proof.* There is a path $\gamma(s) \in \mathbb{R}^n \setminus \varphi(\partial\Omega)$, $s \in [0, 1]$ such that $\gamma(0) = p_1$, $\gamma(1) = p_2$. Take $\psi \in C^1(\bar\Omega, \mathbb{R}^n)$ such that $\|\psi - \varphi\|_\infty$ is less than the distance from $\gamma$ to $\varphi(\partial\Omega)$. Then $p_1$ and $p_2$ are in the same component of $\mathbb{R}^n \setminus \psi(\partial\Omega)$. Consequently, $d(\psi, \Omega, p_1) = d(\psi, \Omega, p_2)$ (Lemma 6.32). But by definition, $d(\varphi, \Omega, p_j) = d(\psi, \Omega, p_j)$, $j = 1, 2$. This gives the desired result. $\qquad\square$

We also have

**Theorem 6.40.** *Let* $\varphi, \psi \in C(\bar{\Omega}, \mathbb{R}^n)$ *be such that* $\varphi = \psi$ *on* $\partial\Omega$. *If* $p \notin \varphi(\partial\Omega) = \psi(\partial\Omega)$, *then* $d(\varphi, \Omega, p) = d(\psi, \Omega, p)$.

*Proof.* Consider the function

$$H(x, t) = t\varphi(x) + (1-t)\psi(x), \quad x \in \bar{\Omega}, \ t \in [0, 1].$$

Then $h_t(x) = H(x, t)$ is a continuous map of $[0, 1] \times \bar{\Omega}$ into $\mathbb{R}^n$. Moreover,

$$h_t(x) = \varphi(x), \quad x \in \partial\Omega, \ t \in [0, 1].$$

Hence, $p \notin h_t(\partial\Omega)$ for $t \in [0, 1]$. Consequently, by Theorem 6.38,

$$d(\varphi, \Omega, p) = d(h_1, \Omega, p) = d(h_0, \Omega, p) = d(\psi, \Omega, p).$$

The proof is complete. $\qquad\square$

**Theorem 6.41.** *If* $\varphi \in C(\bar{\Omega}, \mathbb{R}^n)$ *and* $p \notin \varphi(\partial\Omega)$, *then* $d(\varphi - q, \Omega, p - q)$ *is defined for all* $q \in \mathbb{R}^n$, *and* $d(\varphi - q, \Omega, p - q) = d(\varphi, \Omega, p)$.

*Proof.* Let

$$h_t(x) = \varphi(x) - tq, \ p_t = p - tq, \quad x \in \bar{\Omega}, \ t \in [0, 1].$$

Note that $p_t \in h_t(\partial\Omega)$ if and only if $p \in \varphi(\partial\Omega)$. Hence, $p_t \notin h_t(\partial\Omega)$ for $t \in [0, 1]$. By Theorems 6.38 and 6.39, $d(h_t, \Omega, p_t)$ is a continuous function of $t \in [0, 1]$. Hence,

$$d(\varphi - q, \Omega, p - q) = d(h_1, \Omega, p_1) = d(h_0, \Omega, p_0) = d(\varphi, \Omega, p).$$

This completes the proof. $\qquad\square$

**Theorem 6.42.** *Let* $h_t(x) = H(x, t)$ *be a continuous map of* $[0, 1] \times \bar{\Omega}$ *into* $\mathbb{R}^n$, *and let* $p_t = p(t)$ *be a continuous map of* $[0, 1]$ *into* $\mathbb{R}^n$ *such that* $p_t \notin h_t(\partial\Omega)$ *for* $t \in [0, 1]$. *Then* $d(h_t, \Omega, p_t)$ *does not depend on* $t \in [0, 1]$.

*Proof.* Let

$$k_t(x) = h_t(x) - p_t, \quad t \in [0, 1], \ x \in \bar{\Omega}.$$

Then $d(k_t, \Omega, 0) = d(h_t, \Omega, p_t)$ by Theorem 6.41. But $d(k_t, \Omega, 0)$ is independent of $t$ by Theorem 6.38. $\qquad\square$

**Lemma 6.43.** *If*

$$\Omega = \bigcup_{k=1}^{\infty} \Omega_k,$$

*where the $\Omega_k$ are disjoint open sets, then*

$$\partial \Omega_k \subset \partial \Omega, \quad k = 1, 2, \ldots$$

*Proof.* If $x \in \partial \Omega_k \subset \bar{\Omega}_k \subset \bar{\Omega}$ but $x \notin \partial \Omega$, then we must have $x \in \Omega$. This means that $x \in \Omega_j$ for some $j \neq k$. Since $\Omega_j$ is open and $\Omega_j \cap \Omega_k = \phi$, the point $x$ cannot be on $\partial \Omega_k$. This contradiction proves the lemma. □

**Theorem 6.44.** *Under the same hypotheses, if $\varphi \in C(\bar{\Omega}, \mathbb{R}^n)$ and $p \in \mathbb{R}^n \setminus \varphi(\partial \Omega)$, then*

$$d(\varphi, \Omega, p) = \sum_k d(\varphi, \Omega_k, p).$$

*Proof.* Take $\phi \in C^1(\bar{\Omega}, \mathbb{R}^n)$ such that $\|\psi - \varphi\|_\infty$ is so small that no boundary or critical point satisfies $\psi(x) = p$. Since $\partial \Omega_k \subset \partial \Omega$ for each $k$, we see that $p \notin \psi(\partial \Omega_k)$ and $d(\varphi, \Omega_k, p) = d(\psi, \Omega_k, p)$. Thus,

$$d(\varphi, \Omega, p) = d(\psi, \Omega, p) = \sum_{\psi(x)=p, \, x \in \Omega} \text{sign} \det \psi'(x)$$

$$= \sum_k \sum_{\psi(x)=p, \, x \in \Omega_k} \text{sign} \det \psi'(x)$$

$$= \sum_k d(\psi, \Omega_k, p)$$

$$= \sum_k d(\varphi, \Omega_k, p).$$

This gives the desired result. □

**Theorem 6.45.** *Assume that $\varphi \in C(\bar{\Omega}, \mathbb{R}^n)$ and that $p \in \mathbb{R}^n \setminus \varphi(\partial \Omega)$. If $Q$ is a closed subset of $\Omega$, and $p \notin \varphi(Q)$, then*

$$d(\varphi, \Omega, p) = d(\varphi, \Omega \setminus Q, p).$$

*Proof.* Since $Q$ is compact, we can choose $\psi \in C^1(\bar{\Omega}, \mathbb{R}^n)$ so that $\|\psi - \varphi\|_\infty$ is sufficiently small to guarantee that no boundary or critical point satisfies $\psi(x) = p$ and $p \notin \psi(Q)$. Then

$$d(\varphi, \Omega, p) = d(\psi, \Omega, p) = \sum_{\psi(x)=p,\, x \in \Omega} \text{sign det } \psi'(x)$$

$$= \sum_{\psi(x)=p,\, x \in \Omega \setminus Q} \text{sign det } \psi'(x)$$

$$= d(\psi, \Omega \setminus Q, p)$$

$$= d(\varphi, \Omega \setminus Q, p),$$

since no point $x \in Q$ satisfies $\psi(x) = p$. This completes the proof. $\quad\square$

## 6.7    The Leray–Schauder degree

Suppose that $D$ is an open, bounded subset of $\mathbb{R}^n$ and $\varphi \in C(\bar{D}, \mathbb{R}^m)$, where $m \leq n$. Define $\psi(x) \in C(\bar{D}, \mathbb{R}^n)$ by

$$\psi(x) = x + \varphi(x), \quad x \in \bar{D}.$$

We have

**Lemma 6.46.** *If $D_m = \mathbb{R}^m \cap D$ and $\psi_m$ is the restriction of $\psi$ to $R^m \cap \bar{D}$, then*

$$d(\psi, D, p) = d(\psi_m, D_m, p), \tag{6.23}$$

*where $d$ represents the Brouwer degree and $p \in \mathbb{R}^m \setminus \psi(\partial D)$.*

*Proof.* If $D_m = \phi$, then $p \notin \psi(\bar{D})$, and both sides of (6.23) are 0. Otherwise, $\psi_m$ maps $\bar{D}_m$ into $\mathbb{R}^m$. Since $\partial(\mathbb{R}^m \cap D) \subset \mathbb{R}^m \cap \partial D$, we have $p \notin \psi_m(\partial D_m)$. If $\psi(x) = p$, then $x = p - \varphi(x) \in \mathbb{R}^m$. Thus, $\psi^{-1}(p) \subset D_m$ and $\psi^{-1}(p) = \psi_m^{-1}(p)$.

First, assume that $\varphi \in C^1(\bar{D}, \mathbb{R}^m)$ and that $J_{\psi_m}(x) \neq 0$ when $\psi_m(x) = p$. Then

$$d(\psi, D, p) = \sum_{x \in \psi^{-1}(p)} \text{sign } J_\psi(x)$$

$$= \sum_{x \in \psi_m^{-1}(p)} \text{sign det} \begin{pmatrix} J_{\psi_m}(x) & 0 \\ 0 & I_{n-m} \end{pmatrix}$$

$$= \sum_{x \in \psi_m^{-1}(p)} \text{sign } J_{\psi_m}(x)$$

$$= d(\psi_m, D_m, p).$$

If $\varphi$ and $p$ do not have the added restrictions, choose $\hat{\varphi}_j \in C^1(\bar{D}, \mathbb{R})$ so that $\hat{\varphi} = (\hat{\varphi}_1, \ldots, \hat{\varphi}_n)$ satisfies

$$\hat{\varphi}_j = 0, \quad j = m+1, \ldots, n$$

and

$$|\hat{\varphi}(x) - \varphi(x)| < \rho(p, \psi(\partial D)), \quad x \in \bar{D},$$

where $\rho(p, \psi(\partial D))$ is the distance between $p$ and $\psi(\partial D)$. If $\hat{\psi}(x) = x + \hat{\varphi}(x)$, then

$$|\hat{\psi}(x) - \psi(x)| < \rho(p, \psi(\partial D)), \quad x \in \bar{D}. \tag{6.24}$$

Let $\hat{\psi}_m$ be the restriction of $\hat{\psi}$ to $\bar{D}_m$. In view of Sard's theorem (Theorem 6.25), we can adjust $\hat{\varphi}$ slightly to insure that $J_{\hat{\psi}_m}(x) \neq 0$ when $\hat{\psi}_m(x) = p$. Then $d(\psi, D, p) = d(\hat{\psi}, D, p)$, and we can apply the reasoning above. $\qquad \square$

Let $X$ be a normed vector space, $D$ an open, bounded subset of $X$, and $p$ a point in $X$. Let $T$ be a compact map from $\bar{D}$ to $X$. We define $\varphi = I - T$ and assume that $p \notin \varphi(\partial D)$. We have

**Lemma 6.47.** *If*

$$r = \rho(p, \varphi(\partial D)) = \inf_{x \in \partial D} \|p - \varphi(x)\|,$$

*then $r > 0$.*

*Proof.* Suppose not. Then there is a sequence $\{x_n\} \subset \partial D$ such that $\varphi(x_n) \to p$ as $n \to \infty$. Since the sequence is bounded, there is a renamed subsequence such that $T(x_n)$ converges in $X$ to some element $y$. Then $y \in \overline{T(D)}$ and

$$x_n = T(x_n) + \varphi(x_n) \to y + p \quad \text{as } n \to \infty.$$

Since $\partial D$ is closed, $y + p \in \partial D$. But

$$y = \lim_{n \to \infty} T(x_n) = T(y + p)$$

by continuity. Hence,

$$\varphi(y + p) = y + p - T(y + p) = p.$$

Since $y + p \in \partial D$, this means that $p \in \varphi(\partial D)$, contrary to assumption. Thus, $r > 0$. $\qquad \square$

**Lemma 6.48.** *For each $\varepsilon > 0$ such that $\varepsilon < r$, there is a mapping $T_\varepsilon \in C(\bar{D}, X)$ with finite dimensional range such that*

$$\|T(x) - T_\varepsilon(x)\| < \varepsilon, \quad x \in \bar{D},$$

*and $d(\varphi_\varepsilon, D_\varepsilon, p)$ is defined and independent of $\varepsilon$, where $D_\varepsilon$ is the intersection of $D$ with the finite dimensional subspace spanned by $T_\varepsilon(\bar{D})$ and $p$, and $\varphi_\varepsilon(x)$ is the restriction of*

$$\tilde{\varphi}_\varepsilon(x) = x - T_\varepsilon(x), \quad x \in \bar{D}$$

*to $D_\varepsilon$.*

*Proof.* Let $\eta$ also satisfy $0 < \eta < r$. By Corollary 6.17 there are mappings $T_\varepsilon, T_\eta$ satisfying the conclusion of that corollary for $\varepsilon$ and $\eta$, respectively. Let $\hat{S}$ be the subspace of $X$ spanned by $T_\varepsilon(\bar{D}), T_\eta(\bar{D})$ and $p$. (We have to include them !!!) Let $\hat{D} = D \cap \hat{S}$. Then by Lemma 6.46 we have

$$d(\varphi_\varepsilon, D_\varepsilon, p) = d(\varphi_\varepsilon, \hat{D}, p) \tag{6.25}$$

and

$$d(\varphi_\eta, D_\eta, p) = d(\varphi_\eta, \hat{D}, p). \tag{6.26}$$

Let $S_\varepsilon$ be the finite dimensional subspace of $X$ spanned by the $T_\varepsilon(\bar{D})$ and $p$. Let $D_\varepsilon = D \cap S_\varepsilon$. Then $D_\varepsilon$ is a bounded, open subset of $S_\varepsilon$, and $\partial_\varepsilon D_\varepsilon \subset \partial D$, where $\partial_\varepsilon D_\varepsilon$ is the boundary of $D_\varepsilon$ in $S_\varepsilon$. Note that $\varphi_\varepsilon(\bar{D}_\varepsilon) \subset S_\varepsilon$ and that

$$\|x - T_\varepsilon(x) - p\| \geq \|x - T(x) - p\| - \|T(x) - T_\varepsilon(x)\| > r - \varepsilon > 0,$$
$$x \in \partial D.$$

Thus, $d(\varphi_\varepsilon, D_\varepsilon, p)$ is defined. If $D_\varepsilon = \phi$, then $d(\varphi_\varepsilon, D_\varepsilon, p) = 0$.

Consider the homotopy

$$h_t(x) = t\varphi_\varepsilon(x) + (1 - t)\varphi_\eta(x), \quad x \in \overline{\hat{D}}, \, 0 \leq t \leq 1.$$

Then

$$\|h_t(x) - \varphi(x)\| \leq t\|\varphi_\varepsilon(x) - \varphi(x)\| + (1 - t)\|\varphi_\eta(x) - \varphi(x)\|$$
$$< t\varepsilon + (1 - t)\eta < r.$$

Consequently, we have

$$\|h_t(x) - p\| \geq \|\varphi(x) - p\| - \|\varphi(x) - h_t(x)\| > 0, \quad x \in \partial\hat{D}.$$

By Property 3 of the Brouwer degree, we see that

$$d(\varphi_\varepsilon, \hat{D}, p) = d(\varphi_\eta, \hat{D}, p).$$

We can now apply (6.25) and (6.26) to obtain the desired result.    □

To define the **Leray–Schauder degree**, let $D$ be an open, bounded subset of $X$, and let $T$ be a compact map from $\bar{D}$ to $X$. Take $\varphi = I - T$ and suppose that $p \in X \backslash \varphi(\partial D)$. We can find a continuous map $T_1$ of $\bar{D}$ to $X$ such that its range is finite dimensional and

$$\|T(x) - T_1(x)\| < \rho(p, \varphi(\partial D)), \quad x \in \bar{D}$$

(Corollary 6.17). Let $S_1$ be a finite dimensional subspace of $X$ containing $T_1(\bar{D})$ and $p$. We then define

$$d(\varphi, D, p) = d(\varphi_1, D_1, p),$$

where $\varphi_1 = I - T_1$ and $D_1 = D \cap S_1$. By Lemma 6.48, this definition is independent of $\varphi_1$ and $D_1$.

## 6.8    Properties of the Leray–Schauder degree

We now discuss some of the properties of the Leray–Schauder degree defined in the preceding section. We take $D, T, \varphi$ and $p$ as described there. We have

**Lemma 6.49.** *If $p \in D$, then $d(I, D, p) = 1$. If $p \notin \bar{D}$, then $d(I, D, p) = 0$.*

*Proof.* Take $T_1 = 0$ in the definition and let $S_1$ be the one dimensional subspace containing $p$. Apply Theorems 6.36 and 6.39.    □

**Lemma 6.50.** *If $d(\varphi, D, p) \neq 0$, then there is an $x \in D$ such that $\varphi(x) = p$.*

*Proof.* For each integer $k = 1, 2, \ldots$, there is an operator $T_k$ with finite dimensional range satisfying Lemma 6.48 for $\varepsilon = \frac{1}{k}$. Then

$$\|T_k(x) - T(x)\| < \frac{1}{k}, \quad x \in \bar{D}$$

and

$$d(\varphi, D, p) = d(\varphi_k, D_k, p),$$

where $\varphi_k = I - T_k$, and $D_k$ is the intersection of $D$ with the finite dimensional subspace spanning $T_k(\bar{D})$ and $p$. By Theorem 6.36, for each $k$ sufficiently large there is an $x_k \in D$ such that

$$x_k - T_k(x_k) = p.$$

Since $D$ is bounded, there is a renamed subsequence such that $T(x_k)$ converges to an element $y \in X$. But

$$\|x_k - T(x_k) - p\| = \|T_k(x_k) - T(x_k)\| < \frac{1}{k} \to 0.$$

Hence $x_k \to y + p$. By continuity, $T(x_k) \to T(y + p)$. But $T(x_k) \to y$. Hence, $y = T(y + p)$. This means that $\varphi(y + p) = y + p - T(y + p) = p$, and the proof is complete. $\qquad\square$

## 6.9 Peano's theorem

Picard's theorem (Theorem 2.13) requires $g(t, x)$ to satisfy a Lipschitz condition in $x$. It obtains both existence and uniqueness in an interval. If we do not have the Lipschitz condition, we may lose both existence and uniqueness. However, if $X$ is finite dimensional, we can retain the existence. There is an existence theorem due to Peano which only requires continuity but gives up the important element of uniqueness. This theorem is useful when one does not require uniqueness (this was the case in the proof of Theorem 5.13). Then we do not have to be concerned with verifying Lipschitz continuity (which can sometimes be a pain). We shall describe such a result here. It is known as **Peano's theorem.** We have

**Theorem 6.51.** *Let $X$ be a finite dimensional Banach space, and let*

$$B_0 = \{x \in X : \|x - x_0\| \le R_0\}$$

*and*

$$I_0 = \{t \in \mathbb{R} : |t - t_0| \le T_0\}.$$

*Assume that $g(t, x)$ is a continuous map of $I_0 \times B_0$ into $X$ such that*

$$\|g(t, x)\| \le M_0, \quad x \in B_0, \ t \in I_0. \tag{6.27}$$

*Let $T_1$ be such that*

$$T_1 \le \min(T_0, R_0/M_0), \quad K_0 T_1 < 1. \tag{6.28}$$

*Then there is a solution $x(t)$ of*

$$\frac{dx(t)}{dt} = g(t, x(t)), \quad |t - t_0| \leq T_1, \quad x(t_0) = x_0. \qquad (6.29)$$

*Proof.* As in the case of Picard's theorem we first note that $x(t)$ is a solution of (6.29) if and only if it is a solution of

$$x(t) = x_0 + \int_{t_0}^{t} g(s, x(s))ds, \quad t \in I_1 = \{t \in \mathbb{R} : |t - t_0| \leq T_1\}. \qquad (6.30)$$

For if $x(t)$ is a solution of (6.29), we can integrate to obtain (6.30). Note that $x(t)$ will be in $B$ as long as $t$ is in $I_1$ by (6.28). Conversely, if $x(t)$ satisfies (6.30), it is continuous in $t$ since

$$x(t + h) - x(t) = \int_{t}^{t+h} g(s, x(s))ds,$$

and consequently,

$$\|x(t + h) - x(t)\| \leq M_0|h| \rightarrow 0 \text{ as } h \rightarrow 0.$$

It is also differentiable since $g(s, x(s))$ is continuous and

$$[x(t + h) - x(t)]/h = \frac{1}{h} \int_{t}^{t+h} g(s, x(s))ds \rightarrow g(t, x(t)) \text{ as } h \rightarrow 0.$$

Following the proof of Theorem 2.13, we let $Y$ be the Banach space of all continuous functions $x(t)$ from $I_1$ to $X$ with norm

$$|||x||| = \max_{t \in I_1} \|x(t)\|. \qquad (6.31)$$

For $x(t) \in Y$, let $f(x(t))$ be the right-hand side of (6.30), and let

$$Q = \{x(t) \in Y : |||x - \hat{x}_0||| \leq R_0\},$$

where $\hat{x}_0(t) \equiv x_0$, $t \in I_1$. If $x(t)$ is in $Q$, then $f(x(t))$ satisfies

$$|||f(x) - \hat{x}_0||| \leq \int_{t_0}^{t_0+T_1} M_0 \, ds \leq R_0.$$

Thus $f$ maps $Q$ into $Q$. Also

$$\|f(x(s)) - f(x(t))\| \leq \left| \int_{s}^{t} g(\sigma, x(\sigma))d\sigma \right|$$

$$\leq \left| \int_{s}^{t} \|g(\sigma, x(\sigma))\| \, d\sigma \right| \leq M_0|s - t|.$$

Thus the mapping $f$ is bounded and equicontinuous. By the Arzelà–Ascoli theorem (Theorem C.6), it is a compact map of $Q$ into itself. We can now apply Schauder's fixed point theorem (Theorem 6.15) to show that $f(x)$ has a fixed point which is a solution of (6.30) and come to the desired conclusion. $\qquad\square$

## 6.10    An application

Let us solve the problem

$$-u'' = g(x, u, u'), \quad x \in I = [0, 1], \ u(0) = u(1) = 0. \tag{6.32}$$

We assume that $g(x, t, \tau)$ is a continuous function on $I \times \mathbb{R} \times \mathbb{R}$ and satisfies

$$g(x, t, \tau)^2 \le c_0(|t|^p + |\tau|^p + 1), \quad x \in I, t, \tau \in \mathbb{R} \tag{6.33}$$

and

$$|g(x, t_1, \tau_1) - g(x, t_2, \tau_2)|^2 \le c_1(|t_1 - t_2|^2 + |\tau_1 - \tau_2|^2), \quad x \in I, t_i, \tau_i \in \mathbb{R}, \tag{6.34}$$

where $p < 2$. Before we tackle this problem we examine the linear problem

$$-u'' = f(x), \quad x \in I, \ u(0) = u(1) = 0. \tag{6.35}$$

Let

$$h(t) = \int_0^t f(s)\, ds.$$

Then any solution of (6.35) satisfies

$$u'(t) = -h(t) + u'(0).$$

Consequently,

$$u(x) = -\int_0^x h(t)\, dt + u'(0)x + u(0).$$

Hence,

$$u(1) = -\int_0^1 h(t)\, dt + u'(0) + u(0).$$

Since $u(0) = u(1) = 0$, this yields

$$u(x) = -\int_0^x h(t)\, dt + x\int_0^1 h(t)\, dt. \tag{6.36}$$

A simple calculation shows that (6.36) is indeed a solution of (6.35) for $f(x) \in L^2(I)$. This shows that (6.35) has a unique solution

$$u \in H_0^1 = \{u \in L^2(I) : u' \in L^2(I), \ u(0) = u(1) = 0\}$$

for each $f \in L^2(I)$. Of course, $u''$ is also in $L^2(I)$, but we do not want to use this fact yet.

We note that

$$|h(t)| \leq \left( \int_0^t f(s)^2 ds \right)^{1/2} \left( \int_0^t ds \right)^{1/2} \leq t^{1/2} \|f\|, \quad t \in I.$$

Consequently,

$$|u'(t)| \leq \left( t^{1/2} + \frac{2}{3} \right) \|f\|, \quad t \in I,$$

and

$$\|u'\| \leq 2\|f\|.$$

If we use the norm $\|u'\|$, then $H_0^1$ becomes a Hilbert space. We designate $L$ as the mapping $f \to u$ from $L^2(I)$ to $H_0^1$ given by (6.36), and we put $v = F(u) = Lg(\cdot, u, u')$. If $u \in H_0^1$, then

$$\|g(\cdot, u, u')\|^2 = \int_I g(x, u, u')^2 \, dx \leq c_0 \int_I (|u|^p + |u'|^p + 1) \, dx$$

$$\leq c_0 \left[ \left( \int_I |u|^2 \, dx \right)^{p/2} + \left( \int_I |u'|^2 \, dx \right)^{p/2} + 1 \right]$$

$$\leq c_0 \left( \|u\|^{p/2} + \|u'\|^{p/2} + 1 \right).$$

Since

$$|u(x)| \leq \|u'\|, \quad |u(x) - u(x')| \leq |x - x'|^{1/2} \|u'\|, \quad x, x' \in I,$$

this gives

$$\|g(\cdot, u, u')\|^2 \leq c_0(2\|u'\|^{p/2} + 1), \quad u \in H_0^1.$$

Thus,

$$\|v'\|^2 \leq 4c_0(2\|u'\|^{p/2} + 1), \quad u \in H_0^1.$$

Let $R$ satisfy

$$R^2 \geq 4c_0(2R^{p/2} + 1),$$

and let

$$B = \{u \in H_0^1 : \|u'\| \leq R\}.$$

Then $B$ is a closed, convex, bounded subset of $H_0^1$. Moreover,

$$u \in B \Rightarrow v \in B.$$

Thus $F(u)$ maps $B$ into itself. It is continuous. For if $v_i = F(u_i)$, then

$$\|v_1' - v_2'\|^2 \leq 4\|g(\cdot, u_1, u_1') - g(\cdot, u_2, u_2')\|^2$$

$$= 4\int_I [g(x, u_1, u_1') - g(x, u_2, u_2')]^2 \, dx$$

$$\leq 4c_1 \int_I (|u_1 - u_2|^2 + |u_1' - u_2'|^2) \, dx$$

$$= 4c_1(\|u_1 - u_2\|^2 + \|u_1' - u_2'\|^2) \leq 8c_1\|u_1' - u_2'\|^2.$$

Thus, $F(u)$ is a continuous mapping. It is also compact. For, if $\{u_k\}$ is a sequence of functions in $B$, then

$$\|g(\cdot, u_k, u_k')\|^2 \leq c_0(2R^{p/2} + 1).$$

Consequently, the $v_k = F(u_k)$ satisfy

$$\|v_k'\| \leq R.$$

Since $v_k'' = g(x, u_k, u_k')$, we also have

$$\|v_k''\| \leq R.$$

Thus, the sequences $\{v_k\}, \{v_k'\}$ are bounded and equicontinuous on $I$. This implies that there is a renamed subsequence such that $v_k \to v$, $v_k' \to h$ uniformly on $I$. Of course, $h = v'$, and it follows that $v_k' \to v'$ uniformly on $I$. Thus $v_k \to v$ in $H_0^1$, and we see that $F(u)$ is a compact mapping on $B$. Hence, by the Schauder fixed point theorem (Theorem 6.15), there is a $u \in B$ such that $F(u) = u$. This means that (6.32) has a solution.

## 6.11    Exercises

1. Prove Theorem 6.1 in $\mathbb{R}^1$ without using the degree.

2. Prove: if

$$\|x - y_1\| = \|x - y_2\| = d, \quad x, y_i \in \mathbb{R}^n, \ y_1 \neq y_2,$$

then

$$\left\|x - \frac{1}{2}(y_1 + y_2)\right\| < d.$$

3. Prove Corollary 6.5.

4. Show that the Hilbert cube is a closed, convex, bounded set.

5. Show that for each $n$, $P_n H_0$ is homeomorphic to a closed, bounded, convex subset of $\mathbb{R}^n$, where $H_0$ is the Hilbert cube and $P_n$ is given by (6.8).

6. Prove Corollary 6.11.

7. For $F$ given by (6.10), show that it is a homeomorphism of $K$ onto a closed, convex subset of $H_0^1$.

8. Show that

$$Px = \frac{\displaystyle\sum_{j=1}^{m} f_j(x)x_j}{\displaystyle\sum_{j=1}^{m} f_j(x)}, \quad x \in K \tag{6.37}$$

is a continuous map of $K$ into $co(Q)$.

9. Show that

$$T_\varepsilon(x) = PT(x), \quad x \in D$$

satisfies the conclusions of Corollary 6.17.

10. Prove Lemma 6.19.

11. Prove Lemma 6.21.

12. Show that the function $r(x)$ given by (6.15) has the properties described in Lemma 6.19.

13. Prove Theorem 6.28.

14. In the proof of Theorem 6.29, show that $\psi'(a)$ is an invertible operator on $\mathbb{R}^n$.

15. Why do we not obtain uniqueness in Peano's theorem? Can you give an example where uniqueness fails?

# 7

# Conditional extrema

## 7.1    Constraints

In Chapter 5 we studied problems in which one searched for a function $y(x)$ which minimized the functional (5.6) on functions in $C^1(\bar{\Omega})$ satisfying (5.5). There are times when one is required to minimize functionals such as (5.6) under more restrictive conditions. Sometimes, these conditions are imposed upon the admissible functions by requiring them to satisfy an additional stipulation of the form

$$\int_a^b g(x, y, y')\, dx = c_0,$$

where $g(x, y, z)$ is a given function. Such problems are called **isoperimetric** because the first and best known problem of this type was that of finding a closed curve having a given perimeter which will enclose the largest area. The following theorem will help us deal with such problems.

First we have

**Theorem 7.1.** *Let $H$ be a Hilbert space, and let $G_0, G_1, \ldots, G_N$ be functionals in $C^1(H, \mathbb{R})$. Let*

$$Q = \{u \in H : G_1(u) = \cdots = G_N(u) = 0\},$$

*and assume that*

$$G_0(u_0) = \min_Q G_0.$$

*Then there are numbers $\lambda_0, \lambda_1, \ldots, \lambda_N$ not all 0 such that*

$$\sum_{j=0}^N \lambda_j G_j'(u_0) = 0.$$

*Proof.* Assume not. Then the elements $v_j = G'_j(u_0)$ are linearly independent. This means that

$$\det[(v_j, v_k)] \neq 0.$$

Otherwise, there would be a vector $\beta = (\beta_0, \beta_1, \ldots, \beta_N) \neq 0$ such that

$$\sum_{i=0}^{N} \beta_i(v_i, v_j) = 0, \quad 0 \leq j \leq N.$$

Multiplying the $j$-th equation by $\beta_j$ and summing, we obtain

$$\left\| \sum_{i=0}^{N} \beta_i v_i \right\|^2 = 0,$$

contradicting the fact that the $v_i$ are linearly independent. For $b \in \mathbb{R}^{n+1}$, let $g(b)$ be the matrix

$$g(b) = (g_{ij}(b)),$$

where

$$g_{ij}(b) = (G'_i(u_0 + \sum_{j=0}^{N} b_j v_j), v_j), \quad 0 \leq i, j \leq N.$$

There is a $\delta > 0$ such that $g(b)$ is invertible for $\|b\| < \delta$. Let $\gamma = (1, 0, \ldots 0)$ and consider the differential equation

$$b'(t) = g^{-1}(b(t))\gamma, \quad |t| \leq t_0, \quad b(0) = 0.$$

By Peano's theorem (Theorem 6.51), this can be solved for some $t_0 > 0$. The solution satisfies

$$g(b(t))b'(t) = \gamma,$$

or

$$\sum_{j=0}^{N}(G'_i(u_0 + \sum_{k=0}^{N} b_k(t)v_k), v_j)b'_j(t) = \gamma_i, \quad 0 \leq i \leq N.$$

But this says

$$\frac{d}{dt}G_i(u_0 + \sum_{k=0}^{N} b_k(t)v_k) = \gamma_i, \quad 0 \leq i \leq N.$$

Consequently,

$$G_i(u_0 + \sum_{k=0}^{N} b_k(t)v_k) = 0, \quad 1 \leq i \leq N,$$

while

$$G_0(u_0 + \sum_{k=0}^{N} b_k(t)v_k) = G_0(u_0) + t, \quad |t| \le t_0.$$

This means that

$$u(t) = u_0 + \sum_{k=0}^{N} b_k(t)v_k \in Q$$

and

$$G_0(u(-t_0)) = G_0(u_0) - t_0,$$

contradicting the hypothesis of the theorem. Hence, the $v_j$ must be linearly dependent. □

As an application of Theorem 7.1, we have

**Theorem 7.2.** *If $\tilde{y}_1(x), \ldots, \tilde{y}_m(x)$ minimize the functional*

$$J_0(y_1, \ldots, y_m) = \int_a^b F_0(x, y_1(x), \ldots, y_m(x), y_1'(x), \ldots, y_m'(x))\, dx$$

*under the constraints*

$$J_k(y_1, \ldots, y_m) = \int_a^b F_k(x, y_1(x), \ldots, y_m(x), y_1'(x), \ldots, y_m'(x))\, dx = c_k,$$
$$k = 1, \ldots, n,$$

*then there are numbers $\lambda_0, \lambda_1, \ldots, \lambda_N$ not all 0 such that*

$$\sum_{k=0}^{N} \lambda_k \left[ \frac{\partial F_k}{\partial y_j} - \frac{d}{dx}\frac{\partial F_k}{\partial y_j'} \right] = 0, \quad j = 1, \ldots, m$$

*when*

$$y_j(x) = \tilde{y}_j(x), \quad j = 1, \ldots, m.$$

*Proof.* We apply Theorem 7.1 and note that

$$J_k' = \left[ \frac{\partial F_k}{\partial y_j} - \frac{d}{dx}\frac{\partial F_k}{\partial y_j'} \right]$$

by Corollary 5.7. □

As a special case we have

**Theorem 7.3.** *Suppose $H(x,y,z)$ is a function on $\bar{\Omega} \times \mathbb{R} \times \mathbb{R}$ satisfying the hypotheses of Theorem 5.3, where $\Omega = (a,b)$, and we want to find a function $y(x) \in C^1(\bar{\Omega})$ satisfying*

$$y(a) = a_1, \quad y(b) = b_1, \tag{7.1}$$

*and minimizing the expression*

$$J(y) = \int_\Omega H(x,y(x),y'(x))dx \tag{7.2}$$

*over functions not only satisfying (7.1) but also a condition such as*

$$J_1(y) = \int_\Omega H_1(x,y(x),y'(x))\,dx = l. \tag{7.3}$$

*If $J(y)$ has a minimum $y = u_0(x)$ over the set of $y(x) \in C^1(\bar{\Omega})$ satisfying (7.1), (7.3), and $J_1'(u_0) \neq 0$, then there is a $\lambda \neq 0$ such that*

$$\tilde{J}'(u_0) = 0,$$

*where*

$$\tilde{J}(y) = \int_\Omega \tilde{H}(x,y(x),y'(x))\,dx \tag{7.4}$$

*and*

$$\tilde{H}(x,y(x),y'(x)) = H(x,y(x),y'(x)) + \lambda H_1(x,y(x),y'(x)). \tag{7.5}$$

*Proof.* By Theorem 7.1 there are constants $\lambda_0, \lambda_1$, not both zero, such that

$$\lambda_0 J'(u_0) + \lambda_1 J_1'(u_0) = 0.$$

Since $J_1'(u_0) \neq 0$, we cannot have $\lambda_0 = 0$. Divide by $\lambda_0$ to obtain the desired statement. $\qquad\square$

We present some well known examples.

**Example 7.4.** *Find a curve of length $l$ given by a positive function $y = y(x)$ satisfying (7.1) having the maximum area under the curve. (We assume that $(b-a)^2 + (b_1 - a_1)^2 < l^2$.)*

We want to maximize

$$J(y) = \int_a^b y(x)\,dx$$

under the conditions (7.1) and

$$J_1(y) = \int_a^b \sqrt{1 + \dot{y}^2}\, dx = l.$$

Applying Theorem 7.3, we take

$$\tilde{H}(x, y, \dot{y}) = y + \lambda\sqrt{1 + \dot{y}^2}$$

and solve

$$\tilde{H}_y(x, y(x), \dot{y}(x)) - \frac{d}{dx}\tilde{H}_z(x, y(x), \dot{y}(x)) = 0, \quad x \in \bar{\Omega}. \tag{7.6}$$

By Lemma 5.12, $\tilde{H}$ satisfies

$$\dot{y}\frac{\partial \tilde{H}}{\partial \dot{y}} - \tilde{H} = \text{constant}. \tag{7.7}$$

Thus,

$$y + \lambda\sqrt{1 + \dot{y}^2} - \frac{\lambda\dot{y}^2}{\sqrt{1 + \dot{y}^2}} = c_1$$

or

$$y + \frac{\lambda}{\sqrt{1 + \dot{y}^2}} = c_1.$$

If we let $\dot{y} = \tan t$, we have

$$y - c_1 = -\lambda\cos t, \quad dy = \lambda\sin t\, dt.$$

Hence,

$$dx = \frac{dy}{\tan t} = \frac{\lambda\sin t\, dt}{\tan t} = \lambda\cos t\, dt$$

and

$$x = \lambda\sin t + c_2.$$

Consequently, we have

$$x - c_2 = \lambda\sin t, \quad y - c_1 = -\lambda\cos t.$$

It is clear that this is a family of circles. The constants $c_1, c_2$, and $\lambda$ can be determined by $l$ and (7.1).

**Example 7.5.** *Given a fixed curve $y = f(x)$ connecting the points $(a, a_1)$ and $(b, b_1)$, find a curve $y = y(x)$ of length $l$ connecting the same points such that the two curves enclose the maximum area.*

In this case, we want to maximize

$$J(y) = \int_a^b [y(x) - f(x)] \, dx$$

under the same conditions as before. Here we have

$$\tilde{H}(x, y, \dot{y}) = y - f(x) + \lambda \sqrt{1 + \dot{y}^2}.$$

We proceed as before. (Can we apply Lemma 5.12?)

**Example 7.6.** *Find the shape of a flexible rope of length l extended between the points $(a, a_1)$ and $(b, b_1)$.*

In this case we find a minimum of the functional

$$J(y) = \int_a^b y \sqrt{1 + \dot{y}^2} \, dx$$

over functions satisfying (7.1) and

$$J_1(y) = \int_a^b \sqrt{1 + \dot{y}^2} \, dx = l.$$

In this case we have

$$\tilde{H}(x, y, \dot{y}) = (y + \lambda) \sqrt{1 + \dot{y}^2}.$$

Again, by Lemma 5.12, $\tilde{H}$ satisfies

$$\dot{y} \frac{\partial \tilde{H}}{\partial \dot{y}} - \tilde{H} = \text{constant}. \tag{7.8}$$

Thus,

$$(y + \lambda) \sqrt{1 + \dot{y}^2} - \frac{(y + \lambda) \dot{y}^2}{\sqrt{1 + \dot{y}^2}} = c_1$$

or

$$\frac{y + \lambda}{\sqrt{1 + \dot{y}^2}} = c_1.$$

This time we set $\dot{y} = \sinh t$. Then

$$\sqrt{1 + \dot{y}^2} = \cosh t, y + \lambda = c_1 \cosh t, \ dx = dy / \sinh t = c_1 dt, x = c_1 t + c_2.$$

Thus,

$$y + \lambda = c_1 \cosh[(x - c_2)/c_1],$$

which is a family of catenaries. Again, the constants $c_1, c_2$, and $\lambda$ can be determined by $l$ and (7.1).

## 7.2     Lagrange multipliers

Suppose we want to minimize the functional

$$J = \int_a^b F(x_1, x_2, \dot{x}_1, \dot{x}_2, t)dt$$

under the constraint

$$g(x_1, x_2, \dot{x}_1, \dot{x}_2, t) = 0.$$

By this we mean that $x_1, x_2$ are not independent but are bound together by a relationship. In this case Lemma 5.11 does not apply. Later we shall show that we can incorporate the constraint into the functional $J$ and consider $x_1, x_2$ independent. In particular, we can replace $J$ by

$$J^* = \int_a^b [F(x_1, x_2, \dot{x}_1, \dot{x}_2, t) - \lambda(t)g(x_1, x_2, \dot{x}_1, \dot{x}_2, t)]dt,$$

where $\lambda(t)$ is a function called a **Lagrange multiplier**. We now proceed as before with $F$ replaced by

$$F^*(x_1, x_2, \dot{x}_1, \dot{x}_2, t) = F(x_1, x_2, \dot{x}_1, \dot{x}_2, t) - \lambda g(x_1, x_2, \dot{x}_1, \dot{x}_2, t).$$

**Example 7.7.** *As an example, let us try to minimize*

$$J = \int_a^b (\ddot{x})^2 dt.$$

This does not fit into the situation covered by Lemma 5.2. However, we can create an additional variable by writing

$$x_1 = x, \ x_2 = \dot{x}_1.$$

Then $J$ becomes

$$J = \int_a^b \dot{x}_2^2 \, dt.$$

Since $g(x_1, x_2, \dot{x}_1, \dot{x}_2, t) = x_2 - \dot{x}_1 = 0$, we can use

$$J^* = \int_a^b [\dot{x}_2^2 - \lambda(x_2 - \dot{x}_1)]dt$$

in place of $J$. The Euler equations give

$$0 - \frac{d}{dt}(-\lambda) = 0, \quad \lambda - \frac{d}{dt}(\dot{x}_2) = 0.$$

Thus

$$\dot{\lambda} = 0, \quad \lambda - \ddot{x}_2 = 0.$$

This implies

$$d\ddot{x}_2/dt = d^3 x_2(t)/dt^3 = 0.$$

Thus,

$$x_2(t) = At^2 + Bt + C,$$

where $A, B, C$ are arbitrary constants. Since $\dot{x}_1 = x_2$, we have

$$x = \frac{1}{3}At^3 + \frac{1}{2}Bt^2 + Ct + D.$$

The arbitrary constants are to be determined by given conditions.

**Example 7.8.** *As another example, consider the functional*

$$J = \int_0^b (1 + \dot{x}\dot{y})^{1/2} t^{-1/2} dt,$$

*which we want to minimize under the constraint*

$$y = x + 1.$$

In this case we can minimize

$$J^* = \int_0^b \left[ \left( \frac{1 + \dot{x}\dot{y}}{t} \right)^{1/2} + \lambda(y - x - 1) \right] dt.$$

Euler's equations become

$$-\lambda - \frac{d}{dt} \frac{\dot{y}}{2t(1 + \dot{x}\dot{y})^{1/2}} = 0$$

and

$$\lambda - \frac{d}{dt} \left( \frac{\dot{x}}{2[t(1 + \dot{x}\dot{y})]^{1/2}} \right) = 0.$$

If we eliminate $\lambda$, we obtain

$$\frac{d}{dt} \left( \frac{\dot{x} + \dot{y}}{2[t(1 + \dot{x}\dot{y})]^{1/2}} \right) = 0.$$

Thus,

$$\frac{\dot{x} + \dot{y}}{[t(1 + \dot{x}\dot{y})]^{1/2}} = \text{const.} = C.$$

Since $\dot{y} = \dot{x}$, this becomes

$$\frac{2\dot{y}}{[t(1 + \dot{y}^2)]^{1/2}} = C.$$

Consequently,

$$\dot{y}^2 = \frac{C^2 t}{4 - C^2 t},$$

from which we can determine $y$. We then use $y = x + 1$ to determine $x$. Actually, we could have eliminated $x$ in the beginning and considered only

$$J = \int_0^b (1 + \dot{y}^2)^{1/2} dt.$$

Then we would not have needed the Lagrange multiplier.

## 7.3    Bang–bang control

Suppose we want to drive a car from a stationary point $a$ to a stationary point $b$ along a horizontal driveway assuming that there is no closed garage door between the points. Assume that the only controls that the driver has are the accelerator and brake. If $h(t)$ represents the acceleration (or deceleration) at the time $t$, we assume that it is subject to the following constraints:

$$-\mu \le h(t) \le \nu. \tag{7.9}$$

The equation of motion is

$$\ddot{x}(t) = h(t). \tag{7.10}$$

If the car arrives at point $b$ at time $T$, we want

$$x(0) = a, \quad \dot{x}(0) = 0, \quad x(T) = b, \quad \dot{x}(T) = 0.$$

We are interested in determining the minimum time for effecting the transfer. Assuming that $\dot{x}(t) \ge 0$ for all $t \in [0, T]$ (i.e., that the driver does not reverse), we may consider $v = \dot{x}$ as a function of $x$. Since

$$h = \ddot{x} = v \frac{dv}{dx} = \frac{d}{dx}\left(\frac{1}{2}v^2\right),$$

we have

$$T = \int_0^T dt = \int_a^b \frac{dx}{v} = \int_a^b \frac{dx}{(2w)^{1/2}},$$

where $w(x) = \frac{1}{2}v^2$. Thus,

$$\frac{dw(x)}{dx} = h, \quad w(a) = w(b) = 0. \tag{7.11}$$

In order to deal with the constraints (7.9), we introduce the additional variable $z$ given by

$$z^2 = (h + \mu)(\nu - h). \tag{7.12}$$

The constraints (7.9) guarantee that $z$ is real.

We are finally ready to proceed (make sure the driver is wearing a seat belt). We wish to minimize $T$ under the conditions that (7.11) and (7.12) hold. Using the method of Lagrange multipliers, we take

$$T^* = \int_a^b \left\{ (2w)^{-\frac{1}{2}} + \lambda_1 \left( \frac{dw}{dx} - h \right) + \lambda_2 [z^2 - (h + \mu)(\nu - h)] \right\} dx.$$

If there is an optimum path $w = w(x)$, it will satisfy

$$\frac{\partial F}{\partial w} - \frac{d}{dx} \left( \frac{\partial F}{\partial w'} \right) = 0, \quad \frac{\partial F}{\partial h} - \frac{d}{dx} \left( \frac{\partial F}{\partial h'} \right) = 0, \quad \frac{\partial F}{\partial z} - \frac{d}{dx} \left( \frac{\partial F}{\partial z'} \right) = 0.$$

Thus, we must have

$$(-2w)^{-3/2} - \frac{d\lambda_1}{dx} = 0, \quad -\lambda_1 + \lambda_2(2h + \mu - \nu) = 0, \quad 2z\lambda_2 = 0.$$

The last equation requires either $z = 0$ or $\lambda_2 = 0$. If $\lambda_2 = 0$ then $\lambda_1 = 0$ and, consequently, $(2w)^{-3/2} = 0$. Since $w(x) \equiv \infty$ clearly violates our guidelines, we must conclude that $z = 0$. But then

$$(h + \mu)(\nu - h) \equiv 0$$

which implies that at each time $t \in [0, T]$ we have either $h(t) = -\mu$ or $h(t) = \nu$. Now, initially we have $h(0) = \nu$ and finally we have $h(T) = -\mu$. Since $h(t)$ must have one of these values in between, all the driver can do is switch from one to the other. Suppose he or she switches only once at the time $t = \alpha$. Then (7.10) becomes

$$\ddot{x}(t) = h(t) = \begin{cases} \nu, & 0 \leq t \leq \alpha, \\ -\mu, & \alpha < t \leq T. \end{cases} \tag{7.13}$$

Integrating, we obtain

$$\dot{x}(t) = \begin{cases} \nu t, & 0 \leq t < \alpha, \\ -\nu(t - T), & \alpha < t \leq T, \end{cases}$$

and

$$x(t) = \begin{cases} \frac{1}{2}\nu t^2 + a, & 0 \leq t < \alpha, \\ -\frac{1}{2}\mu(t - T)^2 + b, & \alpha \leq t \leq T. \end{cases}$$

Since $x(t), \dot{x}(t)$ are required to be continuous in $[0, T]$, this includes the point $t = \alpha$. Hence, we must have

$$\nu\alpha = -\mu(\alpha - T), \quad \frac{1}{2}\nu\alpha^2 + a = -\frac{1}{2}\mu(\alpha - T)^2 + b.$$

From these we obtain

$$\alpha^2 = \frac{2\mu(b - a)}{\nu(\mu + \nu)}$$

and

$$T^2 = \frac{2(b - a)(\mu + \nu)}{\mu\nu}.$$

Thus, the time for the operation is minimized if the driver begins with his or her foot all the way down on the accelerator and then at the time $t = \alpha$ all the way down on the brake until the time $t = T$. (We do **not** recommend this at all.) It is not difficult to understand why this procedure is called bang–bang control.

## 7.4    Rocket in orbit

In order to place a rocket in orbit, it is necessary to obtain a sufficiently high speed at the end of its trajectory. As a simple model, let us assume that it is a single stage rocket launched at an angle $\theta_0$ with the horizon. The thrust is produced by the combustion of fuel and is in the direction of motion. We ignore air resistance (which we should not) and assume that gravity is the only other force acting on the rocket. Since the mass of the rocket diminishes as the propellant is ejected, the force $P$ per unit mass will change with time. The equations of motion are

$$m\ddot{r} = m\dot{v} = m(P\cos\theta, P\sin\theta - g),$$

where $m$ is the mass of the rocket at time $t$, $r = (x_1, x_2)$, $v = \dot{r} = (\dot{x}_1, \dot{x}_2) = (v_1, v_2)$, $\theta$ is the angle the rocket makes with the horizontal direction at time $t$, and $g$ is the gravitational constant. Thus, we have

$$\dot{x}_1 = v_1, \quad \dot{v}_1 = P\cos\theta, \quad \dot{x}_2 = v_2, \quad \dot{v}_2 = P\sin\theta - g. \tag{7.14}$$

If the initial velocity is 0 and the rocket attains its final horizontal speed at time $T$ and height $H$, we wish to maximize

$$v_1(T) = \int_0^T P\cos\theta \, dt.$$

Using the restrictions imposed by (7.14), we try to maximize

$$v_1^*(T) = \int_0^T F(\theta, x_2, \dot{x}_2, v_2, \dot{v}_2)\, dt,$$

where

$$F(\theta, x_2, \dot{x}_2, v_2, \dot{v}_2) = P\cos\theta + \lambda(\dot{x}_2 - v_2) + \mu(\dot{v}_2 - P\sin\theta + g).$$

The Euler equations for this system are

$$\frac{\partial F}{\partial x_2} - \frac{d}{dt}\left(\frac{\partial F}{\partial \dot{x}_2}\right) = 0, \quad \frac{\partial F}{\partial v_2} - \frac{d}{dt}\left(\frac{\partial F}{\partial \dot{v}_2}\right) = 0, \quad \frac{\partial F}{\partial \theta} - \frac{d}{dt}\left(\frac{\partial F}{\partial \dot{\theta}}\right) = 0.$$

Thus, we have

$$\dot{\lambda} = 0, \ \lambda = -\dot{\mu}, \ \mu = -\tan\theta.$$

Consequently, we have $\ddot{\mu} = 0$. This means that $\mu = At + B$. But then, $\tan\theta = -(At + B)$. Since the maximum horizontal speed is achieved at $t = T$, we have $\theta(T) = 0$. Thus, $AT + B = 0$ and $\tan\theta = -B(1 - t/T)$. Since $\theta(0) = \theta_0$ this gives

$$\tan\theta = (1 - t/T)\tan\theta_0. \tag{7.15}$$

We wish to solve for $\theta_0$ and $T$. This is not an easy task. For instance, (7.15) implies

$$T\sec^2\theta\tan\theta\, d\tan\theta = -\tan\theta_0\, dt.$$

This gives

$$v_2 = -\frac{T}{\tan\theta_0}\int P\tan\theta\sec\theta\, dt - gt + A.$$

To simplify matters, we assume that $P$ is a constant. In this case we readily obtain

$$v_2 = -\frac{TP}{\tan\theta_0} - gt + A.$$

Since $v_2(0) = 0$, we see that $A = TP/\sin\theta_0^2$. Thus,

$$v_2 = \frac{TP}{\sin\theta_0}(1 - \cos\theta_0\sec\theta) - gt.$$

Since $\theta(T) = 0$, and $v_2(T) = 0$, we have

$$\frac{TP}{\sin\theta_0}(1 - \cos\theta_0) = gT.$$

Let $\gamma = g/P$. Then

$$\gamma^2 = \frac{(1 - \cos\theta_0)^2}{\sin^2\theta_0}.$$

Thus,

$$\gamma^2 + 1 = \frac{2(1 - \cos\theta_0)}{\sin^2\theta_0}$$

and

$$\gamma^2 - 1 = \frac{2\cos\theta_0(\cos\theta_0 - 1)}{\sin^2\theta_0}.$$

Consequently,

$$\cos\theta_0 = \frac{1 - \gamma^2}{1 + \gamma^2}, \quad \sin\theta_0 = \frac{2\gamma}{1 + \gamma^2}. \tag{7.16}$$

Substituting this into the equation for $v_2$, we obtain

$$v_2 = \frac{TP}{2\gamma}[(1 + \gamma^2) - (1 - \gamma^2)\sec\theta] - gt.$$

Thus,

$$x_2 = \frac{TP}{2\gamma}\left[(1 + \gamma^2)^{1/2} - (1 - \gamma^2)\int\sec\theta dt\right] - \frac{1}{2}gt^2 + B$$

for some constant $B$. As before,

$$\int\sec\theta\, dt = -\frac{T(1 - \gamma^2)}{2\gamma}\int\sec^3\theta\, d\theta$$

$$= -\frac{T(1 - \gamma^2)}{4\gamma}\left[\frac{\sin\theta}{\cos^2\theta} + \log(\sec\theta + \tan\theta)\right].$$

When $\theta = \theta_0$, this expression becomes

$$-\frac{T(1 - \gamma^2)}{4\gamma}\left[\frac{2\gamma(1 + \gamma^2)}{(1 - \gamma^2)^2} + \log\left\{\frac{1 + \gamma}{1 - \gamma}\right\}\right].$$

Since $x_2(0) = 0$, $\theta(0) = \theta_0$, this implies

$$B = \frac{-T^2P}{8\gamma^2}\left[2\gamma(1 + \gamma^2) + (1 - \gamma^2)^2\log\left\{\frac{1 + \gamma}{1 - \gamma}\right\}\right].$$

Since $x_2(T) = H$, $\theta(T) = 0$, we have

$$H = \frac{T^2P(1 + \gamma^2)}{2\gamma} - \frac{1}{2}gT^2 + B.$$

Consequently,

$$\frac{H}{T^2 P} = \frac{1+\gamma^2}{2\gamma} - \frac{1}{2}\gamma + \frac{B}{T^2 P} = \frac{1}{2\gamma} + \frac{B}{T^2 P}$$

$$= \frac{1}{8\gamma^2}\left[4\gamma - 2\gamma(1+\gamma^2) - (1-\gamma^2)^2\log\left\{\frac{1+\gamma}{1-\gamma}\right\}\right]$$

$$= \frac{(1-\gamma^2)}{8t^2}\left[2\gamma - (1-\gamma^2)\log\left\{\frac{1+\gamma}{1-\gamma}\right\}\right].$$

Hence,

$$T^2 = \frac{8\gamma^2 H}{P(1-\gamma^2)\left[2\gamma - (1-\gamma^2)\log\left\{\dfrac{1+\gamma}{1-\gamma}\right\}\right]}. \tag{7.17}$$

## 7.5  A generalized derivative

In our studies of extrema in the calculus of variations, we came across problems such as finding a minimum of

$$F(y) = \int_a^b H(x, y, \dot{y})\, dx \tag{7.18}$$

on functions $y \in H^1(\Omega)$ satisfying

$$y(a) = a_1, \quad y(b) = b_1, \tag{7.19}$$

where $\Omega = (a, b)$. If we want to differentiate $F$ using the Fréchet or Gâteaux derivative, we come across the following problem. If $y, \eta$ both satisfy (7.19), then $y + t\eta$ need not. Therefore, we cannot compute the difference quotient

$$\frac{[F(y+t\eta) - F(y)]}{t} \tag{7.20}$$

We had to adjust the definition to have $\eta$ satisfy

$$\eta(a) = \eta(b) = 0. \tag{7.21}$$

Then $y + t\eta$ will satisfy (7.19) whenever $y$ does. This does not result in any great trauma as long as we realize that we are differentiating with respect to $H_0^1(\Omega)$ in place of $H^1(\Omega)$. However, what do we do when we want to minimize $F(y)$ on the set of those $y \in H^1(\Omega)$ satisfying both (7.19) and

$$\varphi(x, y, \dot{y}) = 0? \tag{7.22}$$

We can still differentiate $F$ with respect to $H_0^1(\Omega)$, but the derivative will not necessarily vanish at a minimum. What can we do?

We are faced with the following situation. Let $X$ be a vector space and let $Q, V, Y$ be Banach spaces such that $Q, V \subset X$. Let $F$ be a mapping from $X$ to $Y$. We are confronted with two problems.

(a) If $u + th$ is not in $V$ for $u \in V$ and $h \in Q$, how can we define the derivative when the difference quotient

$$\frac{[F(u + th) - F(u)]}{t} \qquad (7.23)$$

does not exist?

(b) Even if $V + Q \subset V$, we have difficulty dealing with situations such as

$$G(u) = \min_S G, \qquad (7.24)$$

in which $u + tq$ need not be in $S \subset V$ even though it is in $V$ for $q \in Q$. For then it is not necessarily true that $G'(u) = 0$.

The following approach is intended to deal with these situations.

## 7.6    The definition

Let $X, Y, V, Q$ be as described above. For each $u \in V$ we take $C(V, Q, u)$ to denote the set consisting of those $q \in Q$ for which there exist sequences $\{t_n\} \subset \mathbb{R}$, and $\{q_n\} \subset Q$ such that

$$q_n \to q \text{ in } Q, \quad 0 \neq t_n \to 0, \quad u + t_n q_n \in V. \qquad (7.25)$$

The set of $C(V, Q, u)$ need not be a subspace of $Q$. We let $E(V, Q, u)$ be the smallest subspace of $Q$ containing $C(V, Q, u)$, that is, the set of all finite sums of elements in $C(V, Q, u)$. We have the following definition.

**Definition 7.9.** *A linear operator $A$ from $X$ to $Y$ is called the derivative of $F$ with respect to $Q$ at the point $u$ and denoted by $F_Q'(u)$ if*

*(a) $D(A) = E(V, Q, u)$ and*
*(b) for any sequences $\{t_n\}, \{q_n\}$ satisfying (7.25) we have*

$$t_n^{-1}[F(u + t_n q_n) - F(u)] \to Aq \text{ in } Y \text{ as } n \to \infty. \qquad (7.26)$$

Note that when it exists, $F'_Q(u)$ is uniquely determined on $E(V, Q, u)$. On the other hand, any larger domain might not determine it uniquely. Note that even when $V + Q \subset V$, the requirement (7.26) is stronger than the existence of the limit

$$t^{-1}[F(u + tq) - F(u)] \to Aq \quad \text{as} \quad t \to 0.$$

It is possible for this limit to exist while that of (7.26) does not exist.

## 7.7     The theorem

We now study some of the properties of this derivative.

**Theorem 7.10.** *Assume that $V + Q \subset V$ and that there are $u \in V$ and $T \in B(Q, Y)$ such that*

$$\|T(h_1 - h_2) - [F(u + h_1) - F(u + h_2)]\| \le \delta\|h_1 - h_2\|, \quad h_i \in Q, \|h_i\| < m. \tag{7.27}$$

*Assume further that $R(T) = Y$,*

$$d(h, N(T)) = \inf_{w \in N(T)} \|h - w\| \le C_0\|Th\|, \quad h \in Q \tag{7.28}$$

*and $\delta C_0 < 1$. Then for each $q \in Q$ there is an interval $(-r, r)$ and a mapping $q(t)$ of $(-r, r)$ into $Q$ such that*

$$F(u + tq(t)) = F(u), \quad -r < t < r, \tag{7.29}$$

*and*

$$\|q(t) - q\| \le Ct^{-1}\|F(u + tq) - F(u)\|, \quad -r < t < r. \tag{7.30}$$

Before proving Theorem 7.10, we would like to show how it can be applied. One application is

**Theorem 7.11.** *Under the hypotheses of Theorem 7.10, let*

$$S = \{v \in V : F(v) = F(u)\}. \tag{7.31}$$

*Let $G$ be a functional on $V$ such that*

$$G(u) = \min_S G. \tag{7.32}$$

*Assume that both $F'_Q(u)$ and $G'_Q(u)$ exist. Then*

$$F'_Q(u)q = 0 \implies G'_Q(u)q = 0. \tag{7.33}$$

*Proof.* Let $q$ be any element of $Q$. By Theorem 7.10, there is an interval $(-r, r)$ and a mapping $q(t)$ of $(-r, r)$ into $Q$ such that (7.29) and (7.30) hold. Thus, $u + tq(t) \in S$ for $-r < t < r$. If $F'_Q(u)q = 0$, then $q(t) \to q$ by (7.30). Now

$$G(u + tq(t)) \geq G(u)$$

by (7.32) and

$$t^{-1}[G(u + tq(t)) - G(u)] \to G'_Q(u)q.$$

If $t \to 0$ through positive values, the limit is nonnegative and if it approaches through negative values, the limit is nonpositive. Since the limit is the same for both, we see that $G'_Q(u)q = 0$. □

As a result, we have

**Theorem 7.12.** *Under the same hypotheses, $H = F'_Q(u)$ maps $Q$ onto $Y$ and*

$$G'_Q(u)[1 - H^{-1}F'_Q(u)]q = 0, \quad q \in Q, \tag{7.34}$$

*where $H^{-1}y$ is any element $q \in Q$ such that $Hq = y$.*

In proving Theorem 7.12 we shall make use of

**Lemma 7.13.** *If $X, Y, Z$ are Banach spaces, $H \in B(X, Y), L \in B(X, Z)$, $R(H) = Y$, $N(H) \subset N(L)$, then $LH^{-1} \in B(Y, Z)$, where $H^{-1}y$ is any element $x \in X$ such that $Hx = y$.*

*Proof.* The operator $LH^{-1}$ is well defined. For if $Hx = y$ and $Hx_1 = y$, then $Lx = Lx_1$. To show that it is bounded, it suffices to show that it is continuous at 0 (Theorem A.23). Suppose $y_k \to 0$. Let $Hx_k = y_k$. Then by Theorem A.64,

$$d(x_k, N(H)) \leq C\|y_k\| \to 0.$$

In view of Lemma A.25, there is an $\hat{x}_k \in X$ such that

$$H\hat{x}_k = Hx_k = y_k, \quad \text{and} \quad \|\hat{x}_k\| \leq d(x_k, N(H)) \to 0.$$

Hence

$$Lx_k = L\hat{x}_k \to 0.$$

□

**Theorem 7.14.** *Let $Q, Y$ be Banach spaces, and assume that there are operators $T, H \in B(Q, Y)$ such that (7.28) holds with $\|T - H\| \leq \delta$ and $\delta C_0 < 1$. If $R(T) = Y$, then $R(H) = Y$.*

*Proof.* Let $y$ be any element of $Y$, and let $C_1 > C_0$ be such that $\rho = \delta C_1 < 1$. Then there is an element $z_0 \in Q$ such that

$$T z_0 = y, \quad \|z_0\| \le C_1 \|y\|$$

and inductively, there are elements $z_k \in Q$ such that

$$T z_k = (T - H) z_{k-1}, \quad \|z_k\| \le C_1 \|T z_k\|, \quad k = 1, 2, \ldots$$

Then

$$\|z_k\| \le C_1 \|(T - H) z_{k-1}\| \le \rho \|x_{k-1}\| \le \rho^k \|z_0\|.$$

Thus,

$$h_k \equiv \sum_0^k z_j \to h \text{ in } Q$$

and

$$
\begin{aligned}
T h_k &= \sum_0^k T z_j \\
&= \sum_1^k (T - H) z_{j-1} + y \\
&= (T - H) \sum_{i=0}^{k-1} z_i + y \\
&= (T - H) h_{k-1} + y \\
&\to (T - H) h + y.
\end{aligned}
$$

Thus, $T h = (T - H) h + y$, or $H h = y$. Since $y$ was any element of $Y$, this shows that $R(H) = Y$. $\qquad\square$

We can now give the proof of Theorem 7.12.

*Proof.* First we show that $R(H) = Y$. By (7.27)

$$\|T h - [F(u + h) - F(u)]\| \le \delta \|h\|, \quad h \in Q.$$

Hence $\|T - H\| \le \delta$ and $\delta C_0 < 1$. Since $R(T) = Y$, we see that $R(H) = Y$ by Theorem 7.14. Let $q$ be any fixed element in $Q$, and let $h \in Q$ be any element such that $q - h \in N(H)$. Then $G'_Q(u)(q - h) = 0$ by Theorem 7.11. But $Hh = F'_Q(u)q$. Hence,

$$G'_Q(u)[1 - H^{-1} F'_Q(u)]q = 0.$$

$\qquad\square$

This leads to

**Theorem 7.15.** *Let $F$ be a mapping of $V \subset X$ into $Y$, and $G$ a mapping of $V$ into $\mathbb{R}$. Let $S$ be given by (7.31), and assume that (7.32) holds for some $u \in S$. Assume that $V + Q \subset V$, and that $F'_Q(u+q)$ exists and is in $B(Q, Y)$ for $q \in Q$ small with*

$$\|F'_Q(u+q) - F'_Q(u)\| \to 0 \quad as \quad q \to 0.$$

*Assume further that the range of $F'_Q(u)$ is closed in $Y$ and that $G'_Q(u)$ exists and is in $Q'$. Then there exist $\lambda \in \mathbb{R}$ and $y' \in Y'$ not both vanishing such that*

$$\lambda G'_Q(u) + y' F'_Q(u) = 0. \tag{7.35}$$

In proving Theorem 7.15 we shall use

**Theorem 7.16.** *Let $q \in Q$ be such that $u + sq \in V$ for $0 \le s \le 1$, and let $y' \in Y'$. Assume that*

$$F'_Q(u + sq) \to y' F(u) \quad as \quad s \to 0$$

*and*

$$y' F(u + sq) \to y' F(u+q) \quad as \quad s \to 1.$$

*Then there is a $\theta$ such that $0 < \theta < 1$ and*

$$y'[F(u+q) - F(u)] = y' F'_Q(u + \theta q)q.$$

*Proof.* First we note that $q \in C(V, Q, u + sq)$ for $0 < s < 1$. For if $t_n \to 0$, then $u + sq + t_n q$ is in $V$ for $n$ large. If we put $q_n = q$, then (7.25) holds for $u + sq$. Put $f(s) = y' F(u + sq)$. Then for $0 < s < 1$, we have

$$t_n^{-1}[f(s + t_n) - f(s)]$$
$$= t_n^{-1} y'[F(u + sq + t_n q) - F(u + sq)]$$
$$\to y' G'_Q(u + sq)q \quad as \quad t_n \to 0.$$

Thus, $f(s)$ is differentiable in $(0, 1)$. By hypothesis, it is continuous in $[0, 1]$. Hence, there is a $\theta$ such that $0 < \theta < 1$ and $f(1) - f(0) = f'(\theta)$. This gives the desired result. □

We can now give the proof of Theorem 7.15.

*Proof.* Assume first that $M = R(F'_Q(u)) \ne Y$. Since $M$ is closed in $Y$ there is a $y_0 \in Y$ such that $d = d(y_0, M) > 0$. (Otherwise, for each $y_0 \in Y$ there would be a sequence $\{y_n\} \subset M$ such that $\|y_n - y_0\| \to 0$.

Since $M$ is closed, this would imply that $y_0 \in M$. Thus, we would have $M = Y$.) By Theorem A.17, there is a $y' \in Y'$ such that $\|y'\| = 1$, $y'(y_0) = d > 0$, and $y'(M) = 0$. Hence, (7.35) holds with $\lambda = 0$.

On the other hand, if $M = Y$, we can apply Theorem 7.12. All of the hypotheses of Theorem 7.10 are satisfied. We take $T = F'_Q(u)$. Then by Theorem 7.16 we have

$$y'[T(q_1 - q_2) - F(u + q_1) + F(u + q_2)] = y'[T - F'_Q(u + q_\theta)](q_1 - q_2)$$

for each $y' \in Y'$, where $q_\theta = \theta q_1 + (1 - \theta)q_2$. Thus,

$$\|y'[T(q_1 - q_2) - F(u + q_1) + F(u + q_2)]\|$$
$$\leq \|y'\| \cdot \|T - F'_Q(u + q_\theta)\| \cdot \|q_1 - q_2\|.$$

If the $\|q_i\|$ are sufficiently small, this is bounded by

$$\delta \|y'\| \cdot \|q_1 - q_2\|,$$

where $\delta C_0 < 1$. This is true for each $y' \in Y'$. Thus, by Corollary A.16,

$$\|T(q_1 - q_2) - [F(u + q_1) - F(u + q_2)]\| \leq \delta \|q_1 - q_2\|, \quad \|q_i\| < m$$

for $m$ sufficiently small. Moreover, by Lemma 7.13, the operator $G'_Q(u)H^{-1}$ is in $Y'$. We can now see that (7.34) implies (7.35) with $\lambda = 1$ and $y' = G'_Q(u) \, H^{-1}$. This completes the proof.                    □

Theorem 7.15 is a generalization of the **Lagrange multiplier rule**.

## 7.8     The proof

In this section we give the proof of Theorem 7.10. It will be based on

**Theorem 7.17.** *Let $T$ be a closed operator from a Banach space $Q$ to a Banach space $Y$ satisfying (7.28). Suppose $h_0 \in D(T)$ and $f$ is a mapping of the ball $B = \{h \in Q : \|h - h_0\| < m\}$ into $Y$ such that*

$$\|f(h) - f(h')\| \leq \delta \|h - h'\|, \quad h, h' \in B. \tag{7.36}$$

*Assume that $R(T) = Y$ and $\delta C_0 < 1$. Let $h_1$ be any element of $D(T) \cap B$, and put $y_1 = Th_1 - f(h_1)$. Then for any $\varepsilon > 0$ and any $y \in Y$ such that*

$$\left(\frac{C_0}{(1 - \delta C_0)}\right) \|y - y_1\| + \|h_1 - h_0\| < m, \tag{7.37}$$

*there exists an $h \in D(T) \cap B$ such that*

$$Th - f(h) = y \tag{7.38}$$

*and*

$$\|h - h_1\| \leq \left( \frac{C_0}{(1 - \delta C_0)} + \varepsilon \right) \|y - y_1\|. \tag{7.39}$$

*Proof.* Let $\varepsilon > 0$ be given, and take $C_1 > C_0$ such that (7.37) holds with $C_0$ replaced by $C_1$, with

$$\frac{C_1}{1 - \delta C_1} < \frac{C_0}{1 - \delta C_0} + \varepsilon,$$

and $\rho = \delta C_1 < 1$. From (7.28) and the fact that $R(T) = Y$, we can find an element $z_1 \in D(T)$ such that $Tz_1 = y - y_1$ and $\|z_1\| \leq C_1 \|y - y_1\|$. Define $\{h_k\}, \{z_k\}$ inductively by

$$h_{k+1} = h_1 + \sum_{j=1}^{k} z_j, \quad Tz_k = f(h_k) - f(h_{k-1}), \quad z_k = h_{k+1} - h_k,$$

$$\|z_k\| \leq C_1 \|Tz_k\| = C_1 \|f(h_k) - f(h_{k-1})\| \leq \rho \|z_{k-1}\|.$$

Thus

$$\|z_k\| \leq \rho^{k-1} \|z_1\|.$$

Since $\rho < 1$, we see that $h_k \to h$ in $Q$. Moreover,

$$\|h - h_0\| \leq \|h_1 - h_0\| + \sum_{j=1}^{\infty} \|z_j\| \leq \|h_1 - h_0\| + \sum_{j=1}^{\infty} \rho^{j-1} \|z_1\|$$

$$\leq \|h_1 - h_0\| + \frac{C_1}{1 - \rho} \|y - y_1\| < m.$$

Thus $h \in B$. Also

$$\|h - h_1\| \leq \frac{C_1}{1 - \rho} \|y - y_1\|,$$

showing that (7.39) holds. Finally, we note that

$$Th_{k+1} = Th_1 + \sum_{j=1}^{k} Tz_j = Th_1 + f(h_k) - f(h_1) + y - y_1$$

$$= f(h_k) + y \to f(h) + y \quad \text{as} \quad k \to \infty.$$

Since $T$ is a closed operator, we see that $Th - f(h) = y$ holds. This completes the proof. $\qquad\square$

We can now give the proof of Theorem 7.10.

*Proof.* Put

$$f(h) = Th - F(u + h) + F(u), \quad h \in Q.$$

It maps the ball $\|h\| < m$ into $Y$. Let $q$ be any element in $Q$, and let $\varepsilon > 0$ be given. By (7.27),

$$\|f(h) - f(h')\| \leq \delta\|h - h'\|, \quad h, h' \in B,$$

where $B$ is the ball $\|h\| < m$ in $Q$. We apply Theorem 7.17 with $h_0 = 0$, $h_1 = tq$, $y = 0$. We take $r > 0$ such that

$$\frac{C_0}{1 - \delta C_0}\|F(u + tq) - F(u)\| + \|tq\| < m$$

for $|t| < r$. Since

$$y_1 = T(tq) - f(tq) = F(u + tq) - F(u),$$

this implies

$$\frac{C_0}{1 - \delta C_0}\|y_1\| + \|h_1\| < m$$

for $|t| < r$. Theorem 7.17 now implies that for each $t$ satisfying $|t| < r$ there is an $h \in B$ such that $Th - f(h) = 0$ and

$$\|h - tq\| \leq C\|y_1\|.$$

These translate into

$$F(u + h) - F(u) = 0$$

and

$$\|h - tq\| \leq C\|F(u + tq) - F(u)\|.$$

Put $q(t) = h/t$. Then we have

$$F(u + tq(t)) = F(u), \quad |t| < r,$$

and

$$\|q(t) - q\| \leq Ct^{-1}\|F(u + tq) - F(u)\|, \quad |t| < r.$$

This is precisely what we wanted to prove.                              $\square$

## 7.9    Finite subsidiary conditions

In the isoperimetric problem, the constraints were in the form of integral conditions. We now consider the point-wise constraints of the form

$$\varphi(x, y_1, y_2, \dot{y}_1, \dot{y}_2) = 0.$$

Such constraints arise in many applications as we have just seen. Here we shall show why Lagrange multipliers work. We have

**Theorem 7.18.** *Assume that $F_0(x, y_1, \ldots, y_n, z_1, \ldots, z_n) \in C^1(\bar{\Omega} \times \mathbb{R}^{2n})$. If the functions $\tilde{y}_1(x), \ldots, \tilde{y}_n(x)$ make the functional*

$$J(y) = \int_{\Omega} F_0(x, y_1(x), \ldots, y_n(x), \dot{y}_1(x), \ldots, \dot{y}_n(x))\, dx \qquad (7.40)$$

*have an extremum under the conditions*

$$\varphi_i(x, y_1(x), \ldots, y_n(x), \dot{y}_1(x), \ldots, \dot{y}_n(x)) = 0, \quad j = 1, \ldots, m, \ m < n, \qquad (7.41)$$

*then they satisfy the Euler equations for the functional*

$$\tilde{J}(y) = \int_{\Omega} [\lambda_0 F_0 + \sum_{i=1}^{m} \lambda_i(x)\varphi_i]\, dx \qquad (7.42)$$

*for a suitably chosen constant $\lambda_0$ and functions $\lambda_1(x), \ldots, \lambda_m(x)$ not all zero. If we define*

$$\tilde{F} = \lambda_0 F_0 + \sum_{i=1}^{m} \lambda_i(x)\varphi_i, \qquad (7.43)$$

*then the functions $\tilde{y}_1(x), \ldots, \tilde{y}_n(x), \lambda_1(x), \ldots, \lambda_m(x)$ will satisfy*

$$\frac{\partial \tilde{F}}{\partial y_j} - \frac{d}{dx}\frac{\partial \tilde{F}}{\partial \dot{y}_j} = 0, \quad j = 1, \ldots, n \qquad (7.44)$$

*as well as (7.41). These two sets of equations can be used to determine the $\tilde{y}_j$ and $\lambda_i$. It is assumed that each function*

$$\varphi_k(x, y_1, \ldots, y_n, z_1, \ldots, z_n)$$

*is in $C^1(\bar{\Omega} \times \mathbb{R}^{2n})$ for each $k$, $1 \leq k \leq m$. It is also assumed that the Jacobian*

$$\frac{\partial(\varphi_1, \ldots, \varphi_m)}{\partial(z_1, \ldots, z_m)} \neq 0, \quad x \in \bar{\Omega}, \ y, z \in \mathbb{R}^n.$$

*Proof.* Let

$$V = H^1(\Omega) \times \cdots \times H^1(\Omega), \quad n \text{ times,}$$
$$Q = H_0^1(\Omega) \times \cdots \times H_0^1(\Omega), \quad n \text{ times,}$$

and

$$Y = L^2(\Omega) \times \cdots \times L^2(\Omega), \quad m \text{ times.}$$

Let $F$ be the mapping from $V$ to $Y$ given by

$$F(y_1, \ldots, y_n) = (\varphi_1(x, y_1, \ldots, y_n, \dot{y}_1, \ldots, \dot{y}_n), \ldots,$$
$$\varphi_m(x, y_1, \ldots, y_n, \dot{y}_1, \ldots, \dot{y}_n)), \quad (y_1, \ldots, y_n) \in V.$$

Let

$$u = (\tilde{y}_1(x), \ldots, \tilde{y}_n(x)),$$

and let

$$q = (\eta_1(x), \ldots, \eta_n(x))$$

be any element of $Q$. Since

$$t^{-1}[\varphi_k(x, y_1 + t\eta_1, \ldots, y_n + t\eta_n, \dot{y}_1 + t\dot{\eta}_1, \ldots, \dot{y}_n + t\dot{\eta}_n)$$
$$- \varphi_k(x, y_1, \ldots, y_n, \dot{y}_1, \ldots, \dot{y}_n)]$$

$$\rightarrow \sum_{j=1}^n \left( \frac{\partial \varphi_k}{\partial y_j} \eta_j + \frac{\partial \varphi_k}{\partial \dot{y}_j} \dot{\eta}_j \right), \quad 1 \le k \le m,$$

we have

$$F_Q'(u)q = \left\{ \left( \sum_{j=1}^n [a_{j1}\eta_j + b_{j1}\dot{\eta}_j] \right), \ldots, \left( \sum_{j=1}^n [a_{jm}\eta_j + b_{jm}\dot{\eta}_j] \right) \right\},$$

where

$$a_{jk} = \frac{\partial \varphi_k(x, u)}{\partial y_j}, \quad b_{jk} = \frac{\partial \varphi_k(x, u)}{\partial \dot{y}_j}, \quad 1 \le j \le n, \ 1 \le k \le m.$$

Let

$$M = \{q = (\eta_1, \ldots, \eta_n) \in Q : \eta_j = 0, \ m < j \le n\}$$

and

$$N = \{q = (\eta_1, \ldots, \eta_n) \in Q : \eta_j = 0, \ 1 \le j \le m\}.$$

Then

$$Q = M \oplus N.$$

By hypothesis, the matrix $(b_{jk})$, $1 \leq j, k \leq m$, is invertible. Hence, there is a matrix $(d_{kl})$ such that

$$\sum_{k=1}^{m} b_{jk} d_{kl} = \sum_{k=1}^{m} d_{kl} b_{jk} = \delta_{jl}, \quad 1 \leq j, l \leq m.$$

Let

$$\tilde{T}q = \left( \sum_{j=1}^{m} b_{j1} \dot{\eta}_j, \ldots, \sum_{j=1}^{m} b_{jm} \dot{\eta}_j \right), \quad q \in M.$$

I claim that

$$\tilde{T} \in \Phi(M, Y).$$

To see this, let

$$Uw = \left( \sum_{k=1}^{m} d_{k1} w_k, \ldots, \sum_{k=1}^{m} d_{km} w_m \right), \quad w = (w_1, \ldots, w_m) \in Y.$$

Then $U \in B(Y)$, and

$$U\tilde{T}q = (\dot{\eta}_1, \ldots, \dot{\eta}_m), \quad q \in M.$$

Next we note

**Lemma 7.19.** *A function $w \in L^2(\Omega)$ is the weak derivative of a function $v \in H_0^1(\Omega)$ if and only if $(w, 1) = 0$.*

*Proof.* If $v \in H_0^1(\Omega)$, then there is a sequence $\{v_k\} \subset C_0^1(\Omega)$ such that $\|v_k - v\|_H \to 0$. In particular, $(v'_k, 1) \to (v', 1)$. But

$$(v'_k, 1) = \int_a^b v'_k(x)\, dx = v_k(b) - v_k(a) = 0.$$

Hence $(v'_k, 1) = 0$, and consequently, $(v', 1) = 0$. Conversely, if $(w, 1) = 0$, then there is a sequence $\{w_k\} \in C_0^\infty(\Omega)$ such that $w_k \to w$ in $L^2(\Omega)$. Let

$$\tilde{w}_k = w_k - \frac{(w_k, 1)}{(b - a)}.$$

Then

$$(\tilde{w}_k, 1) = (w_k, 1) - (w_k, 1)\frac{\int_a^b dx}{(b - a)} = 0.$$

Since

$$(w_k, 1) \to (w, 1) = 0,$$

we see that $\tilde{w}_k \to w$ in $L^2(\Omega)$. Set

$$v_k(x) = \int_a^x \tilde{w}_k(y)dy.$$

Then

$$v_k(a) = v_k(b) = 0.$$

Thus, $v_k \in C_0^1(\Omega)$. Moreover, $v_k \to v$ uniformly, where

$$v(x) = \int_a^x w(y)\,dy.$$

Since

$$v_k' = \tilde{w}_k \to w \ \text{ in } \ L^2(\Omega),$$

we see that

$$\|v_k - v\|_H \to 0.$$

Hence, $v \in H_0^1(\Omega)$.                                    $\square$

Let $Z$ be the set of all $w \in Y$ of the form

$$w = (c_1, \dots, c_m),$$

where the $c_k$ are constants. If

$$l_k = (0, \dots, 1, \dots, 0)$$

is the element in $Y$ consisting of functions which are 0 in all positions except in the $k$-th position, where it is the constant 1, then

$$w = (c_1, \dots, c_m) = \sum_{k=1}^m c_k l_k.$$

Hence, $Z$ is a subspace of $Y$ having dimension $m$. By Lemma 7.19

$$Y = R(U\tilde{T}) \oplus Z.$$

Consequently,

$$\beta(U\tilde{T}) = m$$

(Corollary A.72). Since

$$\alpha(U\tilde{T}) = 0$$

we have

$$i(U\tilde{T}) = -m.$$

Thus,

$$UÑ\tilde{T} \in \Phi(M,Y).$$

Since $U \in B(Y)$ is invertible, we see that $\tilde{T} \in \Phi(M,Y)$ and $i(\tilde{T}) = -m.$

Now,

$$T_1 q = (\sum_{j=1}^{m} a_{j1}\eta_j, \ldots, \sum_{j=1}^{m} a_{jm}\eta_j)$$

is a compact operator from $M$ to $Y$. To see this, assume that

$$q_i = (\eta_{1i}, \ldots, \eta_{mi})$$

is a sequence of elements of $M$ such that

$$\|q_i\|_Q \leq C.$$

Thus,

$$\sum_{k=1}^{m} \|\eta_{ki}\|_H^2 \leq C^2.$$

By Lemma 1.21 there is a renamed subsequence such that

$$\eta_{ki} \to \eta_k, \quad 1 \leq k \leq m$$

uniformly in $\bar{\Omega}$. Therefore

$$\sum_{k=1}^{m} a_{jk}\eta_{ki} \to \sum_{k=1}^{m} a_{jk}\eta_k \text{ as } i \to \infty, \quad 1 \leq j \leq m$$

uniformly in $\bar{\Omega}$. Let

$$T_2 = T_1 + \tilde{T}.$$

Since $T_1 \in K(M,Y)$ and $\tilde{T} \in \Phi(M,Y)$ with index $i(\tilde{T}) = -m$, we see that $T_2 \in \Phi(M,Y)$ with $i(T_2) = -m$. Let $P_M, P_N$ be the orthogonal projections of $Q$ onto $M$ and $N$, respectively. We define

$$T = T_2 P_M + T_3 P_N,$$

where

$$T_3 q = \left( \sum_{j=m+1}^{n} [a_{j1}\eta_j + b_{j1}\dot{\eta}_j], \ldots, \sum_{j=m+1}^{n} [a_{jm}\eta_j + b_{jm}\dot{\eta}_j] \right), \quad q \in N.$$

Then $T_3 \in B(N, Y)$ and

$$Tq = \left( \sum_{j=1}^{n} [a_{j1}\eta_j + b_{j1}\dot{\eta}_j], \ldots, \sum_{j=1}^{n} [a_{jm}\eta_j + b_{jm}\dot{\eta}_j] \right) = F_Q'(u)q, \quad q \in Q.$$

Note that $\beta(T_2) < \infty$. Consequently, there is a finite dimensional subspace $W \subset Y$ such that

$$Y = R(T_2 P_M) \oplus W$$

(Lemma A.71). Since $R(T) \supset R(T_2 P_M)$, we see that $R(T)$ is closed in $Y$ (Lemma A.70).

Let

$$J(y) = \int_a^b F_0(x, y_1, \ldots, y_n, \dot{y}_1, \ldots, \dot{y}_n)\, dx, \quad y = (y_1, \ldots, y_n).$$

Then

$$J_Q'(y)q = \int_a^b \sum_{j=1}^{n} \left( \frac{\partial F_0}{\partial y_j} \eta_j + \frac{\partial F_0}{\partial \dot{y}_j} \dot{\eta}_j \right) dx.$$

If

$$u = (\tilde{y}_1, \ldots, \tilde{y}_n),$$

Theorem 7.15 tells us that there are $\lambda_0 \in \mathbb{R}$ and $y' \in Y'$ such that

$$\lambda_0 J_Q'(u) + y' F_Q'(u) = 0.$$

Since $Y$ is a Hilbert space, $Y'$ can be represented by $Y$ (Theorem A.12). Thus, we can take

$$y' = (\lambda_1(x), \ldots, \lambda_m(x)) \in Y,$$

with

$$y'(y) = \int_a^b \sum_{k=1}^{m} \lambda_k(x) y_k(x)\, dx, \quad y \in Y.$$

Consequently,

$$y'(F_Q'(u)q) = \int_a^b \sum_{j=1}^{n} \sum_{k=1}^{m} \lambda_k(x)[a_{jk}\eta_j + b_{jk}\dot{\eta}_j]\, dx.$$

If we put

$$\tilde{F} = \lambda_0 F_0 + \sum_{k=1}^{m} \lambda_k(x)\varphi_k,$$

we obtain

$$y'(F_Q'(u)q) = \int_a^b \sum_{j=1}^n [\frac{\partial \tilde{F}}{\partial y_j} \eta_j + \frac{\partial \tilde{F}}{\partial \dot{y}_j} \dot{\eta}_j] \, dx = \int_a^b \sum_{j=1}^n [\frac{\partial \tilde{F}}{\partial y_j} - \frac{d}{dx} \frac{\partial \tilde{F}}{\partial \dot{y}_j}] \eta_j \, dx.$$

We now apply Lemma 5.1 to obtain the desired conclusion. $\square$

## 7.10 Exercises

1. Show that

$$\det[(v_j, v_k)] = 0$$

implies that there is a vector $\beta = (\beta_0, \beta_1, \ldots, \beta_N) \neq 0$ such that

$$\sum_{i=0}^N \beta_i(v_i, v_j) = 0, \quad 0 \le j \le N.$$

2. Why is the matrix

$$g(b) = (g_{ij}(b)),$$

given by

$$g_{ij}(b) = (G_i'(u_0 + \sum_{j=0}^N b_j v_j), v_j), \quad 0 \le i, j \le N$$

invertible for $\|b\|$ sufficiently small?

3. Show that the system of differential equations

$$b'(t) = g^{-1}(b(t))\gamma, \quad |t| \le t_0, \quad b(0) = 0$$

satisfies the hypotheses of Peano's theorem.

4. Carry out the calculations to solve Example 7.10.5. Can we use Lemma 5.12?

5. Finish solving the problem in Example 7.10.8 using both ways.

6. Fill in the details in the calculations for the bang–bang control problem.

7. Fill in the details in the calculations for the rocket in orbit problem.

8. Show that when it exists, $F_Q'(u)$ is uniquely determined on $E(V, Q, u)$ and that any larger domain might not determine it uniquely.

9. In the proof of Theorem 7.18, show that

$$F_Q'(u)q = \left\{ \left( \sum_{j=1}^{n} [a_{j1}\eta_j + b_{j1}\dot\eta_j] \right), \ldots, \left( \sum_{j=1}^{n} [a_{jm}\eta_j + b_{jm}\dot\eta_j] \right) \right\},$$

where

$$a_{jk} = \frac{\partial \varphi_k(x, u)}{\partial y_j}, \quad b_{jk} = \frac{\partial \varphi_k(x, u)}{\partial \dot y_j}, \quad 1 \le j \le n,\ 1 \le k \le m.$$

10. In that proof, why does

$$U\tilde{T} \in \Phi(M, Y)$$

imply that $\tilde{T} \in \Phi(M, Y)$.

# 8

# Mini-max methods

## 8.1    Mini-max

In this chapter we will show how mini-max methods can also be applied to situations that differ from those that we have considered previously. In order to simplify the arguments, we use convexity in a very significant way.

In this section we will prove the following:

**Theorem 8.1.** *Let $M, N$ be closed subspaces of a Hilbert space $E$ such that $M \cap N = \{0\}$, $E = M \oplus N$, $\dim N < \infty$, and let $Q$ be a closed, bounded, convex subset of $E$ containing $0$. Let $G(u) \in C^1(Q, \mathbb{R})$ be such that for each $w \in Q \cap M$, the functional $G(v + w)$ is concave in $v \in Q \cap N$. For each $w \in Q \cap M$ let $N(w)$ be the set of those $v \in Q \cap N$ such that*

$$G(v + w) = \sup_{g \in Q \cap N} G(g + w), \tag{8.1}$$

*and let*

$$L(w) = \{PG'(v + w) : w \in Q \cap M, \ v \in N(w)\},$$

*where $P$ is the (orthogonal) projection of $E$ onto $M$. Let*

$$S = \{u \in Q : u = v + w, \ w \in M, \ v \in N(w)\}.$$

*Assume*

    *(A) there is an interior point $u_0$ of $Q$ such that $u_0 \in S$ and*

$$G(u_0) = \inf_S G. \tag{8.2}$$

*Then the following hold.*

    *(1) For each $w \in Q \cap M$, the set $N(w)$ is not empty.*
    *(2) For each $w \in Q \cap M$, the set $N(w)$ is convex.*

*(3) For each $w \in Q \cap M$, the set $L(w)$ is convex.*

*(4) For each $u_0 \in S$ satisfying (8.2) which is an interior point of $Q$, we have $0 \in L(Pu_0)$. Therefore, there exists a $u_1 \in S$ such that $PG'(u_1) = 0$.*

*(5) There is a $u_0 \in S$ such that $G'(u_0) = 0$.*

*Proof.* (1) For fixed $w \in Q \cap M$, the functional $G(v + w)$ is concave in $v \in Q \cap N$. Thus, by Lemma 4.3, there is a $v \in Q \cap N$ such that (8.1) holds.

(2) If $v_1, v_2 \in N(w)$ and $v_\theta = (1 - \theta)v_1 + \theta v_2$, $0 \le \theta \le 1$, then

$$G(v_\theta + w) \ge (1 - \theta)G(v_1 + w) + \theta G(v_2, w) = \sup_{g \in Q \cap N} G(g + w).$$

Hence, $v_\theta \in N(w)$ and

$$G(v_\theta + w) = G(v_1 + w) = G(v_2, w).$$

(3) Under the same circumstances,

$$
\begin{aligned}
G(v_\theta + w + th) - G(v_\theta + w) &\ge (1 - \theta)[G(v_1 + w + th) - G(v_\theta + w)] \\
&\quad + \theta[G(v_2 + w + th) - G(v_\theta + w)] \\
&= (1 - \theta)[G(v_1 + w + th) - G(v_1 + w)] \\
&\quad + \theta[G(v_2 + w + th) - G(v_2 + w)]
\end{aligned}
$$

holds for any $t \ne 0$ and $h \in Q \cap M$. If we divide by $t$ and let $0 < t \to 0$, we obtain

$$(G'(v_\theta + w), h) \ge (1 - \theta)(G'(v_1 + w), h) + \theta(G'(v_2 + w), h).$$

If we let $0 > t \to 0$, we obtain the opposite inequality. Since this is true for every $h \in Q \cap M$, we have

$$PG'(v_\theta + w) = (1 - \theta)PG'(v_1 + w) + \theta PG'(v_2 + w). \qquad (8.3)$$

Since $v_\theta \in N(w)$, we see that $PG'(v_\theta + w) \in L(w)$.

(4) Suppose $u_0 = v_0 + w_0$ satisfies hypothesis (A), and $L(w_0)$ does not contain 0. Since $L(w_0)$ is closed, there is an element $h_1 \in L(w_0)$ such that

$$0 \ne \|h_1\| = \inf_{h \in L(w_0)} \|h\|.$$

To see this, let $\{h_k\} \subset L(w_0)$ be any sequence such that

$$\|h_k\| \to \inf_{h \in L(w_0)} \|h\|.$$

Since the $h_k$ are bounded, there is a renamed subsequence converging weakly to an element $h_1 \in E$. Since

$$\|h_1\|^2 = \|h_k\|^2 - 2([h_k - h_1], h_1) - \|h_k - h_1\|^2,$$

this implies

$$\|h_1\|^2 \leq \liminf \|h_k\|^2 = \inf_{h \in L(w_0)} \|h\|.$$

Moreover, there is a renamed subsequence such that $\bar{h}_k = (h_1 + \cdots h_k)/k$ converges to $h_1$ strongly in $E$ (Lemma 4.4). Since $\bar{h}_k \in L(w_0)$ and $L(w_0)$ is closed, we see that $h_1 \in L(w_0)$. Since we are assuming that $L(w_0)$ does not contain 0, this tells us that $h_1 \neq 0$. If $h \in L(w_0)$, the same is true of $h_\theta = (1 - \theta)h_1 + \theta h = h_1 + \theta(h - h_1)$, $0 \leq \theta \leq 1$. Hence,

$$\|h_1\|^2 + 2\theta(h - h_1, h_1) + \theta\|h - h_1\|^2 \geq \|h_1\|^2$$

or

$$2(h - h_1, h_1) + \theta\|h - h_1\|^2 \geq 0.$$

Letting $\theta \to 0$, we have $(h - h_1, h_1) \geq 0$ or

$$(h, h_1) \geq \|h_1\|^2 > 0, \quad h \in L(w_0).$$

This means that

$$(G'(v + w_0), h_1) \geq \|h_1\|^2 > 0, \quad v \in N(w_0). \tag{8.4}$$

For $t$ small and positive, let $v_t$ be any element in $N(w_0 - th_1)$. (Note that $w_0 - th_1 \in Q \cap M$.) Since the $v_t$ are bounded in norm, there is a sequence $t_k \to 0$ such that $v_t \rightharpoonup \tilde{v}_0$ in $Q \cap N$ for $t = t_k$. Since $N$ is finite dimensional, there is a renamed subsequence such that $v_t \to \tilde{v}_0$ in $Q \cap N$. Now,

$$G(v_t + w_0 - th_1) \geq G(v + w_0 - th_1), \quad v \in Q \cap N.$$

Hence, in the limit we have by continuity

$$G(\tilde{u}_0) = G(\tilde{v}_0 + w_0) \geq G(v + w_0), \quad v \in Q \cap N,$$

where $\tilde{u}_0 = \tilde{v}_0 + w_0$. Thus, $\tilde{v}_0 \in N(w_0)$. Since $u_0 \in S$, we have

$G(u_0) = G(\tilde{u}_0)$. Hypothesis (A) tells us that

$$0 \leq G(v_t + w_0 - th_1) - G(u_0) \leq G(v_t + w_0 - th_1) - G(v_t + w_0)$$
$$= -t \int_0^1 (G'(v_t + w_0 - \theta th_1), h_1)d\theta.$$

Dividing by $t$ and taking the limit as $t_k \to 0$, we have in the limit

$$(G'(\tilde{v}_0 + w_0), h_1) \leq 0.$$

This contradicts (8.4), showing that $L(w_0)$ does contain 0. Consequently, there is a $v_1 \in N(w_0)$ such that $PG'(v_1 + w_0) = 0$.

(5) To show that $(1 - P)G'(v_1 + w_0) = 0$, let $g$ be any element in $Q \cap N$. Then for any $t \neq 0$,

$$G(v_1 + w_0 + tg) \leq G(v_1 + w_0).$$

Dividing by $t$ and letting $t \to 0$ through both positive and negative values, we obtain

$$(G'(v_1 + w_0), g) = 0, \quad g \in Q \cap N.$$

Hence $(1 - P)G'(v_1 + w_0) = 0$, and if we combine this with (4), the proof is complete. □

## 8.2    An application

As an application we have

**Theorem 8.2.** *Let $n$ be an integer $\geq 0$, and let $\lambda$ satisfy*

$$1 + n^2 < \lambda < 1 + (n+1)^2. \tag{8.5}$$

*Let $f(x,t)$ be a Carathéodory function satisfying (1.62) such that $g(x,t) = f(x,t) - \lambda t$ is a nondecreasing function in $t$ for each $x \in I$. Assume*

$$C_2 t^2 - W_2(x) \leq 2F(x,t) - \lambda t^2 \leq |V_1 t|^\sigma + W_1(x), \quad x \in I, \ t \in \mathbb{R}, \tag{8.6}$$

*where $W_1, W_2 \in L^1(I)$ and*

$$\int_I |V_1 u|^\sigma dx \leq C_1 \|u\|_H^\sigma, \quad u \in H.$$

*Assume that there are constants $R_0, T_0$ such that*

$$B + C_1 T_0^{\sigma/2} < \left(1 - \frac{\lambda}{(n+1)^2 + 1}\right) T_0, \tag{8.7}$$

*and*

$$B + C_1 T_0^{\sigma/2} < \frac{C_2 + \lambda - n^2 - 1}{n^2 + 1} R_0, \tag{8.8}$$

*where $B = |W_1|_1 + |W_2|_1$.*

*Then problem (1.1),(1.2) has a solution.*

*Proof.* Let $G(u)$ be the functional (1.63), and let $M, N$ be the subspaces of $H$ defined in the proof of the Theorem 2.24. Note that $N$ is finite dimensional. Let $Q$ be the region

$$\|v\|_H^2 \leq R_0, \quad \|w\|_H^2 \leq T_0, \quad v \in N, \ w \in M.$$

We now show that all of the hypotheses of Theorem 8.1 are satisfied. Our hypotheses imply that $G(u) \in C^1(H, \mathbb{R})$ by Theorem 1.20. Next, we note that $G(v + w)$ is concave in $v \in N$ for each $w \in M$. To see this, we note that if we put $u_1 = v_1 + w$, $u_2 = v_2 + w$, then we have

$$
\begin{aligned}
(G'(u_1) - G'(u_2), u_1 - u_2) &= \|u_1 - u_2\|_H^2 - (f(\cdot, u_1) - f(\cdot, u_2), u_1 - u_2) \\
&= \|v_1 - v_2\|_H^2 - (g(\cdot, u_1) - g(\cdot, u_2), u_1 - u_2) \\
&\quad - \lambda \|v_1 - v_2\|^2 \leq 0
\end{aligned}
$$

because of (8.5) and

$$[g(x, t_1) - g(x, t_2)](t_1 - t_2) \geq 0, \quad t_1, t_2 \in \mathbb{R}, \ x \in I.$$

Finally, we show that hypothesis (A) of Theorem 8.1 holds. Let

$$G(u_k) \searrow \alpha = \inf_S G,$$

where $S$ is defined as in Theorem 8.1. Since $S$ is bounded, there is a renamed subsequence such that $w_k \rightharpoonup w_0$ (Theorem A.61). Let $h_k = v_0 + w_k$, $h_0 = w_0 + v_0$, where $v_0 \in N(w_0)$. Then

$$\|h_0\|_H^2 = \|h_k\|_H^2 - 2(h_k - h_0, h_0)_H - \|h_k - h_0\|_H^2.$$

Therefore,

$$G(h_0) = G(h_k) - 2(h_k - h_0, h_0)_H - \|h_k - h_0\|_H^2 + \int_I [F(x, h_k) - F(x, h_0)] dx$$

$$\leq G(u_k) - 2(h_k - h_0, h_0)_H + \int_I [F(x, h_k) - F(x, h_0)] dx,$$

because $v_k \in N(w_k)$. Since the $h_k$ are bounded in $H$, there is a renamed subsequence which converges uniformly in $I$ (Lemma 1.21). Hence,

$$\int_I [F(x, h_k) - F(x, h_0)] dx \to 0.$$

Consequently,

$$G(h_0) \le \liminf G(u_k) = \alpha.$$

Since $v_0 \in N(w_0)$, we have $h_0 \in S$. Thus $G(h_0) = \alpha$. We must now show that $h_0$ is an interior point of $Q$. By (8.6) we have

$$C_2\|u\|^2 - B_2 \le \int_I [2F(x,u) - \lambda u^2] dx \le \int_I |V_1 u|^\sigma dx + B_1$$

where $B_j = |W_j|_1$. Hence

$$\|u\|_H^2 - C_1\|u\|_H^\sigma - B_1 \le 2G(u) \le \|u\|_H^2 - (\lambda - C_2)\|u\|^2 + B_2.$$

If $u = v + w \in Q$, and $\|v\|_H^2 = R$, $\|w\|_H^2 = T$, then

$$R + T - C_1(R+T)^{\sigma/2} - B_1 \le 2G(u) \le R + T - \frac{\lambda + C_2}{n^2 + 1}R + B_2.$$

In particular, we have

$$T - C_1 T^{\sigma/2} - B_1 \le 2G(w),$$

and

$$2G(v) \le R - \frac{\lambda + C_2}{n^2 + 1}R + B_2.$$

This gives

$$2[G(u) - G(w)] \le \left(1 - \frac{\lambda + C_2}{n^2 + 1}\right) R + C_1 T^{\sigma/2} + B,$$

where $B = B_1 + B_2$. Note that this will be negative if $R = R_0$. Now suppose $u_0 = v_0 + w_0 \in S$ and satisfies

$$G(u_0) = \alpha.$$

Since $v_0 \in N(w_0)$, we must have $G(w_0) \le G(u_0)$. This shows that we cannot have $\|v_0\|_H^2 = R_0$ in this case.

Similarly, if $v$ is any function in $Q \cap N$, then

$$2[G(v) - G(w_0)] \le \left(1 - \frac{\lambda + C_2}{n^2 + 1}\right) R - T + C_1 T^{\sigma/2} + B.$$

If $T = T_0$, this is negative by (8.7). But there is a $v \in N(0) \subset S$. We then have

$$\alpha \le G(v) < G(w_0) \le G(u_0) = \alpha.$$

This contradiction shows us that when $u_0 = v_0 + w_0 \in S$ satisfies

$G(u_0) = \alpha$, then $\|w_0\|_H^2 < T_0$. In summary, any such $u_0$ must be an interior point of $Q$.

Thus, all of the hypotheses of Theorem 8.1 are satisfied. We may now conclude that there is a solution of $G'(u) = 0$. This is equivalent to a solution of (1.1),(1.2). $\qquad\square$

## 8.3    Exercises

1. Show that

$$0 \leq G(v_t + w_0 - th_1) - G(u_0) \leq G(v_t + w_0 - th_1) - G(v_t + w_0)$$
$$= -t \int_0^1 (G'(v_t + w_0 - \theta th_1), h_1)d\theta.$$

2. Use the equation in Exercise 8.3.1 to prove

$$(G'(\tilde{v}_0 + w_0), h_1) \leq 0.$$

3. If $u = v + w \in Q$, and $\|v\|_H^2 = R$, $\|w\|_H^2 = T$, show that

$$R + T - C_1(R+T)^{\sigma/2} - B_1 \leq 2G(u) \leq R + T - \frac{\lambda + C_2}{n^2 + 1}R + B_2.$$

4. Prove

$$T - C_1 T^{\sigma/2} - B_1 \leq 2G(w),$$

$$2G(v) \leq R - \frac{\lambda + C_2}{n^2 + 1}R + B_2$$

and

$$2[G(u) - G(w)] \leq \left(1 - \frac{\lambda + C_2}{n^2 + 1}\right)R + C_1 T^{\sigma/2} + B,$$

where $B = B_1 + B_2$.

5. If $v$ is any function in $Q \cap N$, show that

$$2[G(v) - G(w_0)] \leq \left(1 - \frac{\lambda + C_2}{n^2 + 1}\right)R - T + C_1 T^{\sigma/2} + B.$$

# 9

# Jumping nonlinearities

## 9.1    The Dancer–Fučík spectrum

Consider the Dirichlet problem

$$-u'' = bu^+ - au^- \quad \text{on} \ \ \Omega = (0, \pi) \tag{9.1}$$

$$u(0) = u(\pi) = 0, \tag{9.2}$$

where

$$u^\pm(x) = \max\{\pm u(x), 0\}$$

and $a$, $b$ are positive constants. Equation (9.1) is mildly nonlinear, but we shall see that solving it is a rather delicate procedure. Moreover, we shall see that the solvability of (9.1) plays an important role in solving problems of the form

$$-u'' = f(x, u)$$

when

$$\frac{f(x, t)}{t} \to b \ \text{ as } \ t \to +\infty, \quad \frac{f(x, t)}{t} \to a \ \text{ as } \ t \to -\infty.$$

First we note that any solution of (9.1) which is not $\equiv 0$ has at most a finite number of zeros (points of $\Omega$ where $u(x) = 0$). To see this, let $\sigma(x) = \{\sigma_1(x), \sigma_2(x)\}$ be a solution of the system

$$\sigma'(x) = h(x, \sigma)$$

and

$$\sigma(x_0) = 0,$$

where

$$h(x, \sigma) = \begin{bmatrix} -\sigma_2(x) \\ b\sigma_1(x)^+ - a\sigma_1(x)^- \end{bmatrix}.$$

Thus,

$$\begin{cases} \sigma_1'(x) = -\sigma_2(x) \\ \sigma_2'(x) = b\sigma_1(x)^+ - a\sigma_1(x)^-. \end{cases}$$

Note that

$$-\sigma_1'' = b\sigma_1(x)^+ - a\sigma_1(x)^-.$$

Thus, $\sigma_1(x)$ is a solution of (9.1). Note also that for each $x_0 \in \Omega$ the only solution of this system is $\sigma(x) \equiv 0$ near $x_0$. This follows from the uniqueness part of Picard's theorem (Theorem 2.13) and the fact that $\sigma(x) \equiv 0$ is a solution (it is easy to verify that $h(x, \sigma)$ satisfies a Lipschitz condition with respect to $\sigma$).

It follows from this that if $u(x)$ is a solution of (9.1) and satisfies $u(x_0) = 0$ for some $x_0 \in \Omega$, then either $u(x) \equiv 0$ in the neighborhood of $x_0$ or $u'(x_0) \neq 0$. (To see this, just set $\sigma_1(x) = u(x)$, $\sigma_2(x) = -u'(x)$.) Moreover, if $u(x) \equiv 0$ on a subinterval of $\Omega$, then it must vanish on the whole of $\Omega$ since it must vanish identically in the neighborhood of the endpoints of the subinterval. Thus, if $u(x)$ is a solution of (9.1) that does not vanish identically on $\Omega$, then $u'(x) \neq 0$ whenever $u(x) = 0$. This shows that the zeros of $u(x)$ are isolated, and consequently, there can only be a finite number of them.

If $x_1, x_2$ are two consecutive zeros of $u(x)$ satisfying (9.1) in $\bar{\Omega}$, then $u(x)$ satisfies

$$-u'' = bu, \quad x_1 < x < x_2$$

if $u(x) > 0$ in $x_1 < x < x_2$, or it satisfies

$$-u'' = au, \quad x_1 < x < x_2$$

if $u(x) < 0$ in $x_1 < x < x_2$. Consequently, we have

$$u(x) = c \sin b^{1/2}(x - x_1), \quad x_1 \leq x \leq x_2$$

in the former case and

$$u(x) = c \sin a^{1/2}(x - x_1), \quad x_1 \leq x \leq x_2$$

in the latter case. In order that $u(x_2) = 0$, we must have $b^{1/2}(x_2 - x_1) = \pi$ in the first case and $a^{1/2}(x_2 - x_1) = \pi$ in the second. It therefore follows that

$$x_2 - x_1 = \frac{\pi}{b^{1/2}} \quad \text{or} \quad x_2 - x_1 = \frac{\pi}{a^{1/2}}$$

as the case may be. Two such intervals in which $u(x)$ is positive (negative) cannot be adjacent to each other. If there are $k$ intervals where $u(x)$ is positive, and $k$ intervals where it is negative, and $u(x)$ satisfies (9.2), then the sum of the lengths of these intervals must add up to $\pi$. Thus,

$$k\pi \left( \frac{1}{a^{1/2}} + \frac{1}{b^{1/2}} \right) = \pi. \tag{9.3}$$

On the other hand, if there are $k+1$ intervals where $u$ is positive and $k$ intervals where it is negative, then we have

$$\frac{k}{a^{1/2}} + \frac{k+1}{b^{1/2}} = 1. \tag{9.4}$$

If there are $k$ intervals where $u$ is positive and $k+1$ intervals where it is negative, then we have

$$\frac{k+1}{a^{1/2}} + \frac{k}{b^{1/2}} = 1. \tag{9.5}$$

Thus, the system (9.1), (9.2) has a nontrivial solution if, and only if, $a, b$ satisfy one of the equations (9.3), (9.4), or (9.5). These equations describe curves in the $(a, b)$ plane that pass through points of the form $(n^2, n^2)$, where $n = 1, 2, \ldots$ In fact, if $n$ is even, we take $k = n/2$ in (9.3). If $n$ is odd, we take $k = (n-1)/2$ in (9.4) and (9.5). If $n = 1$, these "curves" form two straight lines parallel to the axes and pass through the point $(1, 1)$. If $n > 1$, then the curves are hyperbolas. If $n$ is even, then there is only one curve passing through $(n^2, n^2)$ (given by (9.3)), and if $n$ is odd, there are two (given by (9.4) and (9.5)). Note that curves passing through different points $(n^2, n^2)$ do not intersect. Note also that the nontrivial solutions of (9.1), (9.2) that we pieced together are in $C^2(\Omega)$. This is accomplished by choosing the constants in each subinterval to make the derivatives continuous at each zero.

We call the set of those points $(a, b) \in \mathbb{R}^2$ for which (9.1),(9.2) has a nontrivial solution the **Dancer–Fučík** spectrum and denote it by $\Sigma$. We have found that $\Sigma$ consists of a sequence of curves passing through points $(n^2, n^2) \in \mathbb{R}^2$, with $n = 1, 2, \ldots$ For even $n$, the curves are given by (9.3) with $k = n/2$. For odd $n$, the curves are given by (9.4),(9.5) with $k = (n-1)/2$. Note that for $a > n^2$, the upper curve is given by (9.4), while for $a < n^2$ the upper curve is given by (9.5). We denote the upper curve passing through $(n^2, n^2)$ by $b = \mu_n(a)$. Thus,

$$\mu_n(a) = \begin{cases} \dfrac{(n+1)^2 a}{4a + (n-1)^2 - 4a^{1/2}(n-1)}, & a > n^2, \\[3mm] \dfrac{(n-1)^2 a}{4a + (n+1)^2 - 4a^{1/2}(n+1)}, & a < n^2, \end{cases}$$

when $n$ is odd. When $n$ is even, we have

$$\mu_n(a) = \frac{n^2 a}{4a + n^2 - 4a^{1/2}n}.$$

Note that the curve $b = \mu_n(a)$ is symmetric with respect to the diagonal $a = b$. When $n$ is odd we denote the lower curve by $b = \nu_{n-1}(a)$ (the reason for this index $n-1$ will be made clear later). Hence,

$$\nu_{n-1}(a) = \begin{cases} \dfrac{(n-1)^2 a}{4a + (n+1)^2 - 4a^{1/2}(n+1)}, & a > n^2, \\[3mm] \dfrac{(n+1)^2 a}{4a + (n-1)^2 - 4a^{1/2}(n-1)}, & a < n^2, \end{cases}$$

This curve is also symmetric with respect to the diagonal $a = b$.

## 9.2    An application

We see from (9.3), (9.4), and (9.5) that for each $n > 0$, points $(a, b)$ which satisfy

$$\mu_n(a) < b < \nu_n(a)$$

are not in $\Sigma$. Since

$$(n+1)^2 < \nu_n(a), \quad a < (n+1)^2,$$

and

$$n^2 > \mu_n(a), \quad a > n^2,$$

the square

$$Q_n = \{(a, b) \in \mathbb{R}^2 : n^2 < a, b < (n+1)^2\}$$

is free of the Dancer–Fučík spectrum $\Sigma$. As an application of this fact we shall prove

**Theorem 9.1.** *Let $f(x, t)$ be a Carathéodory function on $\Omega \times \mathbb{R}$ satisfying*

$$|f(x, t)| \le C(|t| + 1), \quad x \in \Omega, \ t \in \mathbb{R} \tag{9.6}$$

*and*

$$\frac{f(x,t)}{t} \rightarrow \begin{cases} a, & \text{as } t \rightarrow -\infty \\ b, & \text{as } t \rightarrow \infty, \end{cases} \qquad (9.7)$$

*where* $\Omega = (0,\pi)$ *and* $(a,b) \in Q_n$. *Then there is a solution* $u \in C^2(\Omega) \cap C(\bar{\Omega})$ *of*

$$-u'' = f(x,u), \qquad x \in \Omega \qquad (9.8)$$

*satisfying (9.2).*

*Proof.* Let $N$ be the subspace of $H_0^1 = H_0^1(\Omega)$ that is spanned by the functions $\sin x, \ldots, \sin nx$, and let $M$ be its orthogonal complement in $H_0^1$. I claim that

$$G(u) = \|u\|_H^2 - 2 \int_\Omega F(x,u)dx, \qquad u \in H_0^1,$$

satisfies

$$m_1 = \sup_N G < \infty, \qquad m_0 = \inf_M G > -\infty.$$

To see this, let $\{v_k\}$ be a sequence in $N$ satisfying $G(v_k) \rightarrow m_1$. If $\|v_k\|_H \leq C$, then there is a renamed subsequence such that $v_k \rightarrow v_0$ in $N$ (recall that $N$ is finite dimensional). Thus,

$$G(v_k) \rightarrow G(v_0) = m_1.$$

Consequently, $m_1 < \infty$. If $\rho_k = \|v_k\|_H \rightarrow \infty$, let $\tilde{v}_k = v_k/\rho_k$. Then $\|\tilde{v}_k\|_H = 1$, and there is a renamed subsequence such that $\tilde{v}_k \rightarrow \tilde{v}_0$. Thus

$$\begin{aligned}
\frac{G(v_k)}{\rho_k^2} &= \|\tilde{v}_k\|_H^2 - 2\int_\Omega [\frac{F(x,v_k)}{v_k^2}]\tilde{v}_k^2 \, dx \\
&\rightarrow \|\tilde{v}_0\|_H^2 - a\|\tilde{v}_0^-\|^2 - b\|\tilde{v}_0^+\|^2 \\
&\leq \|\tilde{v}_0\|_H^2 - n^2\|\tilde{v}_0\|^2 \\
&\quad + (n^2 - a)\|\tilde{v}_0^-\|^2 \\
&\quad + (n^2 - b)\|\tilde{v}_0^+\|^2 \\
&< 0
\end{aligned}$$

since $a, b > n^2$ and $\|\tilde{v}_0\|_H = \lim \|\tilde{v}_k\|_H = 1$. This shows that

$$G(v_k) \rightarrow -\infty \quad \text{as} \quad \|v_k\|_H \rightarrow \infty.$$

Thus, $m_1 < \infty$.

To prove the second inequality, let $\{w_k\} \subset M$ be such that $G(w_k) \to m_0$. If $\|w_k\|_H \leq C$, then there is a renamed subsequence such that $w_k \rightharpoonup w_0$ in $M$ and $w_k \to w_0$ uniformly on $\bar{\Omega}$ (Lemma 3.14). Hence,

$$G(w_0) = \|w_k\|_H^2 - 2([w_k - w_0], w_0)_H - \|w_k - w_0\|_H^2 - 2\int_\Omega F(x, w_0)\, dx$$

$$\leq G(w_k) - ([w_k - w_0], w_0)_H + \int_\Omega [F(x, w_k) - F(x, w_0)]\, dx \to m_0.$$

This tells us that $m_0 > -\infty$. If $\rho_k = \|w_k\|_H \to \infty$, let $\tilde{w}_k = w_k/\rho_k$. Then $\|\tilde{w}_k\|_H = 1$, and there is a renamed subsequence such that $\tilde{w}_k \rightharpoonup \tilde{w}_0$ in $M$ and $\tilde{w}_k \to \tilde{w}_0$ uniformly in $\bar{\Omega}$ (Lemma 3.14). Consequently,

$$\frac{G(w_k)}{\rho_k^2} = 1 - 2\int_\Omega \frac{F(x, w_k)}{\rho_k^2}\, dx$$

$$\to 1 - a\|\tilde{w}_0^-\|^2 - b\|\tilde{w}_0^+\|^2$$

$$= (1 - \|\tilde{w}_0\|_H^2)$$
$$\quad + (\|\tilde{w}_0\|_H^2 - (n+1)^2\|\tilde{w}_0\|^2)$$
$$\quad + [(n+1)^2 - a]\|\tilde{w}_0^-\|^2$$
$$\quad + [(n+1)^2 - b]\|\tilde{w}_0^+\|^2$$

$$> 0$$

no matter what $\tilde{w}_0$ is. Hence,

$$G(w_k) \to \infty \quad \text{as} \quad \|w_k\|_H \to \infty,$$

contrary to assumption. Thus, $m_0 > -\infty$.

We can now apply the sandwich theorem (Theorem 2.5) to conclude that there is a sequence $\{u_k\} \subset H_0^1$ such that

$$G(u_k) \to c, \quad G'(u_k) \to 0.$$

Thus,

$$\|u_k\|_H^2 - 2\int_\Omega F(x, u_k)\, dx \to c,$$

and

$$(G'(u_k), h)_H = (u_k, h)_H - (f(\cdot, u_k), h) = o(\|h\|_H), \quad h \in H_0^1.$$

In particular, we have

$$(G'(u_k), u_k)_H = \|u_k\|_H^2 - (f(\cdot, u_k), u_k) = o(\|u_k\|_H).$$

If $\|u_k\|_H \le C$, then there is a renamed subsequence such that $u_k \rightharpoonup u_0$ in $H_0^1$ and $u_k \to u_0$ uniformly on $\bar{\Omega}$ (Lemma 3.14). In the limit we have

$$(u_0, h)_H - (f(\cdot, u_0), h) = 0, \quad h \in H_0^1.$$

From this it follows that $u_0$ is a solution of (9.8) (Theorem 3.8). It also satisfies (9.2) since each $u_k$ does so. If $\rho_k = \|u_k\|_H \to \infty$, we let $\tilde{u}_k = u_k/\rho_k$ (how did you guess?). Then $\|\tilde{u}_k\|_H = 1$, and there is a renamed subsequence such that $\tilde{u}_k \rightharpoonup \tilde{u}_0$ in $H_0^1$ and $\tilde{u}_k \to \tilde{u}_0$ uniformly on $\bar{\Omega}$ (Lemma 3.14). Thus,

$$\frac{G(u_k)}{\rho_k^2} = 1 - 2 \int_\Omega [\frac{F(x, u_k)}{u_k^2}] \tilde{u}_k^2 dx \to 1 - a\|\tilde{u}_0^-\|^2 - b\|\tilde{u}_0^+\|^2 = 0$$

and

$$\frac{(G'(u_k), h)_H}{\rho_k} = (\tilde{u}_k, h)_H - (\frac{f(\cdot, u_k)}{\rho_k}, h)$$

$$\to (\tilde{u}_0, h)_H + a(\tilde{u}_0^-, h) - b(\tilde{u}_0^+, h) = 0.$$

In particular, we have,

$$\|\tilde{u}_0\|_H^2 - a\|\tilde{u}_0^-\|^2 - b\|\tilde{u}_0^+\|^2 = 0,$$

from which we conclude that $\|\tilde{u}_0\|_H = 1$. We also conclude that $\tilde{u}_0$ is a solution of (9.1) (Theorem 3.8). But $(a, b) \notin \Sigma$. Hence $\tilde{u}_0 \equiv 0$. This contradicts the fact that the $H_0^1$ norm of $\tilde{u}_0$ is one. Consequently, the $\rho_k$ must be bounded, and the theorem follows. □

## 9.3    Exercises

1. Show that

$$h(x, \sigma) = \begin{bmatrix} -\sigma_2(x) \\ b\sigma_1(x)^+ - a\sigma_1(x)^- \end{bmatrix}$$

satisfies a Lipschitz condition with respect to $\sigma$.

2. Prove: If a solution of (9.1) satisfies $u(x) \equiv 0$ on a subinterval of $\Omega$, then it must vanish on the whole of $\Omega$.

3. For a solution $u(x) \not\equiv 0$ of (9.1), can two intervals in which $u(x)$ is positive (negative) and vanishes at the end points be adjacent to each other?

4. Show that the curves of the Dancer–Fučík spectrum passing through different points $(n^2, n^2)$ do not intersect.

5. Show that the nontrivial solutions of (9.1),(9.2) constructed in the text can be made to be in $C^2(\Omega)$.

6. Show that the upper curve of the Dancer–Fučík spectrum going through the point $(n^2, n^2)$ is given by

$$\mu_n(a) = \begin{cases} \dfrac{(n+1)^2 a}{4a + (n-1)^2 - 4a^{1/2}(n-1)}, & a > n^2, \\[2ex] \dfrac{(n-1)^2 a}{4a + (n+1)^2 - 4a^{1/2}(n+1)}, & a < n^2, \end{cases}$$

when $n$ is odd.

7. Show that the lower curve is given by

$$\nu_{n-1}(a) = \begin{cases} \dfrac{(n-1)^2 a}{4a + (n+1)^2 - 4a^{1/2}(n+1)}, & a > n^2, \\[2ex] \dfrac{(n+1)^2 a}{4a + (n-1)^2 - 4a^{1/2}(n-1)}, & a < n^2, \end{cases} \, .$$

8. Show that the square

$$Q_n = \{(a, b) \in \mathbb{R}^2 : n^2 < a, b < (n+1)^2\}$$

is free of the Dancer–Fučík spectrum $\Sigma$.

9. Verify

$$\frac{G(u_k)}{\rho_k^2} = 1 - 2 \int_\Omega [\frac{F(x, u_k)}{u_k^2}] \tilde{u}_k^2 dx \to 1 - a\|\tilde{u}_0^-\|^2 - b\|\tilde{u}_0^+\|^2 = 0$$

in the proof of Theorem 9.1.

10. Verify

$$\frac{(G'(u_k), h)_H}{\rho_k} = (\tilde{u}_k, h)_H - (\frac{f(\cdot, u_k)}{\rho_k}, h)$$

$$\to (\tilde{u}_0, h)_H + a(\tilde{u}_0^-, h) - b(\tilde{u}_0^+, h) = 0.$$

11. Verify

$$\|\tilde{u}_0\|_H^2 - a\|\tilde{u}_0^-\|^2 - b\|\tilde{u}_0^+\|^2 = 0.$$

# 10

# Higher dimensions

## 10.1    Orientation

So far, we have studied one dimensional problems. This was done primarily because the corresponding linear problems in one dimension are more easily solved. (As you know by now, we hate to work.) It saved us the need to develop the tools required to solve the linear problems in higher dimensions. The nonlinear methods are almost the same, but they need basic information from the linear problems in order to function.

We now turn our attention to higher dimensional problems. There is much more to learn about the linear cases than in one dimension, and some of this information can be quite involved and technical. We have attempted to keep the need minimal by considering periodic functions. Even so, the amount of information we shall need is far from trivial. Morover, many of the facts that are true in one dimension are no longer true in higher dimensions. This requires us to demand stronger hypotheses on the nonlinear terms than were needed in the one dimensional case. The higher the dimension, the stronger the assumptions needed. We shall prove everything along the way.

## 10.2    Periodic functions

Let

$$Q = \{x \in \mathbb{R}^n : 0 \le x_j \le 2\pi, \ 1 \le j \le n\} \tag{10.1}$$

be a cube in $\mathbb{R}^n$. By this we mean that $Q$ consists of those points

$$x = (x_1, \ldots, x_n) \in \mathbb{R}^n$$

such that each component $x_j$ satisfies $0 \leq x_j \leq 2\pi$, $1 \leq j \leq n$. We shall consider periodic functions in $Q$, that is, functions having the same values on opposite edges. To this end, we consider $n$-tuples of integers $\mu = (\mu_1, \ldots, \mu_n) \in \mathbb{Z}^n$, where each $\mu_j$ is an integer. We write $\mu x = \mu_1 x_1 + \cdots + \mu_n x_n$ and consider the set $S$ of finite sums of the form

$$u(x) = \sum \alpha_\mu e^{i\mu x}. \tag{10.2}$$

We want these functions to be real valued. In order to accomplish this, we require $\alpha_{-\mu}$ to appear in (10.2) whenever $\alpha_\mu$ appears and to satisfy

$$\alpha_{-\mu} = \overline{\alpha}_\mu. \tag{10.3}$$

If

$$u(x) = \sum \alpha_\mu e^{i\mu x}, \quad v(x) = \sum \beta_\mu e^{i\mu x} \tag{10.4}$$

are any functions in $S$, and $t$ is any real number, we define

$$(u, v)_t = (2\pi)^n \sum (1 + \mu^2)^t \alpha_\mu \beta_{-\mu}, \tag{10.5}$$

where $\mu^2 = \mu_1^2 + \cdots + \mu_n^2$. For each $t \in \mathbb{R}$, this has all of the properties of a scalar product (see Appendix A). The corresponding norm is given by

$$\|u\|_t^2 = (2\pi)^n \sum (1 + \mu^2)^t |\alpha_\mu|^2. \tag{10.6}$$

We are now going to consider the completion of $S$ with respect to the norm (10.6). However, it will be more convenient to start from scratch.

## 10.3    The Hilbert spaces $H_t$

For $t \in \mathbb{R}$ we let $H_t$ be the set of all series of the form

$$u = \sum \alpha_\mu e^{i\mu x}, \tag{10.7}$$

where the $\alpha_\mu$ are complex numbers satisfying

$$\alpha_{-\mu} = \overline{\alpha}_\mu, \tag{10.8}$$

and

$$\|u\|_t^2 = (2\pi)^n \sum (1 + \mu^2)^t |\alpha_\mu|^2 < \infty, \tag{10.9}$$

where $\mu^2 = \mu_1^2 + \cdots + \mu_n^2$. It is not required that the series (10.7) converge in any way, but only that (10.9) hold. If

$$u = \sum \alpha_\mu e^{i\mu x}, \quad v = \sum \beta_\mu e^{i\mu x} \tag{10.10}$$

are members of $H_t$, we can introduce the scalar product

$$(u,v)_t = (2\pi)^n \sum (1+\mu^2)^t \alpha_\mu \beta_{-\mu}. \tag{10.11}$$

With this scalar product, $H_t$ becomes a Hilbert space (see Appendix A). Completeness is the only property that is not obvious. To show this, let

$$u_j = \sum \alpha_\mu^{(j)} e^{i\mu x}$$

be a Cauchy sequence in $H_t$. Then

$$\sum (1+\mu^2)^t |\alpha_\mu^{(j)} - \alpha_\mu^{(k)}|^2 \to 0 \text{ as } j, k \to \infty.$$

Hence $\{\alpha_\mu^{(j)}\}$ is a Cauchy sequence of complex number for each $\mu \in \mathbb{Z}^n$. Thus, there is a subsequence such that

$$\alpha_\mu^{(j)} \to \alpha_\mu \text{ as } j \to \infty$$

for each $\mu \in \mathbb{Z}^n$. Let $\varepsilon > 0$ be given. Take $K$ so large that

$$\sum (1+\mu^2)^t |\alpha_\mu^{(j)} - \alpha_\mu^{(k)}|^2 < \varepsilon^2, \quad j,k > K.$$

Then

$$\sum_{\mu^2 \leq N} (1+\mu^2)^t |\alpha_\mu^{(j)} - \alpha_\mu^{(k)}|^2 < \varepsilon^2, \quad j,k > K,$$

holds for each finite $N$. Letting $k \to \infty$, we have

$$\sum_{\mu^2 \leq N} (1+\mu^2)^t |\alpha_\mu^{(j)} - \alpha_\mu|^2 \leq \varepsilon^2, \quad j > K.$$

Since this is true for any $N$, we have

$$\sum (1+\mu^2)^t |\alpha_\mu^{(j)} - \alpha_\mu|^2 \leq \varepsilon^2, \quad j > K.$$

If we let $u$ represent (10.7), then $u$ is in $H_t$ and $\|u_j - u\|_t \leq \varepsilon$. Thus $H_t$ is complete.

If $u \in S$, then

$$\partial u / \partial x_j = \sum (i\mu_j) \alpha_\mu e^{i\mu x}. \tag{10.12}$$

For each $u \in H_t$ we can define $\partial u/\partial x_j$ by means of (10.12) provided we note that the result is an element of $H_{t-1}$, since

$$\sum (1+\mu^2)^{t-1} \mu_j^2 |\alpha_\mu|^2 \leq \sum (1+\mu^2)^t |\alpha_\mu|^2 < \infty.$$

Thus the "operator" $\partial/\partial x_j$ maps $H_t$ into $H_{t-1}$. Consequently,

$$\Delta u = \sum_{j=1}^{n} \partial^2 u/\partial x_j^2 = -\sum \mu^2 \alpha_\mu e^{i\mu x} \qquad (10.13)$$

maps $H_t$ into $H_{t-2}$ for each $t \in \mathbb{R}$.

If $t \geq 0$, then (10.9) shows that series in $H_t$ converge in $L^2(Q)$. To see this note that

$$\int_Q e^{i(\mu-\nu)x} dx = (2\pi)^n \delta_{\mu\nu}, \quad \mu, \nu \in \mathbb{Z}^n, \qquad (10.14)$$

where

$$\delta_{\mu\nu} = \begin{cases} 1, & \mu = \nu, \\ 0, & \mu \neq \nu. \end{cases}$$

This comes from the fact that

$$\int_0^{2\pi} e^{ikx} dx = \begin{cases} \dfrac{1}{ik}[e^{2\pi i k} - 1] = 0, & k \neq 0, \\ 2\pi, & k = 0, \end{cases}$$

and

$$\int_Q e^{i(\mu-\nu)x} dx = \int_0^{2\pi} e^{i(\mu_1-\nu_1)x_1} dx_1 \cdots \int_0^{2\pi} e^{i(\mu_n-\nu_n)x_n} dx_n.$$

Thus if $u$ is represented by (10.7), let

$$u_N(x) = \sum_{\mu^2 \leq N} \alpha_\mu e^{i\mu x}.$$

Then for $M < N$,

$$\int_Q |u_N(x) - u_M(x)|^2\, dx = \sum_{M < \mu^2, \nu^2 \leq N} \alpha_\mu \alpha_{-\nu} \int_Q e^{i(\mu-\nu)x} dx$$

$$= (2\pi)^n \sum_{M < \mu^2 \leq N} |\alpha_\mu|^2 \to 0 \quad \text{as} \quad M, N \to \infty.$$

Thus the series (10.7) converges in $L^2(Q)$ to a function $u(x)$. Consequently, we have

**Theorem 10.1.** *The Hilbert spaces $H_t \subset L^2(Q)$ for $t \geq 0$. Moreover,*

$$(u, v)_0 = \int_Q uv\, dx, \quad u, v \in H_0. \qquad (10.15)$$

*Proof.* In view of (10.14),

$$\int_Q u(x)e^{-i\nu x}\,dx = \sum \alpha_\mu \int_Q e^{i(\mu-\nu)x}\,dx = (2\pi)^n \alpha_\nu.$$

Hence both sides of (10.15) equal

$$(2\pi)^n \sum \alpha_\mu \beta_{-\mu}$$

when $u, v$ are given by (10.10). This completes the proof.        □

We also have

**Lemma 10.2.** *If $u(x) \in L^2(Q)$ and*

$$\alpha_\mu = (2\pi)^{-n} \int_Q e^{-i\mu x} u(x)\,dx,$$

*then*

$$\left\| u - \sum_{\mu^2 \le N} \alpha_\mu e^{i\mu x} \right\| \to 0 \text{ as } N \to \infty.$$

*Proof.* This is the counterpart of Lemma 1.9 for $n > 1$. To prove it we follow the arguments used in the proof of Lemma 1.9. In particular, we define

$$u_N(x) = \sum_{\mu^2 \le N} \alpha_\mu e^{i\mu x}$$

and verify

$$\|u_N\|^2 = (u_N, u) = (2\pi)^n \sum_{\mu^2 \le N} |\alpha_\mu|^2,$$

$$\|u - u_N\|^2 = \|u\|^2 - (2\pi)^n \sum_{\mu^2 \le N} |\alpha_\mu|^2.$$

Consequently,

$$(2\pi)^n \sum_{\mu^2 \le N} |\alpha_\mu|^2 \le \|u\|^2,$$

which implies

$$(2\pi)^n \sum |\alpha_\mu|^2 \le \|u\|^2.$$

Moreover,

$$\int_Q |u_N(x) - u_M(x)|^2\,dx = \sum_{M < \mu^2, \nu^2 \le N} \alpha_\mu \alpha_{-\nu} \int_Q e^{i(\mu-\nu)x}\,dx$$

$$= (2\pi)^n \sum_{M < \mu^2 \le N} |\alpha_\mu|^2 \to 0 \text{ as } M, N \to \infty.$$

Thus, $u_N$ converges in $L^2(Q)$ to a function $\tilde{u}(x)$. Finally, we make use of the fact that

$$\int_Q e^{i\mu x} v(x) \, dx = 0, \quad \mu \in \mathbb{Z}^n,$$

implies that $v(x) = 0$ a.e. $\qquad \square$

**Corollary 10.3.** *The space $H_0 = L^2(Q)$ with the same norm.*

## 10.4 Compact embeddings

The following is a useful result.

**Lemma 10.4.** *If $s < t$, then every bounded sequence in $H_t$ has a subsequence which converges in $H_s$.*

*Proof.* Let $u_j = \sum \alpha_\mu^{(j)} e^{i\mu x}$ be a bounded sequence in $H_t$, $\|u_j\|_t \le M$. Since $|\alpha_\mu^{(j)}| \le M/(1+\mu^2)^{t/2}$, there is a renamed subsequence such that $\alpha_\mu^{(j)}$ converges as $j \to \infty$ for each $\mu$. For each $N \ge 0$,

$$\sum_{\mu^2 > N} (1+\mu^2)^s |\alpha_\mu^{(j)}|^2 \le \sum_{\mu^2 > N} (1+\mu^2)^t (1+N)^{s-t} |\alpha_\mu^{(j)}|^2 \le M^2/(1+N)^{t-s}.$$

Let $\varepsilon > 0$ be given. Take $N$ so large that

$$(2\pi)^n \sum_{\mu^2 > N} (1+\mu^2)^s |\alpha_\mu^{(j)}|^2 < \varepsilon^2.$$

Take $K$ so large that

$$(2\pi)^n \sum_{\mu^2 \le N} (1+\mu^2)^s |\alpha_\mu^{(j)} - \alpha_\mu^{(k)}|^2 < \varepsilon^2, \quad j, k > K.$$

Thus,

$$\|u_j - u_k\|_s < 3\varepsilon, \quad j, k > K.$$

This proves the lemma. $\qquad \square$

## 10.5 Inequalities

It is clear that

$$|(u, v)_t| \le \|u\|_t \|v\|_t, \quad u, v \in H_t. \tag{10.16}$$

We also have

$$|(u, v)_t| \le \|u\|_{t+s} \|v\|_{t-s}, \quad u \in H_{t+s}, \ v \in H_{t-s}. \tag{10.17}$$

This follows from

$$\left| \sum (1 + \mu^2)^t \alpha_\mu \beta_{-\mu} \right| \leq \left( \sum (1 + \mu^2)^{t+s} |\alpha_\mu|^2 \right)^{1/2}$$
$$\times \left( \sum (1 + \mu^2)^{t-s} |\beta_{-\mu}|^2 \right)^{1/2}.$$

It is also obvious that

$$\|u\|_s \leq \|u\|_t, \quad s < t, \quad u \in H_t. \tag{10.18}$$

Hence,

$$H_t \subset H_s, \quad s < t. \tag{10.19}$$

Let

$$\tau = (\tau_1, \ldots, \tau_n), \ \tau_j \geq 0, \ \tau_j \in \mathbb{Z},$$

and define

$$D^\tau = \left( \frac{\partial}{\partial x_1} \right)^{\tau_1} \cdots \left( \frac{\partial}{\partial x_n} \right)^{\tau_n}.$$

Then $D^\tau$ is a partial derivative of order $|\tau| = \sum \tau_j$. It satisfies

$$D^\tau u = \sum (i\mu_1)^{\tau_1} \cdots (i\mu_n)^{\tau_n} \alpha_\mu e^{i\mu x}$$

for $u$ satisfying (10.7). Hence, $D^\tau$ maps $H_t$ into $H_{t-|\tau|}$ and satisfies

$$\|D^\tau u\|_{t-|\tau|} \leq \|u\|_t, \quad u \in H_t. \tag{10.20}$$

We also note that

**Lemma 10.5.** *For $a > 0$, $t > \frac{1}{2}$, we have*

$$\sum_{m=-\infty}^{\infty} (a + m^2)^{-t} \leq a^{-t} + 2 \int_0^\infty (a + x^2)^{-t} dx. \tag{10.21}$$

*Proof.* We note that

$$(a + m^2)^{-t} \leq \int_{m-1}^m (a + x^2)^{-t} dx \tag{10.22}$$

when $m \geq 1$, and

$$(a + m^2)^{-t} \leq \int_m^{m+1} (a + x^2)^{-t} dx \tag{10.23}$$

when $m \leq -1$. These imply (10.21). □

We can now prove

**Theorem 10.6.** *(Sobolev) If $u \in H_t$ is given by (10.7) and $t > n/2$, then $u$ is continuous on $Q$, and*

$$\max_Q |u(x)| \leq \sum |\alpha_\mu| \leq C_t \|u\|_t, \quad u \in H_t. \tag{10.24}$$

*Proof.* First let us prove (10.24). Note that (10.21) implies

$$\sum_{m=-\infty}^{\infty} (a+m^2)^{-t} \leq a^{-t} + 2a^{(1/2)-t} \int_0^{\infty} (1+y^2)^{-t} dy \leq c_t a^{(1/2)-t} \tag{10.25}$$

when $a \geq 1$, where

$$c_t = 1 + 2 \int_0^{\infty} (1+y^2)^{-t} dy.$$

Hence,

$$\sum (1 + \mu_1^2 + \cdots + \mu_n^2)^{-t} \leq c_t \sum (1 + \mu_1^2 + \cdots + \mu_{n-1}^2)^{-(t-(1/2))}$$
$$\leq c_t c_{t-(1/2)} \sum (1 + \mu_1^2 + \cdots + \mu_{n-2}^2)^{-(t-1)}$$
$$\leq c_t c_{t-(1/2)} c_{t-1} \sum (1 + \mu_1^2 + \cdots$$
$$+ \mu_{n-3}^2)^{-(t-(3/2))}$$
$$\leq c_t c_{t-(1/2)} c_{t-1} c_{t-(3/2)} \cdots c_{t-((n-1)/2)} \equiv C_t^2.$$

Thus,

$$\sum |\alpha_n| = \sum (1+\mu^2)^{-t/2} (1+\mu^2)^{t/2} |\alpha_\mu|$$
$$\leq \left( \sum (1+\mu^2)^{-t} \right)^{1/2} \left( \sum (1+\mu^2)^t |\alpha_\mu|^2 \right)^{1/2}$$
$$\leq C_t \|u\|_t,$$

since $t - ((n-1)/2) > 1/2$. Let $u_N(x) = \sum_{\mu^2 \leq N} \alpha_\mu e^{i\mu x}$. Then $u_N$ is a function in $C^\infty(\bar{Q})$ for each $N$, and for $M < N$,

$$|u_N(x) - u_M(x)| = \left| \sum_{M < \mu^2 \leq N} \alpha_\mu e^{i\mu x} \right| \leq \sum_{M < \mu^2 \leq N} |\alpha_\mu| \to 0, \quad M, N \to \infty.$$

Therefore, $u_N$ converges uniformly to $u$, and

$$|u(x)| = \left| \sum \alpha_\mu e^{i\mu x} \right| \leq \sum |\alpha_\mu| \leq C_t \|u\|_t.$$

$\square$

**Corollary 10.7.** *If $t > \frac{n}{2} + k$, then $H_t \subset C^k(Q)$.*

*Proof.* If $u_N$ is given as in the proof of Theorem 10.6, we have

$$D^\tau u_N = \sum_{\mu^2 \leq N} (i\mu)^\tau \alpha_\mu e^{i\mu x}, \quad |\tau| \leq k,$$

and consequently,

$$
\begin{aligned}
|D^\tau(u_N - u_M)| &\leq \sum_{M < \mu^2 \leq N} |\mu|^{|\tau|}|\alpha_\mu| \\
&\leq \sum (1+\mu^2)^{(k-t/2)}|\mu|^{|\tau|}(1+\mu^2)^{(t-k/2)}|\alpha_\mu| \\
&\leq \left(\sum(1+\mu^2)^{k-t}\right)^{1/2}\left(\sum|\mu|^{2|\tau|}(1+\mu^2)^{t-k}|\alpha_\mu|^2\right)^{1/2} \\
&\leq \left(\sum(1+\mu^2)^{k-t}\right)^{1/2}\left(\sum|\mu|^{2|\tau|}(1+\mu^2)^{t-k}|\alpha_\mu|^2\right)^{1/2} \\
&\leq C_{t-k}\|u\|_t.
\end{aligned}
$$

$\square$

Let

$$H_\infty = \bigcap_t H_t, \quad H_{-\infty} = \bigcup_t H_t. \tag{10.26}$$

We have

**Corollary 10.8.** *The space $H_\infty \subset C^\infty(Q)$.*

Let

$$K = 1 - \Delta = 1 - \left(\frac{\partial}{\partial x_1}\right)^2 - \cdots - \left(\frac{\partial}{\partial x_n}\right)^2. \tag{10.27}$$

Then,

$$K \sum \alpha_\mu e^{i\mu x} = \sum \alpha_\mu (1 - (i\mu)\cdot(i\mu))e^{i\mu x} = \sum \alpha_\mu (1+\mu^2)e^{i\mu x}. \tag{10.28}$$

Define

$$K^t \sum \alpha_\mu e^{i\mu x} = \sum(1+\mu^2)^t \alpha_\mu e^{i\mu x}. \tag{10.29}$$

Then,

$$(K^t u, v)_s = (u, K^t v)_s = (u, v)_{s+t}, \tag{10.30}$$

and

$$\|K^t\|_s = \|u\|_{s+2t}. \tag{10.31}$$

We also have

**Theorem 10.9.** *If $u \in H_{-\infty}$ and*

$$|(u, v)_0| \leq C_1 \|v\|_t, \quad v \in H_\infty, \tag{10.32}$$

*then $u \in H_{-t}$ and*

$$\|u\|_{-t} \leq C_1. \tag{10.33}$$

*Proof.* If $u \in H_{-\infty}$, there is some $s$ such that $u \in H_{-s}$ and $u$ is given by (10.7). Let

$$v_N = \sum_{\mu^2 \leq N} (1 + \mu^2)^{-t} \alpha_\mu e^{i\mu x}. \tag{10.34}$$

Then $v_N \in H_\infty$, and

$$(u, v_N)_0 = (2\pi)^n \sum_{\mu^2 \leq N} (1 + \mu^2)^{-t} |\alpha_\mu|^2$$

$$\leq C_1 \left( (2\pi)^n \sum_{\mu^2 \leq N} (1 + \mu^2)^t (1 + \mu^2)^{-2t} |\alpha_\mu|^2 \right)^{1/2}.$$

Thus,

$$(2\pi)^n \sum_{\mu^2 \leq N} (1 + \mu^2)^{-t} |\alpha_\mu|^2 \leq C_1^2.$$

Since this is true for each $N$,

$$(2\pi)^n \sum (1 + \mu^2)^{-t} |\alpha_\mu|^2 \leq C_1^2.$$

Consequently, $u \in H_{-t}$, and (10.33) holds. $\square$

### 10.6 Linear problems

If

$$f(x) = \sum \gamma_\mu e^{i\mu x}, \tag{10.35}$$

we wish to solve

$$-\Delta u = f. \tag{10.36}$$

In other words, we wish to solve

$$\sum \mu^2 \alpha_\mu e^{i\mu x} = \sum \gamma_\mu e^{i\mu x}.$$

This requires

$$\mu^2 \alpha_\mu = \gamma_\mu, \quad \mu \in \mathbb{Z}^n. \tag{10.37}$$

In order to solve for all of the $\alpha_\mu$, we must have

$$\gamma_0 = 0. \tag{10.38}$$

Hence, we cannot solve (10.35) for all $f$. However, if (10.38) holds, we can solve (10.36) by taking

$$\alpha_\mu = \gamma_\mu/\mu^2 \text{ when } \mu \neq 0. \tag{10.39}$$

On the other hand, we can take $\alpha_0$ to be any number we like, and $u$ will be a solution of (10.36) as long as it satisfies (10.39). Thus we have

**Theorem 10.10.** *If $f$, given by (10.35), is in $H_t$ and satisfies (10.38), then (10.36) has a solution $u \in H_{t+2}$. An arbitrary constant can be added to the solution.*

*Proof.* We take the solution to be of the form (10.7) satisfying (10.39). It is in $H_{t+2}$ since

$$\sum (1 + \mu^2)^{t+2} |\alpha_\mu|^2 = \sum (1 + \mu^2)^{t+2} \mu^{-2} |\gamma_\mu|^2 + |\alpha_0|^2 < \infty$$

when (10.38) holds. Since $\Delta$ acting on any constant gives 0, we can add any constant to the solution. $\square$

Now that we know how to solve (10.36), we turn our attention to the equation

$$-(\Delta + \lambda)u = f, \tag{10.40}$$

where $\lambda \in \mathbb{R}$ is any constant. In this case we want $(\mu^2 - \lambda)\alpha_\mu = \gamma_\mu$, or

$$\alpha_\mu = \gamma_\mu/(\mu^2 - \lambda) \text{ when } \mu^2 \neq \lambda. \tag{10.41}$$

Therefore, we must have

$$\gamma_\mu = 0 \text{ when } \mu^2 = \lambda. \tag{10.42}$$

Not every $\lambda$ can equal some $\mu^2$. This is certainly true if $\lambda < 0$ or if $\lambda \notin \mathbb{Z}$. But even if $\lambda$ is a positive integer, there may be no $\mu$ such that

$$\lambda = \mu^2. \tag{10.43}$$

For instance, if $n = 3$ and $\lambda = 7$ or $15$, there is no $\mu$ satisfying (10.43). Thus there is a subset

$$0 = \lambda_0 < \lambda_1 < \cdots < \lambda_k < \cdots \tag{10.44}$$

of the nonnegative integers for which there are $n$-tuples $\mu$ satisfying (10.43). For $\lambda = \lambda_k$, we can solve (10.40) only if (10.42) holds. In that case, we can solve by taking $\alpha_\mu$ to be given by (10.41) when (10.43) does not hold, and to be given arbitrarily when (10.43) does hold. On the other hand, if $\lambda$ is not equal to any $\lambda_k$, then (10.43) never holds, and we can solve (10.40) by taking the $\alpha_\mu$ to satisfy (10.41). Thus we have

**Theorem 10.11.** *There is a sequence $\{\lambda_k\}$ of nonnegative integers tending to $+\infty$ with the following properties. If $f \in H_t$ and $\lambda \neq \lambda_k$ for every $k$, then there is a unique solution $u \in H_{t+2}$ of (10.40). If $\lambda = \lambda_k$ for some $k$, then one can solve (10.40) only if $f$ satisfies (10.42). In this case, the solution is not unique; there is a finite number of linearly independent periodic solutions of*

$$(\Delta + \lambda_k)u = 0 \tag{10.45}$$

*which can be added to the solution.*

*Proof.* The $\lambda_k \to +\infty$ since the $\mu^2$ are unbounded. Suppose $f \in H_t$ is given by (10.35), and $\lambda \neq \lambda_k$ for any $k$. Define $\alpha_\mu$ by (10.41) and take $u$ to be given by (10.7). Then $u$ is in $H_{t+2}$ and satisfies (10.40), since

$$\sum (1+\mu^2)^{t+2}|\alpha_\mu|^2 = \sum (1+\mu^2)^{t+2}|\mu^2 - \lambda|^{-2}|\gamma_\mu|^2 < \infty,$$

and there is a constant $C$ such that

$$1 + \mu^2 \leq C|\mu^2 - \lambda|, \quad \mu \in \mathbb{Z}^n.$$

The solution is unique since the only solution of

$$(\Delta + \lambda)u = 0 \tag{10.46}$$

is $u = 0$. On the other hand, if $\lambda = \lambda_k$ for some $k$, we can solve (10.40) only if (10.42) holds. In that case, we can solve by taking $\alpha_\mu$ to be given by (10.41) when (10.43) does not hold, and to be given arbitrarily when (10.43) does hold. Again, it follows that $u \in H_{t+2}$. Finally, we note that the functions $\alpha_\mu e^{i\mu x}$, $\mu^2 = \lambda_k$, are linearly independent and are solutions of (10.45). $\qquad\square$

The values $\lambda_k$ for which (10.45) has a nontrivial solution (i.e., a solution which is not $\equiv 0$) are called *eigenvalues*, and the corresponding nontrivial solutions are called *eigenfunctions*.

Let us analyze the situation a bit further. Suppose $\lambda = \lambda_k$ for some $k$, and $f \in H_t$ is given by (10.35) and satisfies (10.42). If $v \in H_t$ is given by

$$v = \sum \beta_\mu e^{i\mu x}, \tag{10.47}$$

then

$$(f, v)_t = (2\pi)^n \sum (1 + \mu^2)^t \gamma_\mu \beta_{-\mu},$$

by (10.11). Hence we have

$$(f, v)_t = 0 \tag{10.48}$$

for all $v \in H_t$ satisfying (10.47) and

$$\beta_\mu = 0 \text{ when } \mu^2 \neq \lambda_k. \tag{10.49}$$

On the other hand,

$$(\Delta + \lambda_k)v = \sum (\lambda_k - \mu^2)\beta_\mu e^{i\mu x}. \tag{10.50}$$

Thus $v$ is a solution of (10.45) iff it satisfies (10.49). Conversely, if $(f, v)_t = 0$ for all $v$ satisfying (10.47) and (10.49), then $f$ satisfies (10.42). Combining these, we have

**Theorem 10.12.** *If $\lambda = \lambda_k$ for some $k$, then there is a solution $u \in H_{t+2}$ of (10.40) iff*

$$(f, v)_t = 0 \tag{10.51}$$

*for all $v \in H_t$ satisfying*

$$(\Delta + \lambda_k)v = 0. \tag{10.52}$$

*Moreover, any solution of (10.52) can be added to the solution of (10.40).*

What are the solutions of (10.52)? By (10.50), they are of the form

$$v = \sum_{\mu^2 = \lambda_k} \beta_\mu e^{i\mu x} = \sum_{\mu^2 = \lambda_k} [a_\mu \cos \mu x + b_\mu \sin \mu x],$$

where we took $\beta_\mu = a_\mu - ib_\mu$. Thus (10.51) becomes

$$(f, \cos \mu x)_t = 0, \ (f, \sin \mu x)_t = 0 \text{ when } \mu^2 = \lambda_k. \tag{10.53}$$

## 10.7    Nonlinear problems

Now we turn the discussion to that of nonlinear multidimensional problems. Let $f(x, t)$ be a continuous function on $Q \times \mathbb{R}$ such that $f(x, t)$ is periodic in the components of $x$ with period $2\pi$ for each fixed $t \in \mathbb{R}$. We would like to solve the equation

$$-\Delta u = f(x, u(x)). \tag{10.54}$$

(For convenience, we are absorbing the function $u$ into $f(x, u)$.) Following the procedures we used in the one dimensional case, we first look for a functional such that $w = G'(u) = \Delta u$. At this point we will have to think seriously about the Hilbert space $H$ in which we will work. (Does this sound familiar?) Since we are looking for a solution of (10.54), it would be natural to take $H = H_t$. But if $u$ is in $H_t$ and $w = \Delta u$, we must have $w \in H_{t-2}$. By (10.17), this suggests taking

$$\langle v, w \rangle = (v, w)_{t-1}. \tag{10.55}$$

Note that this choice of $\langle u, v \rangle$ does satisfy (10.17). Thus, we want

$$[G(u + v) - G(u) - (v, \Delta u)_s]/\|v\|_t \to 0 \text{ as } \|v\|_t \to 0,$$

where $s = t - 1$. In this case we can make use of the identity

$$(v, \Delta u)_s = -(\nabla v, \nabla u)_s, \quad u, v \in S, \tag{10.56}$$

where

$$\nabla v = (\partial v / \partial x_1, \dots, \partial v / \partial x_n)$$

and

$$(\nabla v, \nabla u)_s = (\partial v / \partial x_1, \partial u / \partial x_1)_s + \cdots + (\partial v / \partial x_n, \partial u / \partial x_n)_s$$

$$= \sum_{k=1}^{n} \sum \mu_k^2 (1 + \mu^2)^s \beta_\mu \alpha_{-\mu}$$

$$= \sum \mu^2 (1 + \mu^2)^s \beta_\mu \alpha_{-\mu}$$

$$= -(v, \Delta u)_s$$

$$= -(\Delta v, u)_s.$$

This suggests trying

$$G(u) = \|\nabla u\|_s^2, \quad u \in H_t, \tag{10.57}$$

where $s = t - 1$. It looks good because

$$[G(u + v) - G(u) - 2(\nabla v, \nabla u)_s]/\|v\|_t = \|\nabla v\|_s^2/\|v\|_t \to 0 \text{ as } \|v\|_t \to 0,$$

since

$$\|\nabla v\|_s^2 + \|v\|_s^2 = \|v\|_t^2, \quad v \in H_t.$$

Hence, we have

**Theorem 10.13.** *If $G$ is given by (10.57), then*

$$G'(u) = -2\,\Delta u, \quad u \in H_t. \tag{10.58}$$

Next we turn to finding a functional $G$ which will satisfy

$$G'(u) = f(x, u), \quad u \in H_t. \tag{10.59}$$

Here it is not obvious how to proceed, so let us consider the simple case when $t = 1$. Going back to the definition, we want to find a functional $G$ such that $[G(u + v) - G(u) - (v, f(\cdot, u))_0]/\|v\|_1$ converges to 0 as $\|v\|_1 \to 0$. As before, we want to find a function $F(t)$ such that

$$F'(t) = f(x, t).$$

This would suggest that we try something of the form

$$G(u) = \int_Q F(x, u(x)) \, dx = (F(x, u), 1)_0. \tag{10.60}$$

If we now apply our definition, we want

$$[G(u + v) - G(u) - (v, f(\cdot, u))_0]/\|v\|_1 \tag{10.61}$$
$$= \int_Q [F(x, u + v) - F(x, u) - vf(x, u)]/\|v\|_1.$$

But

$$F(x, u + v) - F(x, u) = \int_0^1 [dF(x, u + \theta v)/d\theta] \, d\theta \tag{10.62}$$
$$= \int_0^1 F_t(x, u + \theta v) v \, d\theta.$$

Thus the left-hand side of (10.61) equals

$$\int_Q \int_0^1 [F_t(x, u + \theta v) - f(x, u)] v \, d\theta \, dx/\|v\|_1.$$

We want this expression to converge to 0 as $\|v\|_1 \to 0$. This would suggest that we take $F_t(x, t) = f(x, t)$. As we did in the one dimensional case, we try

$$F(x, t) = \int_0^t f(x, s) \, ds. \tag{10.63}$$

Then the expression above is bounded in absolute value by the square root of

$$\int_Q \int_0^1 |f(x, u + \theta v) - f(x, u)|^2 d\theta \, dx. \tag{10.64}$$

In order to proceed, we must make an assumption on $f(x, t)$. We are now hampered by the fact that for $n > 1$, functions in $H_1$ need not be continuous or even bounded. In fact, the best we can do is

**Theorem 10.14.** *If* $1 \leq q \leq 2^* := 2n/(n-2)$, *then*

$$|u|_q \leq C_q \|u\|_1, \quad u \in H_1,$$

*where*

$$|u|_q := \left( \int_Q |u|^q dx \right)^{1/q}. \tag{10.65}$$

*If* $1 \leq q < 2^*$, *then every bounded sequence in* $H_1$ *has a subsequence which converges in* $L^q(Q)$.

These statements will be proved later (Corollaries 10.45 and 10.46). As a consequence, we have

**Corollary 10.15.** *If* $1 \leq q < 2^*$, *then a bounded sequence in* $H_1$ *has a subsequence that converges weakly in* $H_1$, *strongly in* $L^q(Q)$ *and a.e. in* $Q$.

*Proof.* First we find a subsequence that converges weakly in $H_1$. We can do this by Theorem A.61. By Theorem 10.14, there is a subsequence of this subsequence that converges in $L^q(Q)$. We then find a subsequence of this subsequence that converges a.e. in $Q$ (Theorem B.25). $\qquad\square$

If we want to use the functional $G(u)$ given by (10.60) and we want this functional to be continuous on $H_1$ and have a derivative on this space, we will have to make assumptions on $f(x,t)$ which are stronger than those made in the one dimensional case because we are restricted by Theorem 10.14. For this purpose we will have to assume

$$|f(x,t)| \leq C(|t|^{q-1} + 1), \quad x \in Q, \ t \in \mathbb{R}, \tag{10.66}$$

*where*

$$1 \leq q < 2^* = \frac{2n}{n-2}, \quad 2^* - 1 = \frac{n+2}{n-2}. \tag{10.67}$$

The first thing we must check is that

$$G_1(u) = \int_Q F(x,u) \, dx$$

is differentiable on $H_1$. The fact that $G_1$ maps $H_1$ into $\mathbb{R}$ follows from Theorem 10.14. For (10.66) implies

$$|F(x,t)| \leq C(|t|^q + |t|),$$

and therefore,

$$\int_Q |F(x, u)|\, dx \le C\|u\|_1^q, \quad u \in H_1$$

in view of Theorem 10.14. Moreover,

$$G_1(u+v) - G_1(u) = \int_Q \int_0^1 f(x, u + \theta v) v\, d\theta\, dx.$$

Thus,

$$\left| G_1(u+v) - G_1(u) - \int_Q f(x, u) v\, dx \right|$$

$$\le \int_Q \int_0^1 |f(x, u + \theta v) - f(x, u)| v\, d\theta\, dx$$

$$\le \left( \int_Q \int_0^1 |f(x, u + \theta v) - f(x, u)|^\rho\, d\theta\, dx \right)^{1/\rho} |v|_{2*},$$

where $\rho = 2n/(n+2)$. By (10.66), the integrand on the right is bounded by

$$C(|u|^q + |v|^q + 1),$$

which is in $L^1(Q)$ for $u, v \in H_1$ in view of Theorem 10.14. Since the integrand converges to 0 point-wise, we see that $G_1'(u)$ exists.

If we now want to solve

$$-\Delta u = f(x, u), \qquad (10.68)$$

our arguments suggest that we try

$$2G(u) = \|\nabla u\|_0^2 - 2 \int_Q F(x, u)\, dx, \quad u \in H_1. \qquad (10.69)$$

We have

**Proposition 10.16.** *The functional $G(u)$ has a Fréchet derivative $G'(u)$ on $H_1$ given by*

$$(G'(u), v)_1 = (\nabla u, \nabla v) - (f(\cdot, u), v), \quad u, v \in H_1. \qquad (10.70)$$

*Proof.* We have by (10.69)

$$G(u+v) - G(u) - (\nabla u, \nabla v) + (f(\cdot, u), v)$$

$$= \frac{1}{2}\|\nabla v\|^2 - \int_Q [F(x, u+v) - F(x, u) - v f(x, u)]\, dx. \qquad (10.71)$$

The first term on the right-hand side of (10.71) is clearly $o(\|v\|_1)$ as $\|v\|_1 \to 0$. Since

$$F(x, u+v) - F(x, u) = \int_0^1 [dF(x, u+\theta v)/d\theta] \, d\theta$$

$$= \int_0^1 f(x, u+\theta v)v \, d\theta,$$

the integral in (10.71) equals

$$\int_Q \int_0^1 [f(x, u+\theta v) - f(x, u)]v \, d\theta \, dx.$$

By Hölder's inequality (Theorem B.23), this is bounded by

$$\left( \int_Q \int_0^1 |[f(x, u+\theta v) - f(x, u)]|^{q'} \, d\theta \, dx \right)^{1/q'} \|v\|_q. \tag{10.72}$$

In view of (10.71), the proposition will be proved if we can show that the expression (10.72) is $o(\|v\|_1)$. By (10.67), the second factor is $O(\|v\|_1)$. Hence it suffices to show that the first factor in (10.72) is $o(1)$. The integrand is bounded by

$$(|(u+\theta v)|^{q-1} + |u|^{q-1} + 1)^{q'} \leq C(|u|^q + |v|^q + 1). \tag{10.73}$$

If the first factor in (10.72) did not converge to 0 with $\|v\|_1$, then there would be a sequence $\{v_k\} \subset H_1$ such that $\|v_k\|_1 \to 0$ while

$$\int_Q \int_0^1 |[f(x, u+\theta v_k) - f(x, u)]|^{q'} \, d\theta \, dx \geq \varepsilon > 0. \tag{10.74}$$

In view of Corollary 10.15, there is a renamed subsequence such that $\|v_k\|_q \to 0$ and $v_k \to 0$ a.e. But by (10.73), the integrand of (10.74) is majorized by

$$C(|u|^q + |v_k|^q + 1)$$

which converges in $L^1(Q)$ to

$$C(|u|^q + 1).$$

Moreover, the integrand converges to 0 a.e. Hence the left-hand side of (10.74) converges to 0, contradicting (10.74). This proves the proposition. □

**Proposition 10.17.** *The derivative $G'(u)$ given by (10.70) is continuous in $u$.*

*Proof.* By (10.70), we have

$$(G'(u_1) - G'(u_2), v)_1 = (\nabla u_1 - \nabla u_2, \nabla v)$$
$$- \int_Q v[f(x, u_1) - f(x, u_2)]\, dx$$
$$\leq \|\nabla u_1 - \nabla u_2\| \cdot \|\nabla v\|$$
$$+ \|v\|_q \left( \int_Q |[f(x, u_1) - f(x, u_2)]|^{q'}\, dx \right)^{1/q'}.$$

Thus,

$$\|G'(u_1) - G'(u_2)\|_1 \leq \|u_1 - u_2\|_1$$
$$+ C \left( \int_Q |[f(x, u_1) - f(x, u_2)]|^{q'}\, dx \right)^{1/q'}. \quad (10.75)$$

Reasoning as in the proof of Proposition 10.16, we show that the right-hand side of (10.75) converges to 0 as $u_1 \to u_2$ in $H_1$. $\qquad\square$

It now follows under the given hypotheses that

$$G'(u) = -\Delta u - f(x, u), \quad u \in H_1. \quad (10.76)$$

Thus, if we can find a $u \in H_1$ such that

$$G'(u) = 0, \quad (10.77)$$

we will have found a solution of (10.68). So far we have

**Theorem 10.18.** *If $f(x, t)$ satisfies (10.66) and (10.67), then the functional $G(u)$ given by (10.69) is continuously differentiable and satisfies (10.76).*

In the next section we shall give conditions on $f(x, t)$ which will guarantee the $G$ has a minimum on $H_1$.

## 10.8    Obtaining a minimum

The next step is to find a $u \in H_1$ such that $G'(u) = 0$. The simplest situation is when $G(u)$ has an extremum. We now give a condition on $f(x, t)$ that will guarantee that $G(u)$ has a minimum. We assume that there is a function $W(x) \in L^1(Q)$ such that

$$W(x) \geq F(x, t) \to -\infty \text{ a.e. as } |t| \to \infty. \quad (10.78)$$

We let $N$ be the subspace of constant functions in $H_1$. Let $M$ be the subspace of those functions in $H_1$ which are orthogonal to $N$, that is, functions $w \in H_1$ which satisfy

$$(w, 1)_1 = \int_Q w(x)\, dx = 0.$$

We shall need

**Theorem 10.19.** *The expression* $\|u\|_0 \leq \|\nabla u\|_0, \quad u \in M.$

*Proof.* If $u = \sum \alpha_\mu e^{i\mu x}$, then $(u, 1)_0 = (2\pi)^n \alpha_0$. Hence $u \in M$ iff $\alpha_0 = 0$. This implies

$$\|\nabla u\|_0^2 = \sum_{k=1}^n \|\partial u/\partial x_k\|_0^2 = \sum_{k=1}^n (2\pi)^n \sum \mu_k^2 |\alpha_\mu|^2$$

$$= (2\pi)^n \sum_{\mu^2 \geq 1} \mu^2 |\alpha_\mu|^2 \geq (2\pi)^n \sum |\alpha_\mu|^2 = \|u\|_0^2.$$

This establishes the theorem. $\qquad\qquad\qquad\qquad\qquad\qquad\qquad\qquad\square$

**Theorem 10.20.** *Under hypotheses (10.66) and (10.78) there is a u in $H_1$ such that*

$$G(u) = \min_{H_1} G.$$

*Proof.* Let

$$\alpha = \inf_{H_1} G.$$

Let $\{u_k\}$ be a minimizing sequence, that is, a sequence satisfying

$$G(u_k) \searrow \alpha.$$

Assume first that

$$\rho_k = \|u_k\|_1 \to \infty.$$

Note that

$$2G(u) \geq \|\nabla u\|_0^2 - 2\int_Q W(x)\, dx \geq \frac{1}{2}\|w\|_1^2 - 2\int_Q W(x)\, dx, \quad u \in H_1,$$

where $u = v + w$, $v \in N$, $w \in M$. We see from this that if $\{u_k\}$ is a minimizing sequence, then

$$\|w_k\|_1 \leq C.$$

By Corollary 10.15, there is a renamed subsequence such that $w_k$ converges in $L^2(Q)$ and a.e. in $Q$. Since

$$|u_k(x)| \geq |v_k| - |w_k(x)|,$$

we see that

$$|u_k(x)| \to \infty \text{ a.e.}$$

Since

$$W(x) \geq F(x, u_k(x)) \to -\infty \text{ a.e.},$$

we have

$$\int_Q F(x, u_k(x))\, dx \to -\infty$$

(Theorem B.16). Thus,

$$G(u_k) \to \infty,$$

contrary to assumption. Consequently,

$$\rho_k = \|u_k\|_1 \leq C.$$

Then, by Corollary 10.15, there is a renamed subsequence such that $u_k$ converges weakly to a function $u_0$ in $H_1$, strongly in $L^q(Q)$ and a.e. in $Q$. Since

$$|F(x, u_k)| \leq C(|u_k|^q + |u_k|),$$

and

$$F(x, u_k) \to F(x, u_0) \text{ a.e.},$$

we have

$$\int_Q F(x, u_k)dx \to \int_Q F(x, u_0)dx$$

(Theorem B.24). Moreover, by the weak convergence,

$$(u_k, v)_1 \to (u_0, v)_1, \quad v \in H_1.$$

Since $\|u_0\|_1^2 = \|u_k\|_1^2 - 2([u_k - u], u_0)_1 - \|u_k - u_0\|_1^2$, we have

$$G(u_0) \leq \|u_k\|_1^2 - 2([u_k - u], u_0)_1 - \|u_0\|_0^2 - 2\int_Q F(x, u_0)dx$$

$$= G(u_k) - 2([u_k - u], u_0)_1 + \|u_k\|_0^2$$

$$- \|u_0\|_0^2 + 2\int_Q [F(x, u_k) - F(x, u_0)]dx$$

$$\to \alpha.$$

Thus,

$$\alpha \leq G(u_0) \leq \alpha,$$

and the theorem is proved. □

As a consequence we have

**Corollary 10.21.** *Under hypotheses (10.66) and (10.78), the equation (10.54) has a solution $u \in H_1$.*

*Proof.* By Theorem 10.20, $G$ has a minimum on $H_1$. It now follows from (10.76) that (10.54) holds. □

## 10.9     Another condition

We were able to obtain a minimum for the functional (10.69) and consequently a solution of (10.54) under the assumptions (10.66) and (10.78). In particular, this requires that $F(x,t)$ be bounded from above by a function $W(x) \in L^1(Q)$. We now discuss a situation which does not require such a bound. To begin, we study the case in which

$$|f(x,t)| \leq C(|t| + 1), \quad x \in Q,\ t \in \mathbb{R}. \tag{10.79}$$

This implies

$$|F(x,t)| \leq C(|t|^2 + |t|), \quad x \in Q,\ t \in \mathbb{R}. \tag{10.80}$$

We shall see that solvability depends upon the asymptotic behavior of $f(x,t)$ as $|t| \to \infty$. First we consider the assumption

$$f(x,t)/t \to \alpha(x) \text{ as } |t| \to \infty, \tag{10.81}$$

which implies

$$2F(x,t)/t^2 \to \alpha(x) \text{ as } |t| \to \infty, \tag{10.82}$$

where

$$\alpha(x) \leq 0,\ \alpha(x) \not\equiv 0. \tag{10.83}$$

In this case we have

**Lemma 10.22.**

$$\inf_{H_1} G > -\infty.$$

*Proof.* If the lemma were not true, there would be a sequence $\{u_k\} \subset H_1$ such that $G(u_k) \searrow -\infty$. Assume first that $\rho_k = \|u_k\|_1 \to \infty$. Let $\tilde{u}_k = u_k/\rho_k$. Then $\|\tilde{u}_k\|_1 = 1$, and there is a renamed subsequence such that $\tilde{u}_k \to \tilde{u}$ weakly in $H_1$, strongly in $L^2(Q)$ and a.e. in $Q$(Corollary 10.15). For those $x \in Q$ such that $\{u_k(x)\}$ is bounded, we have

$$2F(x, u_k(x))/\rho_k^2 \to 0.$$

In this case, $\tilde{u} = 0$. For those $x \in Q$ such that $|u_k(x)| \to \infty$ we have

$$2F(x, u_k(x))/\rho_k^2 = [2F(x, u_k(x))/u_k(x)^2]\tilde{u}_k^2 \to \alpha(x)\tilde{u}(x)^2.$$

Hence, this limit holds for all $x \in Q$. Since

$$|F(x, u_k)|/\rho_k^2 \le C(\tilde{u}_k(x)^2 + |\tilde{u}_k(x)|/\rho_k)$$

and $C(\tilde{u}_k(x)^2 + |\tilde{u}_k(x)|/\rho_k)$ converges in $L^1(Q)$, this implies

$$2\int_Q F(x, u_k)\, dx/\rho_k^2 \to \int_Q \alpha(x)\tilde{u}(x)^2\, dx$$

(Theorem B.20). Consequently,

$$2G(u_k)/\rho_k^2 = \|\nabla \tilde{u}_k\|_0^2 + \|\tilde{u}_k\|_0^2 - \|\tilde{u}_k\|_0^2 - 2\int_Q F(x, u_k)\, dx/\rho_k^2$$

$$= 1 - \|\tilde{u}_k\|_0^2 - 2\int_Q F(x, u_k)\, dx/\rho_k^2$$

$$\to (1 - \|\tilde{u}\|_0^2) - \int_Q \alpha(x)\tilde{u}(x)^2 dx \ge 0.$$

The only way the right-hand side can vanish is if

$$\|\tilde{u}\|_0 = 1, \quad \int_Q \alpha(x)\tilde{u}(x)^2\, dx = 0.$$

Since $\|\tilde{u}\|_1 \le 1$, we have $\nabla \tilde{u} \equiv 0$. Then $\tilde{u} \equiv$ a constant. Since $\alpha(x) \le 0$ and $\alpha(x) \not\equiv 0$, this constant must be 0. But this contradicts $\|\tilde{u}\|_0 = 1$. Thus

$$G(u_k) \to \infty,$$

showing that the $\rho_k$ must be bounded. But then

$$2G(u_k) = \|\nabla u_k\|_0^2 - 2 \int_Q F(x, u_k)\, dx$$
$$\geq -C \int_Q \{|u_k|^2 + |u_k|\}dx$$
$$\geq -C'\|u_k\|_0^2$$
$$> -K.$$

This completes the proof. □

Because of Lemma 10.4 we have hopes of finding a minimum for $G$. In fact we have

**Theorem 10.23.** *Under the hypotheses (10.80)–(10.83), there is a function $u_0$ in $H_1$ such that*

$$G(u_0) = \min_{H_1} G. \tag{10.84}$$

*Proof.* Let

$$m_0 = \inf_{H_1} G,$$

and let $u_k$ be a sequence such that $G(u_k) \searrow m_0$. As we saw in the proof of Lemma 10.22, the norms $\|u_k\|_1$ are bounded. Consequently, by Corollary 10.15 there is a renamed subsequence such that $u_k \to u_0$ weakly in $H_1$, strongly in $L^2(Q)$ and a.e. in $Q$. Since

$$|F(x, u_k)| \leq C(|u_k|^2 + |u_k|),$$

and

$$F(x, u_k) \to F(x, u_0) \quad \text{a.e.},$$

we have

$$\int_Q F(x, u_k)\, dx \to \int_Q F(x, u_0)\, dx.$$

Moreover, by weak convergence

$$(u_k, v)_1 \to (u_0, v)_1, \quad v \in H_1.$$

Since $\|u_0\|_1^2 = \|u_k\|_1^2 - 2([u_k - u], u_0)_1 - \|u_k - u_0\|_1^2$, we have

$$2G(u_0) \le \|u_k\|_1^2 - 2([u_k - u], u_0)_1 - \|u_0\|_0^2 - 2\int_Q F(x, u_0)\, dx$$

$$= 2G(u_k) - 2([u_k - u], u_0)_1 + \|u_k\|_0^2 - \|u_0\|_0^2$$

$$+ 2\int_Q [F(x, u_k) - F(x, u_0)]\, dx$$

$$\to 2m_0.$$

Thus

$$m_0 \le G(u_0) \le m_0,$$

and the theorem is proved.                    □

As a consequence we have

**Corollary 10.24.** *If (10.80)–(10.83) hold, then the equation (10.54) has a solution $u \in H_1$.*

*Proof.* By Theorem 10.23, $G$ has a minimum on $H_1$. It now follows from (10.76) that (10.54) holds.                    □

## 10.10    Nontrivial solutions

Corollaries 10.21 and 10.24 guarantee us that solutions of (10.68) exist, but as far as we know the solutions may be identically 0. Such a solution is called "trivial" because it usually has no significance in applications. We have the same situation that we had in the one dimensional case. If $f(x, 0) \equiv 0$, we know that $u \equiv 0$ is a solution of (10.68), and any method of solving it is an exercise in futility unless we know that the solution we get is not trivial. On the other hand, if $f(x, 0) \not\equiv 0$, then we do not have to worry about trivial solutions. We now consider the problem of insuring that the solutions provided by Corollaries 10.21 and 10.24 are indeed nontrivial even when $f(x, 0) \equiv 0$. We have

**Theorem 10.25.** *In addition to the hypotheses of either Theorem 10.20 or Theorem 10.23, assume that there is a $t_0 \in \mathbb{R}$ such that*

$$\int_Q F(x, t_0)\, dx > 0. \tag{10.85}$$

*Then the solutions of (10.68) provided by Corollary 10.21 or Corollary 10.24 are nontrivial.*

*Proof.* We show that the minima provided by Theorem 10.20 or Theorem 10.23 are negative. If this is the case, then the solution $u_0$ satisfies $G(u_0) < 0$. But $G(0) = 0$. This shows that $u_0 \neq 0$. To prove that $G(u_0) < 0$, let $v \equiv t_0$. Then

$$2G(v) = \|\nabla v\|_0^2 - 2 \int_Q F(x, v)\, dx < 0$$

by (10.85). Hence, $G(u_0) < 0$, and the proof is complete.                        □

## 10.11    Another disappointment

We showed that (10.68) has a solution if (10.80)–(10.83) hold. We did this by showing that $G(u)$ has a minimum on $H_1$. However, this does not work if

$$\int_Q \alpha(x)\, dx > 0. \tag{10.86}$$

To see this, let $u = c$, a constant. Then

$$2G(c)/c^2 = -2 \int_Q F(x, c)\, dx/c^2 \to \int_Q \alpha(x)\, dx < 0 \quad \text{as} \ |c| \to \infty.$$

Consequently

$$G(c) \to -\infty \quad \text{as} \ c^2 \to \infty, \tag{10.87}$$

showing that $G$ is unbounded from below. We shall learn how to deal with this situation in the next section.

## 10.12    The next eigenvalue

In proving Theorem 10.11, we showed that $\lambda$ is an eigenvalue of $-\Delta$ if and only if it satisfies (10.43). It follows that the next eigenvalue after $\lambda_0 = 0$ is $\lambda_1 = 1$. This suggests that we try to replace (10.83) with (10.86) and

$$\alpha(x) \leq 1, \alpha(x) \not\equiv 1. \tag{10.88}$$

Moreover, we know that in this case, (10.87) holds, and $G$ is unbounded from below. However, we suspect that the unboundedness from below is due only to (10.87), and if one considers the set $W$ consisting of those $u \in H_1$ which are "orthogonal" to the constants, that is, satisfy

$$(u, 1)_1 = (u, 1)_0 = 0, \tag{10.89}$$

one should find that $G$ is bounded below on $W$.

**Theorem 10.26.** *If (10.66), (10.80), (10.82), (10.86), and (10.88) hold, then*

$$G(w) \to \infty \quad \text{as} \quad \|w\|_1 \to \infty, \quad w \in W.$$

*Proof.* Let $w_k$ be any sequence in $W$ such that $\rho_k = \|w_k\|_1 \to \infty$. Let $\tilde{w}_k = w_k/\rho_k$. Then

$$2G(w_k)/\rho_k^2 = \|\nabla \tilde{w}_k\|_0^2 - 2\int_Q F(x, w_k)/\rho_k^2 \, dx.$$

Since $\|\tilde{w}_k\|_1 = 1$, there is a renamed subsequence such that $\tilde{w}_k \to \tilde{w}$ weakly in $H_1$, strongly in $H_0$, and a.e. in $Q$ (Corollary 10.15). Moreover

$$\|\nabla \tilde{w}_k\|_0^2 = \|\tilde{w}_k\|_1^2 - \|\tilde{w}_k\|_0^2 = 1 - \|\tilde{w}_k\|_0^2 \to 1 - \|\tilde{w}\|_0^2.$$

Consequently,

$$2G(w_k)/\rho_k^2 \to 1 - \|\tilde{w}\|_0^2 - \int_Q \alpha(x)\tilde{w}^2 \, dx$$
$$= (1 - \|\tilde{w}\|_1^2) + (\|\nabla \tilde{w}\|_0^2 - \|\tilde{w}\|_0^2)$$
$$+ \int_Q [1 - \alpha(x)]\tilde{w}^2 \, dx \geq 0.$$

The only way this can vanish is if

$$\|\tilde{w}\|_1 = 1 \tag{10.90}$$

$$\|\nabla \tilde{w}\|_0 = \|\tilde{w}\|_0 \tag{10.91}$$

and

$$\int_Q [1 - \alpha(x)]\tilde{w}^2 \, dx = 0. \tag{10.92}$$

If $\tilde{w} = \sum \gamma_\mu e^{i\mu x}$, then (10.91) implies

$$\sum \mu^2 |\gamma_\mu|^2 = \sum |\gamma_\mu|^2.$$

Since $\tilde{w} \in W$, $\gamma_0 = 0$. Thus $\gamma_\mu = 0$ whenever $\mu^2 > 1$. This means that

$$\tilde{w} = \sum_{\mu^2 = 1} \gamma_\mu e^{i\mu x}.$$

This is a finite sum. Let

$$h(z) = \sum_{\mu^2 = 1} \gamma_\mu e^{i\mu z}, \quad z_j = x_j + iy_j.$$

Then $h(z)$ is analytic in the whole of $\mathbb{C}^n$, and it satisfies

$$\int_Q [1 - \alpha(x)] |h(x)|^2 dx = 0.$$

In view of (10.88), this implies that $h(x) = 0$ on a set of positive measure in $Q$. From this we can conclude that $h(x) \equiv 0$. There are several ways of showing this. One way is to notice that

$$h(x) = \sum_{k=1}^{n} (a_k \cos x_k + b_k \sin x_k) = \sum_{k=1}^{n} h_k(x_k).$$

We note that

**Lemma 10.27.** *If*

$$h(x) = a \cos x + b \sin x, \quad 0 \le x \le 2\pi,$$

*and $h(x) = c$ on a set $e$ of positive measure, then $c = 0$ and*

$$h(x) \equiv 0, \quad 0 \le x \le 2\pi.$$

*Proof.* Since $e$ has a limit point (a finite set of points has measure 0) and $h(x)$ has a convergent Taylor series, it must be identically equal to $c$ on the whole interval $I = [0, 2\pi]$. The only way this can happen is when $c = 0$. $\qquad\square$

For the $n$ dimensional case we have

**Lemma 10.28.** *If $E \subset Q$ is a set of positive measure and*

$$h(x) = c, \quad x \in E,$$

*then $c = 0$ and*

$$h_k(x_k) \equiv 0, \quad x_k \in I = [0, 2\pi], \ 1 \le k \le n.$$

*Proof.* We use induction. The lemma is true for $n = 1$ (Lemma 10.27). Assume it is true for $n - 1$, where $n > 1$. Let $y = (x_1, \ldots, x_{n-1}) \in \mathbb{R}^{n-1}$ be any point in

$$Q_{n-1} = \{x \in \mathbb{R}^{n-1} : 0 \le x_j \le 2\pi, \ j = 1, \ldots, n-1\}.$$

For each $x_n \in \mathbb{R}$, let

$$E(x_n) = \{y \in Q_{n-1} : (y, x_n) \in E\},$$

and let $e \subset \mathbb{R}$ be the set of those $x_n \in \mathbb{R}$ such that

$$m(E(x_n)) > 0.$$

Then $m(e) > 0$ by Fubini's theorem (Theorem B.26) since

$$\int_Q \mu(E)\,dx = \int_e \left[ \int_{Q_{n-1}} \mu(E(x_n))dy \right]\,dx_n,$$

where $\mu(E)$ is the characteristic function of $E$. For each $x_n \in e$ we have

$$\sum_{k=1}^{n-1} h_k(x_k) = c - h_n(x_n), \quad y = (x_1, \ldots, x_{n-1}) \in E(x_n).$$

Since $m(E(x_n)) > 0$ for $x_n \in e$, we have by the induction hypothesis

$$c - h_n(x_n) = 0, \quad x_n \in e$$

and

$$h_k(x_k) \equiv 0, \quad x_k \in I, \ 1 \le k \le n-1.$$

Since this is true for every $x_n \in e$, we have by Lemma 10.27 that

$$h_k(x_k) \equiv 0, \quad x_k \in I, \ 1 \le k \le n.$$

$\square$

From this we see that $\tilde{w} \equiv 0$. But this is impossible by (10.90). Hence

$$\lim G(w_k)/\rho_k^2 > 0.$$

Since this is true for any sequence in $W$, the theorem follows. $\square$

**Theorem 10.29.** *The functional $G$ is bounded from below on $W$.*

*Proof.* Suppose $\{w_k\} \subset W$ is such that $G(w_k) \searrow -\infty$. By Theorem 10.26, the sequence must be bounded. But then

$$\begin{aligned}
2G(w_k) &= \|\nabla w_k\|_0^2 - 2\int_Q F(x, w_k)\,dx \\
&\ge -C\int_Q (|w_k|^2 + |w_k|)\,dx \\
&\ge -C'(\|w_k\|_0^2 + 1) \\
&\ge -C'',
\end{aligned}$$

providing a contradiction. $\square$

### 10.13    A Lipschitz condition

In order to solve (10.77) when we do not have a minimum, we must verify that $G'(u)$ given by (10.76) satisfies a Lipschitz condition of the form (2.54). Initially, we showed that this will be true if $f(x, t)$ satisfies (2.36). We can do better, but we cannot do as well as we did in the one dimensional case. The degree of improvement gets worse as $n$ increases. In fact, we have

**Lemma 10.30.** *If* $2 \leq q \leq 2^*$, *and* $f(x, t)$ *satisfies*

$$|f(x, t) - f(x, s)| \leq C(|t|^{q-2} + |s|^{q-2} + 1)|t - s|, \qquad (10.93)$$

*then* $G'(u)$ *satisfies*

$$\|G'(u) - G'(v)\|_{-1} \leq C(\|u\|_1^{q-2} + \|v\|_1^{q-2} + 1)\|u - v\|_1, \quad u, v \in H_1. \qquad (10.94)$$

*Proof.* It suffices to show that

$$\int_Q |f(x, u) - f(x, v)| \cdot |h|\, dx \leq C(|u|_{2^*}^{q-2} + |v|_{2^*}^{q-2} + 1)|u - v|_{2^*}|h|_{2^*}. \qquad (10.95)$$

By Corollary 10.46, this implies

$$|(G'(u) - G'(v), h)_0| \leq \|\nabla(u - v)\|_0 \|\nabla h\|_0$$
$$+ C(\|u\|_1^{q-2} + \|v\|_1^{q-2} + 1)\|u - v\|_1 \|h\|_1,$$

which implies (10.94). To prove(10.95), we note that the left-hand side of (10.95) is bounded by

$$\left( \int_Q |f(x, u) - f(x, v)|^{2^{*'}} dx \right)^{1/2^{*'}} \left( \int_Q |h|^{2^*} dx \right)^{1/2^*}$$

$$\leq \left( \int_Q (|u|^{q-2} + |v|^{q-2} + 1)^{2^{*'}} |u - v|^{2^{*'}} dx \right)^{1/2^{*'}} |h|_{2^*}$$

$$\leq \left( \int_Q (|u|^{q-2} + |v|^{q-2} + 1)^{2^{*'}\rho} dx \right)^{1/\rho 2^{*'}}$$

$$\times \left( \int_Q |u - v|^{2^{*'}\rho'} dx \right)^{1/\rho' 2^{*'}} |h|_{2^*}$$

$$\leq C \left( \int_Q (|u|^{2^{*'}\rho(q-2)} + |v|^{2^{*'}\rho(q-2)} + 1) dx \right)^{1/\rho 2^{*'}} |u - v|_{2^{*'}\rho'} |h|_{2^*}$$

$$\leq C(|u|_{2^*\rho(q-2)}^{q-2} + |v|_{2^*\rho(q-2)}^{q-2} + 1) |u - v|_{2^{*'}\rho'} |h|_{2^*},$$

where $p' = p/(p-1)$. Take $\rho = (2^* - 1)/(2^* - 2)$. Then $2^{*\prime}\rho' = 2^*$ and $2^{*\prime}\rho(q-2) \le 2^*$. This gives (10.95) and completes the proof. □

**Remark 10.31.** *As we mentioned before, we can dispense with the requirement that $G'(u)$ satisfy a local Lipschitz condition as long as it is continuous. This is accomplished by replacing $G'(u)$ with a locally Lipschitz "pseudo-gradient." The difficulty lies in showing that such a pseudo-gradient exists. We shall present the argument later in Appendix D.*

## 10.14    Splitting subspaces

It should come as no surprise that we have the same problems in higher dimensions as we had in one dimension. We now have the following situation. The functional $G$ is bounded above on the constants (the subspace $V$) and below on the subspace $W$ orthogonal to the constants. We can now apply Theorem 2.5 to obtain

**Theorem 10.32.** *Under the hypotheses of Theorem 10.26 and Lemma 10.30, there is a PS sequence satisfying*

$$G(u_k) \to a, \quad G'(u_k) \to 0.$$

*Proof.* By (10.86) and Theorem 10.26, $G$ is bounded above on $V$ and bounded below on $W$. By Lemma 10.30, $G'$ satisfies a local Lipschitz condition. We may now apply Theorem 2.5 to reach the desired conclusion. □

This theorem raises hope of finding a stationary point of $G$, that is, a solution of (10.77). A step in this direction is

**Theorem 10.33.** *Under the hypotheses of Theorem 10.26, if for some $a \in \mathbb{R}$, $G'(u) \ne 0$ for all $u \in H_1$ such that $G(u) = a$, then there is a $\delta > 0$ such that*

$$\|G'(u)\|_{-1} \ge \delta \text{ when } a - \delta \le G(u) \le a + \delta. \tag{10.96}$$

*Proof.* Suppose not. Then there would be a sequence $\{u_k\} \subset H_1$ such that

$$G(u_k) \to a, \quad G'(u_k) \to 0.$$

Consequently, by Proposition 10.16,

$$(\nabla u_k, \nabla v)_0 - (f(u_k), v)_0 = o(\|v\|_1), \quad v \in H_1. \tag{10.97}$$

If $\rho_k = \|u_k\|_1 \to \infty$, let $\tilde{u}_k = u_k/\rho_k$. Then $\|\tilde{u}_k\|_1 = 1$, and hence there is a renamed subsequence such that $\tilde{u}_k \to \tilde{u}$ weakly in $H_1$, strongly in $L^2(Q)$, and a.e. in $Q$ (Corollary 10.15). For this subsequence (10.97) implies

$$(\nabla \tilde{u}_k, \nabla v)_0 - (f(u_k)/\rho_k, v)_0 \to 0, \quad v \in H_1.$$

Thus in the limit

$$(\nabla \tilde{u}, \nabla v)_0 - \int_Q \alpha(x)\tilde{u}v \, dx = 0, \quad v \in H_1. \tag{10.98}$$

Here we used the fact that

$$f(x, u_k)/\rho_k = [f(x, u_k)/u_k]\tilde{u}_k \to \alpha\tilde{u} \quad \text{a.e.}$$

and

$$|f(x, u_k)|/\rho_k \leq C(|\tilde{u}_k| + 1/\rho_k).$$

This implies

$$-\Delta\tilde{u} = \alpha(x)\tilde{u}.$$

Set $\tilde{u} = \gamma + \tilde{w}$, where $\gamma$ is a constant and $\tilde{w} \in W$. Take $v = \gamma - \tilde{w}$ in (10.98). Then

$$-\|\nabla\tilde{w}\|_0^2 = \int_Q \alpha[\gamma + \tilde{w}][\gamma - \tilde{w}]dx = \int_Q \alpha(\gamma^2 - \tilde{w}^2)dx.$$

Consequently,

$$(\|\nabla\tilde{w}\|_0^2 - \|\tilde{w}\|_0^2) + \int_Q [1 - \alpha]\tilde{w}^2 dx + \gamma^2 \int_Q \alpha(x)dx = 0.$$

As we saw above, this implies $\gamma = 0$, $\tilde{w} = 0$. Hence $\tilde{u} = 0$. But (10.97) also implies

$$\|\nabla\tilde{u}_k\|_0^2 - (f(\cdot, u_k)/\rho_k, \tilde{u}_k)_0 \to 0,$$

or

$$1 - \|\tilde{u}_k\|_0^2 - (f(\cdot, u_k)/\rho_k, \tilde{u}_k)_0 \to 0,$$

which gives

$$1 - \|\tilde{u}\|_0^2 - \int_Q \alpha\tilde{u}^2 \, dx = 0.$$

This shows that $\tilde{u} \not\equiv 0$. Hence the $\rho_k$ are bounded. But then there is a renamed subsequence such that $u_k \to u$ weakly in $H_1$, strongly in $H_0$, and a.e. in $Q$ (Corollary 10.15). Now (10.97) implies that

$$(\nabla u, \nabla v)_0 - (f(\cdot, u), v)_0 = 0, \quad v \in H_1. \tag{10.99}$$

Hence $G'(u) = 0$, and

$$\|\nabla u\|_0^2 - (f(\cdot, u), u)_0 = 0.$$

Moreover, by (10.97),

$$\|\nabla u_k\|_0^2 = (f(\cdot, u_k), u_k)_0 + o(1) \to (f(\cdot, u), u)_0 = \|\nabla u\|_0^2.$$

Thus $\nabla u_k$ converges strongly to $\nabla u$ in $L^2(Q)$. This implies that

$$2G(u_k) = \|\nabla u_k\|_0^2 - 2 \int_Q F(x, u_k)dx \to \|\nabla u\|_0^2 - 2 \int_Q F(x, u)dx = 2G(u).$$

Since $G(u_k) \to a$, we have $G(u) = a$, contradicting the hypothesis. This proves the theorem.      $\square$

We can conclude from Theorems 10.32 and 10.33 that

**Theorem 10.34.** *Under the hypotheses of Theorem 10.26 and Lemma 10.30, $G$ has a critical point which is a solution of (10.54).*

**Remark 10.35.** *As we noted previously, a sequence satisfying*

$$G(u_k) \to a, \ G'(u_k) \to 0, \tag{10.100}$$

*is called a Palais–Smale sequence (PS sequence) at level a. A functional $G$ is said to satisfy the Palais–Smale condition (PS condition) at level a if every PS sequence at level a has a convergent subsequence. It is said to satisfy the Palais–Smale condition if it satisfies the PS condition at all levels. Theorem 10.33 states that under the hypotheses of Theorem 10.26, if $G$ has a PS sequence at level a, then it has a critical point at that level.*

## 10.15    The question of nontriviality

We essentially follow the arguments used in the one dimensional periodic problems. In the case of Theorem 10.34, we have the same problem that faced us when we proved Corollaries 10.21 and 10.24, namely, if $f(x, 0) \equiv 0$, what guarantee do we have that the solution provided is not $u \equiv 0$. In the case of these corollaries, we were able to find a criterion (namely (10.85)) which provides such a guarantee (cf. Theorem 10.25). This was accomplished by showing that the minimum obtained in (10.84) was less than 0. Since $G(0) = 0$, this shows that the solutions obtained in Corollaries 10.21 and 10.24 were not 0. However, this does not work in the case of Theorem 10.34 since $m_0 \leq G(0) \leq m_1$ (cf. (2.41)–(2.40)). Thus in order to guarantee that our solutions are not $\equiv 0$, we must devise another means of attack. A method that worked in the one dimensional case is as follows. Suppose, in addition to the hypotheses of Theorem 10.34, we have assumptions which will provide positive constants $\varepsilon, \rho$ so that

$$G(u) \geq \varepsilon \quad \text{when} \quad \|u\|_1 = \rho.$$

Since $G(0) = 0$, this creates the image of 0 being in a valley surrounded by mountains of minimum height $\varepsilon$. Now we maintain that there is a solution of (10.77) which satisfies

$$\varepsilon \leq G(u) \leq m_1$$

in place of (2.42). To see this we follow the proof of Theorem 2.8 and claim that if (10.77) does not have such a solution, then there is a $\delta > 0$ such that (2.41) holds whenever $u$ satisfies

$$\varepsilon - \delta \leq G(u) \leq m_1 + \delta.$$

We draw our curves as before from each point of $V$ along which $G$ decreases at a rate of at least $\delta$. Again we arrive at a curve $S = \{\sigma(T)v : v \in V\}$. The claim now is that $S$ has points both inside and outside the sphere $\|u\|_1 = \rho$. To see this note that there are points $v \in V$ so far away from the origin that $\sigma(T)v$ never reaches the sphere. On the other hand, $G(\sigma(t)0)$ is nonincreasing in $t$. Consequently, $\sigma(t)0$ cannot exit the sphere. Thus, $\sigma(T)0$ is inside the sphere while there are points $\sigma(T)v$ outside. By continuity, there must be a point $v_1 \in V$ such that $G(\sigma(T)v_1) = \varepsilon$. But

$$G(\sigma(T)v) \leq \varepsilon - \delta, \quad v \in V.$$

This contradiction provides the desired solution. This procedure is summarized by

**Lemma 10.36.** *In addition to the hypotheses of Theorem 10.34, assume that there are positive constants $\varepsilon, \rho$ such that*

$$G(u) \geq \varepsilon \qquad\qquad\qquad (10.101)$$

*when*

$$\|u\|_1 = \rho. \qquad\qquad\qquad (10.102)$$

*Then there is a solution $u$ of (10.68) satisfying (10.101).*

The proof of Lemma 10.36 is similar to that of Lemma 2.20 and is omitted.

In the next section we shall give simple conditions under which the "mountains" can be constructed.

## 10.16    The mountains revisited

We now want to give sufficient conditions on $F(x, t)$ which will imply that the origin is surrounded by mountains. By this we mean the following.

**Theorem 10.37.** *Assume that (10.66) holds and that there is a $\delta > 0$ such that*

$$F(x, t) \geq 0, \quad |t| \leq \delta. \tag{10.103}$$

*Then for each positive $\rho \leq \delta/2$, we have either*

(a) *there is an $\varepsilon > 0$ such that*

$$G(u) \geq \varepsilon, \quad \|u\|_1 = \rho, \tag{10.104}$$

*or*

(b) *there is a constant function $v_0 \in V$ such that $|v_0| = \rho/(2\pi)^{n/2} \leq \delta/2$, and*

$$f(x, v_0) \equiv 0. \tag{10.105}$$

*Moreover, the constant function $v_0 \in V$ satisfying (10.105) is a solution of (10.54).*

*Proof.* For each $u \in H_1$ write $u = v + w$, where $v \in V$, $w \in W$. Then

$$2G(u) = \|\nabla w\|_0^2 - 2\int_Q F(x, u)\, dx \geq \|\nabla w\|_0^2 - 2\int_{|u|>\delta} |F(x, u)|\, dx.$$

Now

$$\|u\|_1 \leq \rho \Rightarrow \|v\|_0^2 + \|w\|_1^2 \leq \rho^2 \Rightarrow (2\pi)^{n/2}|v| \leq \rho.$$

Thus if $\rho \leq (2\pi)^{n/2}\delta/2$, then $|v| < \delta/2$. Hence, if

$$\|u\|_1 \leq \rho, \quad |u(x)| \geq \delta,$$

then

$$\delta \leq |u(x)| \leq |v| + |w(x)| \leq \delta/2 + |w(x)|.$$

Consequently,

$$\delta \le |u(x)| \le 2|w(x)|.$$

We may assume that $q$ satisfies $2 < q \le 2^*$. Then

$$2G(u) \ge \|\nabla w\|_0^2 - C \int_{|u|>\delta} (|u|^q + |u|)\, dx$$

$$\ge \|\nabla w\|_0^2 - C(1 + \delta^{1-q}) \int_{|u|>\delta} |u|^q\, dx$$

$$\ge \|\nabla w\|_0^2 - C' \int_{2|w|>\delta} |w|^q dx$$

$$\ge \frac{1}{2}\|w\|_1^2 - C'|w|_q^q$$

$$\ge \frac{1}{2}\|w\|_1^2 - C''\|w\|_1^q$$

$$\ge \left(\frac{1}{2} - C''\|w\|_1^{q-2}\right)\|w\|_1^2.$$

Hence,

$$G(u) \ge \frac{1}{5}\|w\|_1^2, \quad \|u\|_1 \le \rho, \tag{10.106}$$

for $\rho > 0$ sufficiently small. For each such $\rho$ assume that there is no $\varepsilon > 0$ for which (10.104) holds. Then there is a $\{u_k\} \subset H_1$ such that $\|u_k\|_1 = \rho$ and $G(u_k) \to 0$. Write $u_k = v_k + w_k$, where $v_k \in V$, $w_k \in W$. Then $\|w_k\|_1 \to 0$ by (10.106). This means that $\|v_k\|_1 \to \rho$. Thus, $|v_k| \to \rho/(2\pi)^{n/2}$. Since the constants $\{v_k\}$ are bounded, there is a renamed subsequence such that $v_k \to v_0$. Clearly $|v_0| = \rho/(2\pi)^{n/2} \le \delta/2$, and

$$\int_Q F(x, v_0)\, dx = G(v_0) = 0.$$

In view of (10.103), $F(x, v_0) \equiv 0$, and $v_0$ is a minimum point of $F(x, t)$ in $|t| < \delta$. Hence, the derivative of $F(x, t)$ with respect to $t$ must vanish at $t = v_0$. This gives $f(x, v_0) \equiv 0$. Since $\rho$ was any sufficiently small constant, we see that (b) holds. This completes the proof. □

We note that (b) implies that every constant function $v_0 \in V$ satisfying (10.105) is a solution of $G'(v) = 0$. We therefore have

**Corollary 10.38.** *Under the hypotheses of Theorem 10.37, either (a) holds for all $\rho > 0$ sufficiently small, or (10.68) has an infinite number of solutions.*

Combining our results so far, we have

**Theorem 10.39.** *Under the hypotheses of Theorems 10.26, 10.37, and Lemma 10.30, G has a nontrivial critical point which is a solution of (10.54).*

*Proof.* We combine Lemma 10.36 and Theorem 10.37. $\qquad\square$

## 10.17    Other intervals between eigenvalues

Suppose $f(x,t)$ satisfies (10.80) and (10.82), but $\alpha(x)$ does not satisfy (10.88). Are there other intervals $(a,b)$ such that a solution of (10.54) for $u \in H_1$ can be found when $a \leq \alpha(x) \leq b$? We are going to show that this is indeed the case. In fact we have

**Theorem 10.40.** *Assume that (10.66), (10.80), and (10.93) hold. Let $\lambda_m$ and $\lambda_{m+1}$ be consecutive eigenvalues of $-\Delta$. Assume that (10.82) holds with $\alpha(x)$ satisfying*

$$\lambda_m \leq \alpha \leq \lambda_{m+1}, \quad \lambda_m \not\equiv \alpha \not\equiv \lambda_{m+1}. \tag{10.107}$$

*Then (10.54) has a solution. If, in addition, (10.103) holds, then we are assured that it has a nontrivial solution.*

*Proof.* First, we note by (10.9), that

$$\|u\|_1^2 = (2\pi)^n \sum (1+\mu^2)|\alpha_\mu|^2 < \infty, \tag{10.108}$$

where the $\alpha_\mu$ are given by Lemma 10.2. Let

$$N = \{u \in H_1 : \alpha_\mu = 0 \ for \ |\mu| > \lambda_m\}.$$

Thus,

$$\|u\|_1^2 = \sum_{|\mu| \leq \lambda_m} \mu^2 |\alpha_\mu|^2 \leq \lambda_m \|u\|^2, \quad u \in N. \tag{10.109}$$

Let

$$M = \{u \in H : \alpha_\mu = 0 \ for \ |\mu| \leq \lambda_m\}.$$

In this case,

$$\|u\|_1^2 = \sum_{|\mu| \geq \lambda_{m+1}} \mu^2 |\alpha_\mu|^2 \geq \lambda_{m+1} \|u\|^2, \quad u \in M. \tag{10.110}$$

Note that $M, N$ are closed subspaces of $H_1$ and that $M = N^\perp$. Note also that $N$ is finite dimensional. Next, we consider the functional (10.69) and show that

$$G(v) \to -\infty \text{ as } \|v\|_1 \to \infty, \quad v \in N, \tag{10.111}$$

and

$$G(w) \to \infty \text{ as } \|w\|_1 \to \infty, \quad w \in M. \tag{10.112}$$

Assuming these for the moment, we note that they imply

$$\inf_M G > -\infty; \quad \sup_N G < \infty. \tag{10.113}$$

This is easily seen from the fact that (10.112) implies that there is an $R > 0$ such that

$$G(w) > 0, \quad \|w\|_1 > R, \ w \in M.$$

Consequently, if the first statement in (10.113) were false, there would be a sequence satisfying

$$G(w_k) \to -\infty, \quad \|w_k\|_1 \leq R, \ w_k \in H_1.$$

But this would imply that there is a renamed subsequence such that $w_k \to w_0$ weakly in $H_1$, strongly in $L^2(Q)$, and a.e. in $Q$ to a limit $w_0 \in H_1$ (Corollary 10.15). Since

$$|F(x, w_k)| \leq C(|w_k|^2 + |w_k|),$$

and

$$F(x, w_k) \to F(x, w_0) \text{ a.e.},$$

we have

$$\int_Q F(x, w_k)\, dx \to \int_Q F(x, w_0)\, dx.$$

Thus,

$$G(w_k) \geq -\int_Q F(x, w_k)\, dx \to -\int_Q F(x, w_0)\, dx > -\infty.$$

This contradiction verifies the first statement in (10.113). The second is verified similarly by (10.111).

We are now in a position to apply Theorem 2.5. This produces a sequence in $H_1$ satisfying

$$G(u_k) \to c, \quad G'(u_k) \to 0, \tag{10.114}$$

where $c$ is finite. In particular, this implies

$$(G'(u_k), v)_1 = (u_k, v)_1 - (f(\cdot, u_k), v) = o(\|v\|_1), \quad \|v\|_1 \to 0. \quad (10.115)$$

Assume first that

$$\rho_k = \|u_k\|_1 \to \infty. \quad (10.116)$$

Set $\tilde{u}_k = u_k/\rho_k$. Then $\|u_k\|_1 = 1$, and consequently, by Corollary 10.15, there is a renamed subsequence that converges to a limit $\tilde{u} \in H_1$ weakly in $H_1$, strongly in $L^2(Q)$, and a.e. in $Q$. Thus,

$$(\tilde{u}_k, v)_H - (f(\cdot, u_k)/\rho_k, v) \to 0, \quad v \in H. \quad (10.117)$$

As we saw before, this implies in the limit that

$$(\tilde{u}, v)_1 = (\alpha \tilde{u}, v), \quad v \in H_1. \quad (10.118)$$

Let

$$\tilde{u} = \tilde{w} + \tilde{v}, \quad \hat{u} = \tilde{w} - \tilde{v}. \quad (10.119)$$

Then

$$(\tilde{u}, \hat{u})_1 = (\alpha \tilde{u}, \hat{u}).$$

This implies

$$\|\tilde{w}\|_1^2 - \|\tilde{v}\|_1^2 = (\alpha[\tilde{w} + \tilde{v}], \tilde{w} - \tilde{v}) = (\alpha \tilde{w}, \tilde{w}) - (\alpha \tilde{v}, \tilde{v}),$$

since

$$(\alpha \tilde{v}, \tilde{w}) = (\alpha \tilde{w}, \tilde{v}) = \int_Q \alpha(x) \tilde{v}(x) \tilde{w}(x)\, dx.$$

Thus,

$$\|\tilde{w}\|_1^2 - (\alpha \tilde{w}, \tilde{w}) = \|\tilde{v}\|_1^2 - (\alpha \tilde{v}, \tilde{v}).$$

This says

$$(\|\tilde{w}\|_1^2 - \lambda_{m+1}\|\tilde{w}\|^2) + \int_Q [\lambda_{m+1} - \alpha(x)]\tilde{w}^2\, dx$$
$$= (\|\tilde{v}\|_1^2 - \lambda_m\|\tilde{v}\|^2) + \int_Q [\lambda_m^2 - \alpha(x)]\tilde{v}^2\, dx.$$

We write this as $A + B = C + D$. In view of (10.107), (10.109), and (10.110), $A \geq 0$, $B \geq 0$, $C \leq 0$, $D \leq 0$. But this implies $A = B = C = D = 0$. If

$$\tilde{u} = \sum \tilde{\alpha}_\mu \varphi_\mu,$$

then in view of (10.110) the only way $A$ can vanish is if

$$\tilde{w} = \sum_{k=1}^{n} (a_k \cos \lambda_{m+1} x_k + b_k \sin \lambda_{m+1} x_k).$$

But then, $B$ cannot vanish unless $\tilde{w} \equiv 0$ (Lemma 10.28). Similar reasoning shows that $C = D = 0$ implies that $\tilde{v} \equiv 0$. On the other hand, (10.82) implies

$$2G(u_k)/\rho_k^2 = \|\tilde{u}\|_1^2 - 2 \int_Q F(x, \tilde{u}) \, dx/\rho_k^2 \to 1 - 2 \int_Q \alpha(x)\tilde{u}^2 \, dx = 0,$$

from which we conclude that $\tilde{u} \not\equiv 0$. This contradiction shows that the assumption (10.116) is incorrect. Once this is known, we can conclude that by Corollary 10.15, there is a renamed subsequence that converges to a limit $u \in H_1$ weakly in $H_1$, strongly in $L^2(Q)$, and a.e. in $Q$. It then follows from (10.115) that

$$(u, v)_1 - (f(\cdot, u), v) = 0, \quad v \in H_1. \tag{10.120}$$

It remains to prove (10.111) and (10.112). Let $\{w_k\} \subset M$ be any sequence such that $\rho_k = \|w_k\|_1 \to \infty$. Let $\tilde{w}_k = w_k/\rho_k$. Then $\|\tilde{w}_k\|_1 = 1$. Thus, there is a renamed subsequence such that (2.30) and (2.31) hold. This implies

$$2G(w_k)/\rho_k^2 = 1 - 2 \int_Q \frac{f(x, w_k)}{w_k^2} \tilde{w}_k^2 \, dx \to 1 - \int_Q \alpha(x)\tilde{w}^2(x) \, dx$$

$$= (1 - \|\tilde{w}\|_1^2) + (\|\tilde{w}\|_1^2 - \lambda_{m+1}\|\tilde{w}\|^2)$$

$$+ \int_I [\lambda_{m+1} - \alpha(x)]\tilde{w}^2(x) \, dx$$

$$= A + B + C.$$

As before, we note that $A \geq 0$, $B \geq 0$, $C \geq 0$. The only way $G(w_k)$ can fail to become infinite is if $A = B = C = 0$. As before, $B = C = 0$ implies that $\tilde{w} \equiv 0$. But this contradicts the fact that $A = 0$. Thus, $G(w_k) \to \infty$ for each such sequence. This proves (10.112). The limit

(10.111) is proved in a similar fashion. This completes the proof of the first part of Theorem 10.40. The last statement follows from Lemma 10.36 and Theorem 10.37. $\qquad\square$

## 10.18 An example

Before we become overjoyed by our success in finding Palais–Smale sequences and obtaining solutions, we want to stress the fact that not all Palais–Smale sequences are created equal. Here is another example of a simple situation in which no Palais–Smale sequence has a convergent subsequence.

Let us consider the Hilbert space $H = \mathbb{R}^2$. For $u = (x, y) \in H$ define

$$G(u) = x^2 - (x - 1)^3 y^2.$$

Then,

$$G'(u) = \nabla G(u) = \{2x - 3(x - 1)^2 y^2, -2(x - 1)^3 y\}.$$

Note that $G$ satisfies all of the hypotheses of Theorem 2.26. Clearly, $G(0) = 0$. Moreover, there are positive constants $\varepsilon, \rho$ such that

$$G(u) \geq \varepsilon \|u\|^2, \quad \|u\| \leq \rho. \tag{10.121}$$

For if (10.121) did not hold, there would be a sequence $\{u_k\} \subset H$ such that $\rho_k = \|u_k\| \to 0$ and

$$G(u_k)/\|u_k\|^2 \to 0. \tag{10.122}$$

Let $\tilde{u}_k = u_k/\rho_k$. Then

$$G(u_k)/\rho_k^2 = \tilde{x}_k^2 - (x_k - 1)^3 \tilde{y}_k^2.$$

Since $\tilde{x}_k^2 + \tilde{y}_k^2 = 1$, there is a renamed subsequence such that $\tilde{x}_k \to \tilde{x}$, $\tilde{y}_k \to \tilde{y}$, $\tilde{x}^2 + \tilde{y}^2 = 1$. Consequently,

$$G(u_k)/\rho_k^2 \to \tilde{x}^2 + \tilde{y}^2 = 1,$$

contradicting (10.122). Thus (10.121) holds, and this implies (2.99). If we take $\varphi_0 = (1, 1)$, then

$$G(r\varphi_0) = r^2 - (r - 1)^3 r^2 \to -\infty \ as \ r \to \infty.$$

Thus (2.100) holds for some constant $C_0$. Since $G$ is a polynomial, all

of the hypotheses of Theorem 2.26 are satisfied. Thus, there is a Palais–
Smale sequence satisfying (2.101). Hence,

$$x_k^2 - (x_k - 1)^3 y_k^2 \to c, \quad \varepsilon \le c \le C_0.$$

$$2x_k - 3(x_k - 1)^2 y_k^2 \to 0,$$

and

$$2(x_k - 1)^3 y_k \to 0.$$

Either there is a renamed subsequence such that $x_k \to 1$, or the whole
sequence is such that $x_k$ is bounded away from 1. In the latter case,
$y_k \to 0$, and

$$|(x_k - 1)y_k|^2 = |(x_k - 1)^3 y_k|^{2/3} |y_k|^{4/3} \to 0.$$

Consequently, $x_k \to 0$. But then, $G(u_k) \to 0$, and this is not the sequence
(2.101). In the former case,

$$(x_k - 1)^2 y_k^2 \to 2/3.$$

Hence, $|y_k| \to \infty$, and there is no convergent subsequence. Thus, $G$ does
not satisfy the PS condition. Note that $G$ has no critical point other
than $u = 0$. For, in order that $u$ be a critical point, we need either $x = 1$
or $y = 0$. But either of these implies that $x = 0$. This shows that the
only critical point of $G$ is 0.

## 10.19    Satisfying the PS condition

In solving the problem (10.68), our approach has been to find a sequence
$\{u_k\}$ such that (10.100) holds and then show that this implies that $\{u_k\}$
has a convergent subsequence. So far we have shown this only when
$f(x,t)$ satisfies (10.83). In this section we allow $f(x,t)$ to satisfy (10.66)
with $q < 2^*$ and give sufficient conditions which will guarantee that the
PS condition holds for $G(u)$ given by (10.69). We have

**Theorem 10.41.** *If there are constants $\gamma > 2, C$ such that*

$$H_\gamma(x,t) := \gamma F(x,t) - tf(x,t) \le C(t^2 + 1) \tag{10.123}$$

*and there is a function $W(x) \in L^1(\Omega)$ such that*

$$-W(x) \le F(x,t)/t^2 \to \infty \text{ as } |t| \to \infty, \quad x \in \Omega, \tag{10.124}$$

*then (10.100) implies that $\{u_k\}$ has a convergent subsequence.*

*Proof.* If (10.100) holds, then

$$\|\nabla u_k\|_0^2 - 2\int_Q F(x, u_k)\, dx \to a, \qquad (10.125)$$

$$(\nabla u_k, \nabla v)_0 - (f(u_k), v)_0 = o(\|v\|_1), \quad v \in H_1, \qquad (10.126)$$

and

$$\|\nabla u_k\|_0^2 - (f(u_k), u_k) = o(\|u_k\|_1). \qquad (10.127)$$

If we multiply (10.127) by 2, (10.125) by $\gamma$ and subtract, we obtain

$$(\gamma - 2)\|\nabla u_k\|_0^2 - 2\int_Q H_\gamma(x, u_k)\, dx = o(\|u_k\|_1). \qquad (10.128)$$

Assume that $\rho_k = \|u_k\|_1 \to \infty$, and let $\tilde u_k = u_k/\rho_k$. Then $\|\tilde u_k\|_1 = 1$, and (10.128) implies

$$(\gamma - 2)(1 - \|\tilde u_k\|_0^2) - 2\int_Q H_\gamma(x, u_k)\, dx/\rho_k^2 \to 0.$$

However,

$$H_\gamma(x, u_k)/\rho_k^2 = [H_\gamma(x, u_k)/u_k^2]\tilde u_k^2.$$

There is a renamed subsequence such that $\tilde u_k \to \tilde u$ weakly in $H_1$, strongly in $H_0$, and a.e. in $Q$ (Corollary 10.15). By (10.125),

$$\int_\Omega \frac{2F(x, u_k) + u_k^2}{u_k^2} \tilde u_k^2\, dx \to 1,$$

and by (10.127)

$$\int_\Omega \frac{u_k f(x, u_k) + u_k^2}{u_k^2} \tilde u_k^2\, dx \to 1.$$

Let

$$\Omega_1 = \{x \in \Omega : \tilde u(x) \neq 0\}, \quad \Omega_2 = \Omega \setminus \Omega_1.$$

Then

$$\frac{2F(x, u_k)}{u_k^2}\tilde u_k^2 \to \infty, \quad x \in \Omega_1$$

by (10.124). If $\Omega_1$ has positive measure, then

$$\int_\Omega \frac{2F(x, u_k)}{u_k^2}\tilde u_k^2\, dx \geq \int_{\Omega_1} \frac{2F(x, u_k)}{u_k^2}\tilde u_k^2\, dx + \int_{\Omega_2} [-W(x)]\, dx \to \infty.$$

Thus, the measure of $\Omega_1$ must be 0, that is, we must have $\tilde{u} \equiv 0$ a.e. Thus, $\tilde{u}_k \to 0$ in $L^2(Q)$ implying

$$\int_\Omega \frac{2F(x, u_k)}{u_k^2} \tilde{u}_k^2 \, dx \to 1$$

and

$$\int_\Omega \frac{u_k f(x, u_k)}{u_k^2} \tilde{u}_k^2 \, dx \to 1.$$

Consequently,

$$\int_\Omega \frac{\gamma F(x, u_k) - u_k f(x, u_k)}{u_k^2} \tilde{u}_k^2 \, dx \to \frac{\gamma}{2} - 1.$$

But by (10.123),

$$\limsup \frac{\gamma F(x, u_k) - u_k f(x, u_k)}{u_k^2} \tilde{u}_k^2 \le \limsup C \frac{u_k^2 + 1}{u_k^2} \tilde{u}_k^2 = 0,$$

which implies that $\gamma/2 - 1 \le 0$, contrary to assumption. Thus, the $\rho_k$ must be bounded. Consequently, there is a renamed subsequence such that $u_k \to u$ weakly in $H_1$, strongly in $L^q(Q)$, and a.e. in $Q$ (Corollary 10.15). In view of (10.66),

$$|f(x, u_k) - f(x, u)| \le C(|u_k|^{q-1} + |u|^{q-1} + 1),$$

and

$$\int_Q |u_k|^{q-1} |v| \, dx \le |u_k|_q^{q-1} |v|_q,$$

which converges. Therefore (10.126) implies

$$(\nabla u, \nabla v)_0 - (f(u), v)_0 = 0, \quad v \in H_1,$$

which means that $u$ is a solution of (10.68) and satisfies

$$\|\nabla u\|_0^2 - (f(u), u)_0 = 0.$$

Moreover, by (10.127)

$$\|\nabla u_k\|_0^2 = (f(u_k), u_k) + o(1) \to (f(u), u) = \|\nabla u\|_0^2,$$

showing that $u_k$ converges to $u$ in $H_1$. Hence, the PS condition holds, and the proof is complete.                                    □

**Remark 10.42.** *In Theorem 10.41 we could have used hypothesis (1.114) in place of (10.124). The proof would have followed the reasoning of Theorem 1.37.*

### 10.20    More super-linear problems

If $f(x,t)$ does not satisfy (10.79), then the problem (10.54) is called super-linear. In the one dimensional case we proved a theorem concerning super-linear equations (Theorem 2.27). Now we consider a corresponding theorem for higher dimensions. The process is almost the same, but for higher dimensions we must place stronger restrictions on the growth of $f(x,t)$ because of the Sobolev inequalities. We have

**Theorem 10.43.** *Assume (10.66), (10.67), (10.103), and (10.93). Then under the hypotheses of Theorem 10.41, the problem (10.54) has a nontrivial solution.*

*Proof.* Apply Theorems 2.26, 10.37, and 10.41 to the functional $G(u)$ given by (10.69). We have to show that (2.100) holds. This follows from (10.124). In fact, we have

$$G(k\varphi) = -\int_Q F(x, k\varphi)\, dx$$

for any constant function $\varphi$. If we take $\varphi \equiv 1$, we have

$$G(k\varphi)/k^2 = -\int_Q F(x, k)/k^2\, dx \to -\infty \text{ as } k \to \infty.$$

This completes the proof. $\qquad\square$

### 10.21    Sobolev's inequalities

Unlike the one dimensional case, it is not true in higher dimensions that functions having weak derivatives in $L^2$ are bounded and continuous. In fact, they need not be either. The best that can be said for them is that they are in some $L^q$ space with $q$ depending on the dimension. The higher the dimension, the worse the $q$ is; it may be very close to 2. These facts are revealed in the Sobolev inequalities which we now describe.

We know from the definition that

$$\|u\|_0 \le \|u\|_1, \quad u \in H_1.$$

Actually, one can improve upon this. In fact, we have

**Theorem 10.44.** *For each $p \ge 1$, $q \ge 1$ satisfying*

$$\frac{1}{p} \le \frac{1}{q} + \frac{1}{n} \tag{10.129}$$

*there is a constant $C_{pq}$ such that*

$$|u|_q \le C_{pq}(|\nabla u|_p + |u|_p), \quad u \in C^1(Q), \tag{10.130}$$

*where*

$$|u|_q = \left(\int_Q |u|^q dx\right)^{1/q} \tag{10.131}$$

*and*

$$|\nabla u| = \left(\sum_{k=1}^n \left|\frac{\partial u}{\partial x_k}\right|^2\right)^{1/2}. \tag{10.132}$$

*Proof.* First we note that for each integer $r \ge 1$

$$\int (h_1 \cdots h_r)^{1/r} \le \left(\int h_1\right)^{1/r} \cdots \left(\int h_r\right)^{1/r}, \tag{10.133}$$

by Hölder's inequality (Theorem B.23), where the $h_j \ge 0$. Let

$$H_{jk...\ell} = \int |\nabla u| dx_j dx_k \cdots dx_\ell.$$

Assume that there is a $y \in Q$ such that $u(x) = 0$ whenever $x_j = y_j$ for some $j$. Then

$$u(x) = \int_{y_j}^{x_j} \frac{\partial u}{\partial x_j} dx_j.$$

This implies that

$$|u(x)| \le \int_0^{2\pi} \left|\frac{\partial u}{\partial x_j}\right| dx_j$$

$$\le h_j(x) = \int_0^{2\pi} |\nabla u(x)| dx_j.$$

(Note that $h_j(x)$ does not depend on $x_j$.) Then

$$|u(x)|^n \le h_1 \cdots h_n. \tag{10.134}$$

Take $r = n - 1$. We claim that

$$\int |u(x)|^{n/r} dx_1 \cdots dx_k$$

$$\le \begin{cases} H_{1\cdots k}^{k/r}(x)(H_{1\cdots k,(k+1)}(x)H_{1\cdots k,(k+2)}(x) \cdots H_{1\cdots k,n}(x))^{1/r}, & k < n, \\ H_{1\cdots k}^{k/r}(x), & k = n. \end{cases}$$

$$\tag{10.135}$$

We prove (10.135) by induction. We note that it holds for $k = 1$. By (10.133)

$$\int |u(x)|^{n/r}\, dx_1 \le \int (h_1 \cdots h_n)^{1/r}\, dx_1$$

$$\le h_1^{1/r} \int (h_2 \cdots h_n)^{1/r}\, dx_1$$

$$\le H_1^{1/r}(H_{12} \cdots H_{1n})^{1/r},$$

which shows that (10.135) holds for $k = 1$. Assume that it holds for $k$. Then

$$\int |u(x)|^{n/r} dx_1 \cdots dx_k\, dx_{k+1}$$

$$\le H_{1\cdots(k+1)}^{1/r} \int (H_{1\cdots k}^{k/r} H_{1\cdots k,(k+2)} \cdots H_{1\cdots k,n})^{1/r}\, dx_{k+1}$$

$$\le H_{1\cdots(k+1)}^{(k+1)/r}(H_{1\cdots(k+1),(k+2)} H_{1\cdots(k+1),(k+3)} \cdots H_{1\cdots(k+1),n})^{1/r}.$$

Thus (10.135) holds for $k + 1$. In particular, if we take $k = r = n - 1$, we obtain (10.135) for $k = n$. Thus, (10.135) holds for $1 \le k \le n$. If we take $k = n$, we obtain

$$\int |u(x)|^{n/r}\, dx_1 \cdots dx_n \le H_{1\cdots n}^{n/r} = |\nabla u|_1^{n/r}.$$

This implies

$$|u|_{n/(n-1)} \le |\nabla u|_1 \tag{10.136}$$

when $u$ satisfies the hypotheses given above. Hence, (10.130) is proved for the case $p = 1$ and $u(x)$ vanishing when $x_j = y_j$ for some $j$. To derive it without this restriction, let

$$\hat{Q} = [-2\pi, 2\pi]^n = \{x \in \mathbb{R}^n : -2\pi \le x_j \le 2\pi,\ 1 \le j \le n\}.$$

For each $j$, let $x_j'$ be the vector $x$ with the $x_j$ component missing (e.g., $x_3' = \{x_1, x_2, x_4, \ldots, x_n\}$). Thus we can write $x = \{x_j', x_j\}$ for each $j$. In consecutive order, we extend $u \in C^1(Q)$ to a $C^1$ function of $\{x_j', x_j\}$ for $-2\pi \le x_j \le 2\pi, j = 1, \ldots, n$. This is done as follows. For $-2\pi \le x_j \le 0$, we define

$$u(x_j', x_j) = 4u(x_j', -\tfrac{1}{2}x_j) - 3u(x_j', -x_j).$$

It is easily checked that $u$ is a $C^1$ function of $x$ for $-2\pi \le x_j \le 2\pi$.

Moreover,

$$\int_{-2\pi}^0 |u(x_j', x_j)|\, dx_j \le 4 \int_{-2\pi}^0 |u(x_j', -\tfrac{1}{2}x_j)|\, dx_j$$

$$+ 3 \int_{-2\pi}^0 |u(x_j', -x_j)|\, dx_j$$

$$\le 8 \int_0^\pi |u(x_j', z)|\, dz$$

$$+ 3 \int_0^{2\pi} |u(x_j', z)|\, dz$$

$$\le 11 \int_0^{2\pi} |u(x_j', x_j)|\, dx_j$$

and

$$\int_{-2\pi}^0 |D_j u(x_j', x_j)|\, dx_j \le 2 \int_{-2\pi}^0 |D_j u(x_j', -\tfrac{1}{2}x_j)|\, dx_j$$

$$+ 3 \int_{-2\pi}^0 |D_j u(x_j', -x_j)|\, dx_j$$

$$\le 4 \int_0^\pi |D_j u(x_j', z)|\, dz$$

$$+ 3 \int_0^{2\pi} |D_j u(x_j', z)|\, dz$$

$$\le 7 \int_0^{2\pi} |D_j u(x_j', x_j)|\, dx_j.$$

We carry out this procedure in consecutive order from $j = 1$ to $j = n$. Consequently,

$$|u|_{1,\hat{Q}} \le C|u|_{1,Q}, \quad |\nabla u|_{1,\hat{Q}} \le C|\nabla u|_{1,Q}. \tag{10.137}$$

Let $\psi(t) \in C^\infty(\mathbb{R})$ be such that

$$0 \le \psi(t) \le 1$$

and

$$\psi(t) = \begin{cases} 1, & t \ge 0, \\ 0, & t \le -2\pi. \end{cases}$$

We define

$$\hat{u}(x) = \psi(x_1) \cdots \psi(x_n) u(x).$$

Note that $\hat{u}(x) = u(x)$ when $x \in Q$ and $\hat{u}(x) = 0$ if any $x_j = -2\pi$. Thus

$\hat{u}$ satisfies in $\hat{Q}$ the hypotheses that produced (10.136) for the cube $Q$. Now

$$D_j\hat{u}(x) = \psi(x_1)\cdots\psi'(x_j)\cdots\psi(x_n)u(x) + \psi(x_1)\cdots\psi(x_n)D_ju(x).$$

Hence,

$$|D_j\hat{u}(x)| \le C(|D_ju(x)| + |u(x)|),$$

and

$$|\hat{u}|_{n',\hat{Q}} \le C|\nabla\hat{u}|_{1,\hat{Q}}$$

by (10.136), where $n' = n/(n-1)$. Consequently,

$$\begin{aligned}|u|_{n',Q} \le |\hat{u}|_{n',\hat{Q}} &\le C|\nabla\hat{u}|_{1,\hat{Q}} \\ &\le C'(|\nabla u|_{1,\hat{Q}} + |u|_{1,\hat{Q}}) \\ &\le C''(|\nabla u|_{1,Q} + |u|_{1,Q}).\end{aligned}$$

Thus we have

$$|u|_{n'} \le C(|\nabla u|_1 + |u|_1) \qquad (10.138)$$

holding for all $u \in C^1(Q)$ without restriction. Next, assume that $u(x) > 0$ in $Q$, and let $\rho = (n-1)q/n \ge 1$. Then (10.138) implies

$$|u^\rho|_{n/(n-1)} \le C(|\nabla(u^\rho)|_1 + |u^\rho|_1).$$

Thus, by Hölder's inequality (Theorem B.23),

$$|u^\rho|_{n'} \le C'(\rho|u^{\rho-1}\nabla u|_1 + |u^{\rho-1}u|_1) \le C'|u^{\rho-1}|_{p'}\,(\rho|\nabla u|_p + |u|_p), \qquad (10.139)$$

where $p' = p/(p-1), n' = n/(n-1)$. By (10.129),

$$\frac{1}{p'} = 1 - \frac{1}{p} \ge 1 - \frac{1}{n} - \frac{1}{q} = \frac{1}{n'} - \frac{1}{q}.$$

Hence,

$$q \ge \left(\frac{q}{n'} - 1\right)p' = (\rho-1)p',$$

and

$$\left(\int_Q u^{(\rho-1)p'}\right)^{1/p'} \le \left(\int_Q u^{\rho n'}\right)^{(\rho-1)/(\rho p')}\left(\int_Q 1\right)^{1/(\rho p')} \le C|u|_q^{\rho-1}.$$

It therefore follows from (10.139) that there is a constant $C_0$ such that

$$|u|_q \le C_0(\rho|\nabla u|_p + |u|_p). \qquad (10.140)$$

Consequently, we see that (10.130) holds for positive $u$.

In the general case, let

$$u_\varepsilon(x) = (u^2 + \varepsilon^2)^{1/2}, \quad \varepsilon > 0.$$

Then

$$\frac{\partial(u_\varepsilon^\rho)}{\partial x_j} = \rho u_\varepsilon^{\rho-1} \frac{\partial u_\varepsilon}{\partial x_j}.$$

We apply (10.130) to $u_\varepsilon$ and use the fact that

$$|u| \le |u_\varepsilon| \le |u| + \varepsilon, \quad |\nabla u_\varepsilon| \le |\nabla u|$$

to show that (10.130) holds for general $u$. This proves the theorem.    □

As a corollary, we have

**Corollary 10.45.** *If*

$$1 \le q \le 2^* := 2n/(n-2),$$

*then*

$$|u|_q \le C_q \|u\|_1, \quad u \in H_1.$$

*Proof.* Since $S$ is dense in $H_1$, it suffices to prove the corollary for $u \in S$. We take $p = 2$ in (10.129). Then $2^* = 2n/(n-2)$ is the largest value of $q$ which satisfies (10.129). The corollary now follows from Theorem 10.44.    □

We also have

**Corollary 10.46.** *If $1 \le q < 2^*$, then every bounded sequence in $H_1$ has a subsequence which converges in $L^q(Q)$.*

*Proof.* We may assume that $2 < q < 2^*$. We shall prove that

$$|u|_q \le C \|u\|_0^a \cdot \|u\|_1^b, \quad u \in H_1, \tag{10.141}$$

where $a = 2(2^* - q)/q(2^* - 2)$, $b = 2^*(q - 2)/q(2^* - 2)$. Once we have (10.141), we can prove the corollary as follows. If $\|u_k\|_1 \le M$, then there is a renamed subsequence which converges in $H_0$ (Lemma 10.4). Hence,

$$|u_j - u_k|_q \le C\|u_j - u_k\|_0^a \cdot \|u_j - u_k\|_1^b$$

$$\le (2M)^b C\|u_j - u_k\|_0^a \to 0 \text{ as } j, k \to \infty.$$

Therefore, $\{u_k\}$ converges in $L^q(Q)$. We prove (10.141) by noting that

$$\int_Q |u|^q \, dx \le \left( \int_Q u^2 \, dx \right)^{qa/2} \left( \int_Q |u|^{2^*} \, dx \right)^{qb/2^*},$$

by Hölder's inequality (Theorem B.23). (Note that $a/2 + b/2^* = 1$.) Thus

$$|u|_q \le |u|_2^a |u|_{2^*}^b,$$

and this implies (10.141) in view of Corollary 10.45. This completes the proof. $\qquad\square$

Note that our proof required $q < 2^*$, since we needed $a > 0$.

## 10.22    The case $q = \infty$

It would appear from Theorem 10.44 that when $p \ge n$, one should be able to obtain the inequality

$$|u|_\infty \le C(|\nabla u|_p + |u|_p), \quad u \in C^1(Q), \tag{10.142}$$

where

$$|u|_\infty = \operatorname{esssup}_Q |u(x)|.$$

(For discontinuous functions the essential supremum ignores sets of measure 0.) When $p = n$, this is not true, but otherwise we have

**Theorem 10.47.** *If $p > n$, then (10.142) holds.*

*Proof.* Assume that $u(x) > 0$ on $Q$. By (10.138)

$$|u|_{n'} \le C_0(|\nabla u|_1 + |u|_1) \le C_0 |Q|^{1/p'} (|\nabla u|_p + |u|_p).$$

Let

$$\tilde{u} = u / C_0 |Q|^{1/p'} (|\nabla u|_p + |u|_p).$$

Then

$$|\tilde{u}|_{n'} \le 1,$$

and if $\rho > 1$, we have

$$\begin{aligned}
|\tilde{u}^\rho|_{n'} &\le C_0(|\nabla \tilde{u}^\rho|_1 + |\tilde{u}^\rho|_1) \\
&\le C_0(\rho |\tilde{u}^{\rho-1} \nabla \tilde{u}|_1 + |\tilde{u}^{\rho-1} \tilde{u}|_1) \\
&\le C_0 |\tilde{u}^{\rho-1}|_{p'} (\rho |\nabla \tilde{u}|_p + |\tilde{u}|_p).
\end{aligned}$$

Since

$$|\tilde{u}^{\rho-1}|_{p'} = \left(\int_Q \tilde{u}^{(\rho-1)p'}\,dx\right)^{1/p'} \leq \left(\int_Q \tilde{u}^{\rho p'}\,dx\right)^{(\rho-1)/(\rho p')} |Q|^{1/\rho p'},$$

we have

$$|\tilde{u}^{\rho}|_{n'} \leq C_0 |\tilde{u}|_{\rho p'}^{\rho-1}|Q|^{1/\rho p'}(\rho|\nabla\tilde{u}|_p + |\tilde{u}|_p).$$

But

$$C_0|Q|^{1/\rho p'}(\rho|\nabla\tilde{u}|_p + |\tilde{u}|_p) \leq \rho,$$

since $\rho > 1$ and $|Q| \geq 1$. Hence,

$$|\tilde{u}^{\rho}|_{n'} = |\tilde{u}|_{\rho n'}^{\rho} \leq \rho|\tilde{u}|_{\rho p'}^{\rho-1}.$$

This is also true for $\rho = 1$. Let $\sigma = n'/p' > 1$, and take $\rho = \sigma^k$, $k = 0, 1, 2, \ldots$ Then we have

$$|\tilde{u}|_{\sigma^k n'} \leq \sigma^{k/\sigma^k}|\tilde{u}|_{\sigma^{k-1}n'}^{1-(1/\sigma^k)}.$$

This implies

$$|\tilde{u}|_{\sigma^N n'} \leq \sigma^{\sum_{k=0}^{N}(k/\sigma^k)}.$$

We saw before that this is true for $N = 0$. If it is true for $N - 1$, then

$$|\tilde{u}|_{\sigma^N n'} \leq \sigma^{N/\sigma^N}\sigma^{\sum_{k=0}^{N-1}(k/\sigma^k)} = \sigma^{\sum_{k=0}^{N}(k/\sigma^k)},$$

since $\sigma > 1$. Thus,

$$|\tilde{u}|_{\sigma^N n'} \leq \sigma^{\sum_{k=0}^{\infty}(k/\sigma^k)} \equiv C_1$$

for each $N$. Let $C_2$ be any number greater than $C_1$, and let $\Omega$ be the set on which $|\tilde{u}(s)| \geq C_2$. Then

$$C_2|\Omega|^{1/(\sigma^N n')} \leq C_1, \quad N = 0, 1, 2, \ldots,$$

where

$$|\Omega| = \int_\Omega 1\,dx.$$

If $|\Omega| \neq 0$, then $|\Omega|^{1/(\sigma^N n')} \to 1$ as $N \to \infty$. This implies that $C_2 \leq C_1$, contrary to assumption. Hence, $|\Omega| = 0$. This shows that

$$|\tilde{u}|_\infty \leq C_1.$$

If we now use the definition of $\tilde{u}$, we obtain (10.142).                    □

## 10.23    Sobolev spaces

For $m$ a nonnegative integer and $p \geq 1$, we consider the norm

$$\|u\|_{m,p} = \sum_{|\tau| \leq m} |D^\tau u|_p, \quad u \in C^m(Q).$$

An equivalent norm is

$$\left( \sum_{|\tau| \leq m} \int_Q |D^\tau u|^p \, dx \right)^{1/p}.$$

We note

**Theorem 10.48.** *If*

$$p \geq 1, q \geq 1, \frac{1}{p} \leq \frac{1}{q} + \frac{m}{n},$$

*then*

$$|u|_q \leq C\|u\|_{m,p}, \quad u \in C^m(Q).$$

*Proof.* The theorem is true for $m = 1$ in view of Theorem 10.44. Assume it is true for $m - 1$. Let $q_1$ satisfy

$$\frac{1}{q_1} = \frac{1}{p} - \frac{m-1}{n}$$

if $(m-1) < n/p$ and $q_1 = 1$, otherwise. In either case,

$$\frac{1}{p} \leq \frac{1}{q_1} + \frac{m-1}{n}.$$

By the induction hypothesis,

$$|u|_{q_1} \leq C\|u\|_{m-1,p}, \quad u \in C^{m-1}(Q).$$

Thus,

$$|D_j u|_{q_1} \leq C\|D_j u\|_{m-1,p} \leq C\|u\|_{m,p}, \quad u \in C^m(Q).$$

Hence,

$$|\nabla u|_{q_1} + |u|_{q_1} \leq C\|u\|_{m,p}, \quad u \in C^m(Q).$$

Moreover,

$$\frac{1}{q_1} \leq \frac{1}{q} + \frac{1}{n},$$

since

$$\frac{1}{p} - \frac{m-1}{n} \leq \frac{1}{q} + \frac{1}{n}.$$

Thus,

$$|u|_q \leq C(|\nabla u|_{q_1} + |u|_{q_1}) \leq C\|u\|_{m,p},$$

and the theorem is proved. $\qquad\square$

We let $W^{m,p}(Q)$ be the completion of $C^m(Q)$ with respect to the norm $\|u\|_{m,p}$. What kind of functions are in $W^{m,p}(Q)$? If $u \in W^{m,p}(Q)$, then there is a sequence $\{u_k\} \subset C^m(Q)$ such that

$$\|u_k - u\|_{m,p} \to 0.$$

Thus

$$\|u_j - u_k\|_{m,p} \to 0.$$

This means that

$$\sum_{|\tau| \leq m} |D^\tau u_j - D^\tau u_k|_p \to 0,$$

as $j, k \to \infty$. Consequently, for each $\tau$ such that $|\tau| \leq m$ there is a function $u_\tau \in L^p(Q)$ such that

$$|D^\tau u_k - u_\tau|_p \to 0.$$

The function $u_\tau$ does not depend on the sequence $\{u_k\}$, for if $\{\hat{u}_k\}$ is another sequence converging to $\hat{u}$ in $W^{m,p}(Q)$, then

$$\|\hat{u}_k - u_k\|_{m,p} \to 0.$$

This implies that $\hat{u}_\tau = u_\tau$ for each $\tau$. We call $u_\tau$ the generalized strong $D^\tau$ derivative of $u$ in $L^p(Q)$, and denote it by $D^\tau u$. We have

**Theorem 10.49.** *Under the hypotheses of Theorem 10.48,*

$$|u|_q \leq C\|u\|_{m,p}, \quad u \in W^{m,p}(Q). \tag{10.143}$$

*Proof.* For a sequence $\{u_k\}$ in $C^m(Q)$ converging to $u$ in $W^{m,p}(Q)$, we have by Theorem 10.48,

$$|u_j - u_k|_q \leq C\|u_j - u_k\|_{m,p}.$$

Thus $u_k \to \hat{u}$ in $L^q(Q)$. Since $u_k \to u$ in $L^p(Q)$, we must have $\hat{u} = u$ a.e. Since

$$|u_k|_q \leq C\|u_k\|_{m,p},$$

we have (10.143). $\qquad\square$

We also have

**Theorem 10.50.** *If $m > n/p$, then*

$$|u|_\infty \le C\|u\|_{m,p}, \quad u \in C^m(Q).$$

*Proof.* If $m - 1 < n/p$, let

$$\frac{1}{q} = \frac{1}{p} - \frac{m-1}{n}.$$

Otherwise, take $q > n$. Then

$$|u|_q \le C\|u\|_{m-1,p}, \quad u \in C^{m-1}(Q),$$

in view of Theorem 10.48. Hence,

$$|D_j u|_q \le C\|D_j u\|_{m-1,p} \le C\|u\|_{m,p},$$

implying

$$|\nabla u|_q \le C\|u\|_{m,p}.$$

Since $q > n$, we have

$$|u|_\infty \le C(|\nabla u|_q + |u|_q) \le C'\|u\|_{m,p}$$

(Theorem 10.47). This proves the theorem. $\qquad\square$

We also have

**Theorem 10.51.** *If $m > n/p$ and $u \in W^{m,p}(Q)$, then $u \in C(Q)$, and*

$$\max_Q |u| \le C\|u\|_{m,p}. \tag{10.144}$$

*Proof.* If $\{u_k\}$ is a sequence in $C^m(Q)$ converging to $u$ in $W^{m,p}(Q)$, then

$$|u_j - u_k|_\infty \le C\|u_j - u_k\|_{m,p} \to 0.$$

Hence, $u_k$ converges uniformly on $Q$ to a continuous function $\hat{u}$. Since $u_k \to u$ in $L^p(Q)$, we must have $\hat{u} = u$ a.e. $\qquad\square$

**Corollary 10.52.** *If $m - \ell > n/p$, then $W^{m,p}(Q) \subset C^\ell(Q)$, and*

$$\max_{|\tau| \le \ell} \max_Q |D^\tau u| \le C\|u\|_{m,p}, \quad u \in W^{m,p}(Q).$$

*Proof.* We apply Theorem 2.26 to the derivatives of $u$ up to order $\ell$. $\qquad\square$

## 10.24     Exercises

1. Show that

$$u(x) = \sum \alpha_\mu e^{i\mu x} \tag{10.145}$$

   is real valued when $\alpha_{-\mu}$ appears whenever $\alpha_\mu$ appears and satisfies

$$\alpha_{-\mu} = \overline{\alpha}_\mu. \tag{10.146}$$

2. Show that $H_t$ is a Hilbert space for each real $t$.

3. Show that

$$\sum (1 + \mu^2)^t |\alpha_\mu^{(j)} - \alpha_\mu^{(k)}|^2 \to 0 \text{ as } j, k \to \infty$$

   implies that there is a subsequence such that

$$\alpha_\mu^{(j)} \to \alpha_\mu \text{ as } j \to \infty$$

   for each $\mu \in \mathbb{Z}^n$.

4. Why does letting $k \to \infty$ in

$$\sum_{\mu^2 \leq N} (1 + \mu^2)^t |\alpha_\mu^{(j)} - \alpha_\mu^{(k)}|^2 < \varepsilon^2, \quad j, k > K,$$

   imply

$$\sum_{\mu^2 \leq N} (1 + \mu^2)^t |\alpha_\mu^{(j)} - \alpha_\mu|^2 \leq \varepsilon^2, \quad j > K.$$

5. Why does this imply

$$\sum (1 + \mu^2)^t |\alpha_\mu^{(j)} - \alpha_\mu|^2 \leq \varepsilon^2, \quad j > K.$$

6. Show that both sides of (10.15) equal

$$(2\pi)^n \sum \alpha_\mu \beta_{-\mu}$$

   when $u, v$ are given by (10.10).

7. Prove Lemma 10.2.

8. Prove (10.22) and (10.23). Show that they imply (10.21).

9. Prove (10.25).

10. In the proof of Theorem 10.6, how do we know that $u_N$ converges uniformly to $u$ and not some other function?

11. Show that there is a constant $C$ such that

$$1 + \mu^2 \le C|\mu^2 - \lambda|, \quad \mu \in \mathbb{Z}^n,$$

when $\lambda$ is not an eigenvalue of $-\Delta$.

12. List the first 20 eigenvalues of $-\Delta$ when $n = 3$.

13. Prove (10.56).

14. Show that

$$\|\nabla v\|_s^2 + \|v\|_s^2 = \|v\|_t^2, \quad v \in H_t,$$

when $s = t - 1$.

15. Show that

$$|F(x,t)| \le C(|t|^q + |t|), \quad x \in Q,\ t \in \mathbb{R},$$

implies

$$\int_Q |F(x,u)|\, dx \le C\|u\|_1^q, \quad u \in H_1.$$

16. Carry out the details of the proof of Lemma 10.27.

17. Prove:

$$\int_Q \mu(E)\, dx = \int_e \left[ \int_{Q_{n-1}} \mu(E(x_n))dy \right] dx_n,$$

where $\mu(E)$ is the characteristic function of $E$.

18. Prove (10.133).

19. Let

$$\hat{Q} = [-2\pi, 2\pi]^n = \{x \in \mathbb{R}^n : -2\pi \le x_j \le 2\pi,\ 1 \le j \le n\}.$$

For each $j$, let $x'_j$ be the vector $x$ with the $x_j$ component missing (e.g., $x'_3 = \{x_1, x_2, x_4, \ldots, x_n\}$). Thus we can write $x = \{x'_j, x_j\}$ for each $j$. For $-2\pi \le x_j \le 0$, define

$$u(x'_j, x_j) = 4u(x'_j, -\tfrac{1}{2}x_j) - 3u(x'_j, -x_j).$$

Show that

$$\int_{-2\pi}^{0} |u(x_j', x_j)|\, dx_j \le 4 \int_{-2\pi}^{0} |u(x_j', -\tfrac{1}{2}x_j)|\, dx_j$$

$$+ 3 \int_{-2\pi}^{0} |u(x_j', -x_j)|\, dx_j$$

$$\le 8 \int_{0}^{\pi} |u(x_j', z)|\, dz$$

$$+ 3 \int_{0}^{2\pi} |u(x_j', z)|\, dz$$

$$\le 11 \int_{0}^{2\pi} |u(x_j', x_j)|\, dx_j.$$

20. Under the same circumstances, show that

$$\int_{-2\pi}^{0} |D_j u(x_j', x_j)|\, dx_j \le 2 \int_{-2\pi}^{0} |D_j u(x_j', -\tfrac{1}{2}x_j)|\, dx_j$$

$$+ 3 \int_{-2\pi}^{0} |D_j u(x_j', -x_j)|\, dx_j$$

$$\le 4 \int_{0}^{\pi} |D_j u(x_j', z)|\, dz$$

$$+ 3 \int_{0}^{2\pi} |D_j u(x_j', z)|\, dz$$

$$\le 7 \int_{0}^{2\pi} |D_j u(x_j', x_j)|\, dx_j.$$

21. Conclude that

$$|u|_{1,\hat{Q}} \le C|u|_{1,Q}, \quad |\nabla u|_{1,\hat{Q}} \le C|\nabla u|_{1,Q}. \qquad (10.147)$$

22. Find an estimate for the constant $C_{pq}$ in (10.130).

23. If

$$\tilde{u} = u/C_0 |Q|^{1/p'} (|\nabla u|_p + |u|_p)$$

and $\rho > 1$, show that

$$|\tilde{u}^\rho|_{n'} \le C_0(|\nabla \tilde{u}^\rho|_1 + |\tilde{u}^\rho|_1)$$
$$\le C_0(\rho|\tilde{u}^{\rho-1}\nabla\tilde{u}|_1 + |\tilde{u}^{\rho-1}\tilde{u}|_1)$$
$$\le C_0|\tilde{u}^{\rho-1}|_{p'} (\rho|\nabla\tilde{u}|_p + |\tilde{u}|_p).$$

24. Also, show that

$$|\tilde{u}^p|_{n'} \leq C_0 |\tilde{u}|_{pp'}^{p-1} |Q|^{1/pp'} (\rho|\nabla\tilde{u}|_p + |\tilde{u}|_p).$$

25. Also,

$$C_0 |Q|^{1/pp'} (\rho|\nabla\tilde{u}|_p + |\tilde{u}|_p) \leq \rho.$$

26. Also,

$$|\tilde{u}|_{\sigma^k n'} \leq \sigma^{k/\sigma^k} |\tilde{u}|_{\sigma^{k-1} n'}^{1-(1/\sigma^k)},$$

where $\sigma = n'/p' > 1$.

27. Also,

$$|\tilde{u}|_{\sigma^N n'} \leq \sigma^{\sum_{k=0}^{N} k/\sigma^k}.$$

28. Show that

$$\|u\|_{m,p} = \sum_{|\tau| \leq m} |D^\tau u|_p$$

and

$$\left( \sum_{|\tau| \leq m} \int_Q |D^\tau u|^p dx \right)^{1/p}$$

are equivalent norms.

29. Show that

$$\|\hat{u}_k - u_k\|_{m,p} \to 0$$

implies that $\hat{u}_\tau = u_\tau$ for each $\tau$.

# Appendix A
## Concepts from functional analysis

### A.1    Some basic definitions

Consider a collection $C$ of elements or "vectors" with the following properties:

1. They can be added. If $f$ and $g$ are in $C$, so is $f + g$.
2. $f + (g + h) = (f + g) + h,\quad f,\ g,\ h \in C.$
3. There is an element $0 \in C$ such that $h + 0 = h$ for all $h \in C$.
4. For each $h \in C$ there is an element $-h \in C$ such that $h + (-h) = 0$.
5. $g + h = h + g,\quad g,\ h \in C.$
6. For each real number $\alpha$, $\alpha h \in C$.
7. $\alpha(g + h) = \alpha g + \alpha h.$
8. $(\alpha + \beta)h = \alpha h + \beta h.$
9. $\alpha(\beta h) = (\alpha\beta)h.$
10. To each $h \in C$ there corresponds a real number $\|h\|$ with the following properties:
11. $\|\alpha h\| = |\alpha|\,\|h\|.$
12. $\|h\| = 0$ if, and only if, $h = 0$.
13. $\|g + h\| \leq \|g\| + \|h\|.$
14. If $\{h_n\}$ is a sequence of elements of $C$ such that $\|h_n - h_m\| \to 0$ as $m, n \to \infty$, then there is an element $h \in C$ such that $\|h_n - h\| \to 0$ as $n \to \infty$.

A collection of objects which satisfies statements (1)–(9) and the additional statement

$$1h = h \tag{15}$$

is called a **vector space** (VS) or **linear space.** We will be using real scalars.

A set of objects satisfying statements (1)–(13) is called a **normed vector space** (NVS), and the number $\|h\|$ is called the **norm** of $h$. Although statement (15) is not implied by statements (1)–(9), it is implied by statements (1)–(13). A sequence satisfying

$$\|h_n - h_m\| \to 0 \ \text{ as } \ m, n \to \infty$$

is called a **Cauchy sequence.** Property (14) states that every Cauchy sequence converges in norm to a limit (i.e., satisfies $\|h_n - h\| \to 0$ as $n \to \infty$). Property (14) is called **completeness**, and a normed vector space satisfying it is called a complete normed vector space or a **Banach space.**

We shall write

$$h_n \to h \ \text{ as } \ n \to \infty$$

when we mean

$$\|h_n - h\| \to 0 \ \text{ as } \ n \to \infty.$$

## A.2     Subspaces

**Definition A.1.** *A subset $U$ of a vector space $V$ is called a **subspace** of $V$ if $\alpha_1 x_1 + \alpha_2 x_2$ is in $U$ whenever $x_1$, $x_2$ are in $U$ and $\alpha_1$, $\alpha_2$ are scalars.*

**Definition A.2.** *A subset $U$ of a normed vector space $X$ is called **closed** if for every sequence $\{x_n\}$ of elements in $U$ having a limit in $X$, the limit is actually in $U$.*

A simple consequence of the definitions is

**Lemma A.3.** *A closed subspace of a Banach space is a Banach space with the same norm.*

## A.3     Hilbert spaces

**Lemma A.4.** *Consider a vector space $X$ having a mapping $(f, g)$ from pairs of its elements to the reals such that* *(i)* $(\alpha f, g) = \alpha(f, g)$

*(ii)* $(f + g, h) = (f, h) + (g, h)$

*(iii)* $(f, g) = (g, f)$

*(iv)* $(f, f) > 0$ unless $f = 0$.

*Then*

$$(f, g)^2 \leq (f, f)(g, g), \quad f, g \in X. \tag{A.1}$$

*Proof.* Let $\alpha$ be any scalar. Then

$$(\alpha f + g, \alpha f + g) = \alpha^2(f, f) + 2\alpha(f, g) + (g, g)$$

$$= (f, f)\left[\alpha^2 + 2\alpha\frac{(f, g)}{(f, f)} + \frac{(f, g)^2}{(f, f)^2}\right]$$

$$+ (g, g) - \frac{(f, g)^2}{(f, f)}$$

$$= (f, f)\left[\alpha + \frac{(f, g)}{(f, f)}\right]^2$$

$$+ (g, g) - \frac{(f, g)^2}{(f, f)},$$

where we have completed the square with respect to $\alpha$ and tacitly assumed that $(f, f) \neq 0$. This assumption is justified by the fact that if $(f, f) = 0$, then (A.1) holds vacuously. We now note that the left-hand side of (A.2) is nonnegative by property (iv) listed above. If we now take $\alpha = -(f, g)/(f, f)$, this inequality becomes

$$0 \leq (g, g) - \frac{(f, g)^2}{(f, f)},$$

which is exactly what we want. □

**Definition A.5.** *An expression $(f, g)$ that assigns a real number to each pair of elements of a vector space and satisfies the aforementioned properties is called a* **scalar** *(or* **inner***) product.*

We have essentially proved

**Lemma A.6.** *If a vector space $X$ has a scalar product $(f, g)$, then it is a normed vector space with norm $\|f\| = (f, f)^{1/2}$.*

*Proof.* Again, the only thing that is not immediate is the triangle inequality. This follows from (A.1) since

$$\begin{aligned} \|f+g\|^2 &= \|f\|^2 + \|g\|^2 + 2(f,g) \\ &\leq \|f\|^2 + \|g\|^2 + 2\|f\|\,\|g\| \\ &= (\|f\| + \|g\|)^2. \end{aligned}$$

This gives the desired result.     □

**Definition A.7.** *A vector space which has a scalar product and is complete with respect to the induced norm is called a* **Hilbert space**.

Every Hilbert space is a Banach space, but the converse is not true. Inequality (A.1) is known as the **Cauchy–Schwarz** inequality.

**Proposition A.8.** *The space* $\mathbb{R}^n$ *is a Hilbert space.*

## A.4     Bounded linear functionals

Let $H$ be a Hilbert space and let $(x, y)$ denote its scalar product. If we fix $y$, then the expression $(x, y)$ assigns to each $x \in H$ a number.

**Definition A.9.** *An assignment $F$ of a number to each element $x$ of a vector space is called a* **functional** *and denoted by $F(x)$.*

The scalar product is not the first functional we have encountered. In any normed vector space, the norm is also a functional. The functional $F(x) = (x, y)$ has some very interesting and surprising features. For instance it satisfies

$$F(\alpha_1 x_1 + \alpha_2 x_2) = \alpha_1 F(x_1) + \alpha_2 F(x_2) \tag{A.2}$$

for $\alpha_1$, $\alpha_2$ scalars.

**Definition A.10.** *A functional satisfying (A.2) is called* **linear**.

Another property is

$$|F(x)| \leq M\,\|x\|, \quad x \in H \tag{A.3}$$

which follows immediately from Schwarz's inequality (cf. (A.1)).

**Definition A.11.** *A functional satisfying (A.3) is called* **bounded**. *The norm of such a functional is defined to be*

$$\|F\| = \sup_{x \in H,\; x \neq 0} \frac{|F(x)|}{\|x\|}.$$

Thus for $y$ fixed, $F(x) = (x, y)$ is a bounded linear functional in the Hilbert space $H$. We have

**Theorem A.12.** *For every bounded linear functional $F$ on a Hilbert space $H$ there is a unique element $y \in H$ such that*

$$F(x) = (x, y) \text{ for all } x \in H. \tag{A.4}$$

*Moreover,*

$$\|y\| = \sup_{x \in H, \, x \neq 0} \frac{|F(x)|}{\|x\|} = \|F\|. \tag{A.5}$$

Theorem A.12 is known as the **Riesz representation theorem**.

We also have

**Theorem A.13.** *Let $M$ be a subspace of a normed vector space $X$, and suppose that $f(x)$ is a bounded linear functional on $M$. Set*

$$\|f\| = \sup_{x \in M, \, x \neq 0} \frac{|f(x)|}{\|x\|}.$$

*Then there is a bounded linear functional $F(x)$ on the whole of $X$ such that*

$$F(x) = f(x), \quad x \in M, \tag{A.6}$$

*and*

$$\|F\| = \sup_{x \in X, \, x \neq 0} \frac{|F(x)|}{\|x\|} = \|f\| = \sup_{x \in M, \, x \neq 0} \frac{|f(x)|}{\|x\|}. \tag{A.7}$$

Theorem A.13 is known as the **Hahn–Banach theorem**.

## A.5    The dual space

For any normed vector space $X$, let $X'$ denote the set of bounded linear functionals on $X$. If $f, g \in X'$, we say that $f = g$ if

$$f(x) = g(x) \text{ for all } x \in X.$$

The "zero" functional is the one assigning zero to all $x \in X$. We define $h = f + g$ by

$$h(x) = f(x) + g(x), \quad x \in X,$$

and $g = \alpha f$ by

$$g(x) = \alpha f(x), \quad x \in X.$$

Under these definitions, $X'$ becomes a vector space. We have been employing the expression

$$\|f\| = \sup_{x \neq 0} \frac{|f(x)|}{\|x\|}, \quad f \in X'. \tag{A.8}$$

This is easily seen to be a norm. In fact

$$\sup \frac{|f(x) + g(x)|}{\|x\|} \leq \sup \frac{|f(x)|}{\|x\|} + \sup \frac{|g(x)|}{\|x\|}.$$

Thus $X'$ is a normed vector space. It is therefore natural to ask when $X'$ will be complete. A rather surprising answer is given by

**Theorem A.14.** *The space $X'$ is a Banach space whether or not $X$ is.*

**Theorem A.15.** *Let $X$ be a normed vector space and let $x_0 \neq 0$ be an element of $X$. Then there is a bounded linear functional $F(x)$ on $X$ such that*

$$\|F\| = 1, \quad F(x_0) = \|x_0\|. \tag{A.9}$$

**Corollary A.16.** *For each $x \in X$,*

$$\|x\| = \max_{f \in X', \, f \neq 0} \frac{|f(x)|}{\|f\|}. \tag{A.10}$$

*Consequently, if $x_1$ is an element of $X$ such that $f(x_1) = 0$ for every bounded linear functional $f$ on $X$, then $x_1 = 0$.*

**Theorem A.17.** *Let $M$ be a subspace of a normed vector space $X$, and suppose $x_0$ is an element of $X$ satisfying*

$$d = d(x_0, M) = \inf_{x \in M} \|x_0 - x\| > 0. \tag{A.11}$$

*Then there is a bounded linear functional $F$ on $X$ such that $\|F\| = 1$, $F(x_0) = d$, and $F(x) = 0$ for $x \in M$.*

A subset $U$ of a normed vector space is called **convex** if $\alpha x + (1 - \alpha)y$ is in $U$ for each $x, y \in U$, $0 < \alpha < 1$. Clearly, the closure of a convex set is convex. The following consequence of the Hahn–Banach theorem is sometimes referred to as the "geometric form of the Hahn–Banach Theorem."

**Theorem A.18.** *If $U$ is a closed, convex subset of a normed vector space $X$ and $x_0 \in X$ is not in $U$, then there is an $x' \in X'$ such that*

$$x'(x_0) \geq x'(x), \quad x \in U, \tag{A.12}$$

*and $x'(x_0) \neq x'(x_1)$ for some $x_1 \in U$.*

## A.6    Operators

**Definition A.19.** *Let $X, Y$ be normed vector spaces. A mapping $A$ which assigns to each element $x$ of a set $D(A) \subset X$ a unique element $y \in Y$ is called an* **operator** *(or* **transformation***).*

The set $D(A)$ on which $A$ acts is called the **domain** of $A$.

**Definition A.20.** *The operator $A$ is called* **linear** *if*

(a) $D(A)$ *is a subspace of $X$*
    *and*
(b) $A(\alpha_1 x_1 + \alpha_2 x_2) = \alpha_1 A x_1 + \alpha_2 A x_2$ *for all scalars $\alpha_1$, $\alpha_2$ and all elements $x_1$, $x_2 \in D(A)$.*

**Definition A.21.** *An operator $A$ is called* **bounded** *if there is a constant $M$ such that*

$$\|Ax\| \le M\|x\|, \quad x \in X. \tag{A.13}$$

The norm of such an operator is defined by

$$\|A\| = \sup_{x \ne 0} \frac{\|Ax\|}{\|x\|}. \tag{A.14}$$

Again, it is the smallest $M$ which works in (A.13).

**Definition A.22.** *An operator $A$ is called* **continuous** *at a point $x_0 \in X$ if $x_n \to x$ in $X$ implies $A x_n \to A x$ in $Y$.*

A bounded linear operator is continuous at each point. For if $x_n \to x$ in $X$, then

$$\|A x_n - A x\| \le \|A\| \cdot \|x_n - x\| \longrightarrow 0.$$

We also have

**Theorem A.23.** *If a linear operator $A$ is continuous at one point $x_0 \in X$, then it is bounded, and hence continuous at every point.*

*Proof.* If $A$ were not bounded, then for each $n$ we could find an element $x_n \in X$ such that

$$\|A x_n\| > n\|x_n\|.$$

Set

$$z_n = \frac{x_n}{n\|x_n\|} + x_0.$$

Then $z_n \to x_0$. Since $A$ is continuous at $x_0$, we must have $Az_n \to Ax_0$. But

$$Az_n = \frac{Ax_n}{n\|x_n\|} + Ax_0.$$

Hence,

$$\frac{Ax_n}{n\|x_n\|} \longrightarrow 0.$$

But

$$\frac{\|Ax_n\|}{n\|x_n\|} > 1,$$

providing a contradiction.  □

We let $B(X, Y)$ be the set of bounded linear operators from $X$ to $Y$ having domains equal to the whole of $X$. Under the norm (A.14), one easily checks that $B(X, Y)$ is a normed vector space. As a generalization of Theorem A.14, we have

**Theorem A.24.** *If $Y$ is a Banach space, so is $B(X, Y)$.*

**Lemma A.25.** *Let $X$, $Y$ be normed vector spaces, and let $A$ be a linear operator from $X$ to $Y$. Then for each $x$ in $D(A)$ and $\varepsilon > 0$ there is an element $x_0 \in D(A)$ such that*

$$Ax_0 = Ax, \quad d(x_0, N(A)) = d(x, N(A)),$$

$$d(x, N(A)) \leq \|x_0\| \leq d(x, N(A)) + \varepsilon.$$

*Proof.* There is an $x_1 \in N(A)$ such that $\|x - x_1\| < d(x, N(A)) + \varepsilon$. Set $x_0 = x - x_1$.  □

**Definition A.26.** *Let $X, Y$ be normed vector spaces. A linear operator $K$ from $X$ to $Y$ is called* **compact** *(or* **completely continuous***) if $D(K) = X$ and for every sequence $\{x_n\} \subset X$ such that $\|x_n\| \leq C$, the sequence $\{Kx_n\}$ has a subsequence which converges in $Y$.*

The set of all compact operators from $X$ to $Y$ is denoted by $K(X, Y)$. If $X = Y$, then we write $K(X)$ for $K(X, X)$.

**Definition A.27.** *If $A$ is a linear operator on a normed vector space $X$, a scalar $\lambda$ is called an* **eigenvalue** *if there is a nonzero element $v \in X$ such that*

$$(A - \lambda)v = 0.$$

*Any such $v$ is called a corresponding* **eigenelement** *or* **eigenvector**. *The subspace spanned by the eigenelements corresponding to an eigenvalue $\lambda$ is called its* **eigenspace** *and is denoted by $E(\lambda)$.*

## A.7    Adjoints

Suppose $X, Y$ are normed vector spaces and $A \in B(X, Y)$. For each $y' \in Y'$, the expression $y'(Ax)$ assigns a scalar to each $x \in X$. Thus, it is a functional $F(x)$. Clearly $F$ is linear. It is also bounded since

$$|F(x)| = |y'(Ax)| \le \|y'\| \cdot \|Ax\| \le \|y'\| \cdot \|A\| \cdot \|x\|.$$

Thus, there is an $x' \in X'$ such that

$$y'(Ax) = x'(x), \quad x \in X. \tag{A.15}$$

This functional $x'$ is unique, for any other functional satisfying (A.15) would have to coincide with $x'$ on each $x \in X$. Thus, to each $y' \in Y'$ we have assigned a unique $x' \in X'$. We designate this assignment by $A'$ and note that it is a linear operator from $Y'$ to $X'$. Thus, (A.15) can be written in the form

$$y'(Ax) = A'y'(x). \tag{A.16}$$

The operator $A'$ is called the **adjoint** (or **conjugate**) of $A$.

**Theorem A.28.** *If $A \in B(X, Y)$, then $A' \in B(Y', X')$, and $\|A'\| = \|A\|$.*

The adjoint has the following easily verified properties:

$$(A + B)' = A' + B'. \tag{A.17}$$

$$(\alpha A)' = \alpha A'. \tag{A.18}$$

$$(AB)' = B'A'. \tag{A.19}$$

Many problems in mathematics and its applications can be put in the form: given normed vector spaces $X, Y$ and an operator $A \in B(X, Y)$, one wishes to solve

$$Ax = y. \tag{A.20}$$

The set of all $y$ for which one can solve (A.20) is called the **range** of $A$ and is denoted by $R(A)$. The set of all $x$ for which $Ax = 0$ is called the

**null space** of $A$ and is denoted by $N(A)$. Since $A$ is linear, it is easily checked that $N(A)$ and $R(A)$ are subspaces of $X$ and $Y$, respectively. If $y \in R(A)$, there is an $x \in X$ satisfying (A.20). For any $y' \in Y'$ we have

$$y'(Ax) = y'(y).$$

Taking adjoints we get

$$A'y'(x) = y'(y).$$

If $y' \in N(A')$, this gives $y'(y) = 0$. Thus, a necessary condition that $y \in R(A)$ is that $y'(y) = 0$ for all $y' \in N(A')$.

## A.8      Closed operators

**Definition A.29.** *The operator $A$ is called **closed** if whenever $\{x_n\} \subset D(A)$ is a sequence satisfying*

$$x_n \longrightarrow x \text{ in } X, \quad Ax_n \longrightarrow y \text{ in } Y, \tag{A.21}$$

*then $x \in D(A)$ and $Ax = y$.*

Clearly, we have

**Lemma A.30.** *All operators in $B(X,Y)$ are closed.*

Another obvious statement is

**Lemma A.31.** *If $A$ is closed, then $N(A)$ is a closed subspace of $X$.*

A statement which is not so obvious is

**Theorem A.32.** *If $X, Y$ are Banach spaces, and $A$ is a closed linear operator from $X$ to $Y$ with $D(A) = X$, then*

(a) *there are positive constants $M, r$ such that $\|Ax\| \leq M$ whenever $\|x\| < r$*
   *and*
(b) *$A \in B(X,Y)$.*

Theorem A.32 is called the **closed graph theorem**. As an application of the closed graph theorem, we have an important result known either as the **uniform boundedness principle** or the **Banach–Steinhaus theorem**.

**Theorem A.33.** *Let $X$ be a Banach space, and let $Y$ be a normed vector space. Let $W$ be any subset of $B(X,Y)$ such that for each $x \in X$,*

$$\sup_{A \in W} \|Ax\| < \infty.$$

*Then there is a finite constant $M$ such that $\|A\| \leq M$ for all $A \in W$.*

Another useful consequence of the previous theorems can be stated as follows.

**Theorem A.34.** *Let $A$ be a closed operator from a Banach space $X$ to a Banach space $Y$ such that $R(A) = Y$. If $Q$ is any open subset of $D(A)$, then the image $A(Q)$ of $Q$ is open in $Y$.*

If $A$ is a linear operator from $X$ to $Y$, with $R(A) = Y$ and $N(A) = \{0\}$ (i.e., consists only of the vector 0), we can assign to each $y \in Y$ the unique solution of

$$Ax = y.$$

This assignment is an operator from $Y$ to $X$ and is usually denoted by $A^{-1}$ and called the **inverse operator** of $A$. It is linear because of the linearity of $A$. One can ask: when is $A^{-1}$ continuous? By Theorem A.23, this is equivalent to when is it bounded? A very important answer to this question is given by

**Theorem A.35.** *If $X$, $Y$ are Banach spaces and $A$ is a closed linear operator from $X$ to $Y$ with $R(A) = Y$, $N(A) = \{0\}$, then $A^{-1} \in B(Y,X)$.*

This theorem is sometimes referred to as the **bounded inverse theorem.**

## A.9     Self-adjoint operators

**Definition A.36.** *A linear operator $A$ on a Hilbert space $X$ is called* **self-adjoint** *if it has the property that $x \in D(A)$ and $Ax = f$ if and only if*

$$(x, Ay) = (f, y), \quad y \in D(A).$$

*In particular, it satisfies*

$$(Ax, y) = (x, Ay), \quad x, y \in D(A).$$

**Proposition A.37.** *If A is self-adjoint and*

$$(A - \lambda)x = 0, \quad (A - \mu)y = 0$$

*with* $\lambda \neq \mu$, *then*

$$(x, y) = 0.$$

*Proof.* We have

$$((A - \lambda)x, y) = (x, (A - \mu)y).$$

Thus,

$$(\lambda - \mu)(x, y) = 0.$$

$\square$

**Corollary A.38.** *Eigenelements corresponding to different eigenvalues are orthogonal.*

**Corollary A.39.** *If A has a compact inverse, its eigenvalues cannot have limit points.*

*Proof.* If $x_k$ is an eigenelement corresponding to $\lambda_k \to \lambda_0$ with norm equal to one, then

$$A^{-1}x_k = x_k/\lambda_k$$

is uniformly bounded. Hence, there is a subsequence that converges. But,

$$\|x_k - x_j\|^2 = \|x_k\|^2 - 2(x_k, x_j) + \|x_j\|^2 = 2,$$

showing that no subsequence can converge. $\square$

**Corollary A.40.** *If $A^{-1}$ is compact, then the eigenelements corresponding to the same eigenvalue form a finite dimensional subspace.*

*Proof.* Same proof. Take $\lambda_k = \lambda_0$ for each $k$. $\square$

**Proposition A.41.** *If A is self-adjoint, $A^{-1}$ is compact, and all eigenvalues of A are $\geq \lambda_0$, then*

$$(Ax, x) \geq \lambda_0 \|x\|^2, \quad x \in D(A).$$

**Theorem A.42.** *If $A \in B(X)$ is self-adjoint and satisfies*

$$(Ax, x) \geq 0, \quad x \in X,$$

*then there is a unique self-adjoint operator $B \in B(X)$ such that*

$$(Bx, x) \geq 0$$

*and $B^2 = A$. If A is compact, then B is also compact.*

## A.10     Subsets

**Definition A.43.** *A subset $W$ of a normed vector space is called* **bounded** *if there is a number $b$ such that $\|x\| \leq b$ for all $x \in W$. It is called* **compact** *if each sequence $\{x^{(k)}\}$ of elements of $W$ has a subsequence which converges to an element of $W$.*

**Definition A.44.** *A subset $V \subset W$ is called* **dense** *in $W$ if for every $\varepsilon > 0$ and every $w \in W$ there is a $v \in V$ such that $\|v - w\| < \varepsilon$.*

**Definition A.45.** *A subset $W$ of a normed vector space is called* **separable** *if it has a dense subset that is denumerable. In other words, $W$ is separable if there is a sequence $\{x_k\}$ of elements of $W$ such that for each $x \in W$ and each $\varepsilon > 0$, there is an $x_k$ satisfying $\|x - x_k\| < \varepsilon$.*

**Lemma A.46.** *A compact set is separable.*

**Definition A.47.** *For a subset $W$ of a normed vector space, the* **linear span** *of $W$ is the set of linear combinations of elements of $W$. It is a subspace. Its* **closed linear span** *is the closure of its linear span.*

**Lemma A.48.** *The closed linear span of a separable set is separable.*

**Theorem A.49.** *If $X'$ is separable, so is $X$.*

*Proof.* Let $\{x'_n\}$ be a dense set in $X'$. For each $n$, there is an $x_n \in X$ such that $\|x_n\| = 1$ and

$$|x'_n(x_n)| \geq \|x'_n\|/2$$

by (A.8). Let $M = \overline{[\{x_n\}]}$, the closure of the set of linear combinations of the $x_n$. If $M \neq X$, let $x_0$ be any element of $X$ not in $M$. Then there is an $x'_0 \in M^\circ$ such that $\|x'_0\| = 1$ and $x'_0(x_0) \neq 0$ (Theorem A.17). In particular,

$$x'_0(x_n) = 0, \quad n = 1, 2, \ldots$$

Thus,

$$\|x'_n\|/2 \leq |x'_n(x_n)| = |x'_n(x_n) - x'_0(x_n)|$$

$$\leq \|x'_n - x'_0\| \cdot \|x_n\| = \|x'_n - x'_0\|.$$

Hence,

$$1 = \|x'_0\| \leq \|x'_n - x'_0\| + \|x'_n\| \leq 3\|x'_n - x'_0\|,$$

showing that none of the $x'_n$ can come closer than a distance of $1/3$ from $x'_0$. This contradicts the fact that $\{x'_n\}$ is dense in $X'$. Thus we must have $M = X$. But $M$ is separable. To see this, we note that all linear combinations of the $x_n$ with rational coefficients form a denumerable set. This set is dense in $M$. Hence, $M$ is separable, and the proof is complete. □

## A.11      Finite dimensional subspaces

Let $V$ be a vector space. The elements $v_1, \ldots, v_n$ are called **linearly independent** if the only scalars $\alpha_1, \ldots, \alpha_n$ for which

$$\alpha_1 v_1 + \cdots + \alpha_n v_n = 0 \tag{A.22}$$

are $\alpha_1 = \cdots = \alpha_n = 0$. Otherwise, they are called **linearly dependent.**

**Definition A.50.** *The space $V$ is said to be of* **dimension** $n > 0$ *if*

  *(a) there are n linearly independent vectors in V*
       *and*
  *(b) every set of $n + 1$ elements of V is linearly dependent.*

  *If there are no independent vectors, $V$ consists of just the zero element and is said to be of dimension zero. If $V$ is not of dimension n for any finite n, we say that it is infinite dimensional.*

  Now suppose $\dim V = n$ (i.e., $V$ is of dimension $n$), and let $v_1, \ldots, v_n$ be $n$ linearly independent elements. Then every $v \in V$ can be expressed uniquely in the form

$$v = \alpha_1 v_1 + \cdots + \alpha_n v_n. \tag{A.23}$$

To see this, note that the set $v, v_1, \ldots, v_n$ of $n+1$ vectors must be linearly dependent. Thus, there are scalars $\beta, \beta_1, \ldots, \beta_n$, not all zero, such that

$$\beta v + \beta_1 v_1 + \cdots + \beta_n v_n = 0.$$

Now $\beta$ cannot vanish, for otherwise the $v_1, \ldots, v_n$ would be dependent. Dividing by $\beta$, we get an expression of the form (A.23). This expression is unique. For if

$$v = \alpha'_1 v_1 + \cdots + \alpha'_n v_n,$$

then

$$(\alpha_1 - \alpha'_1)v_1 + \cdots + (\alpha_n - \alpha'_n)v_n = 0,$$

showing that $\alpha_i' = \alpha_i$ for each $i$. If $\dim V = n$, we call any set of $n$ linearly independent vectors in $V$ a **basis** for $V$. Let $X$ be a normed vector space, and suppose that it has two norms $\|\cdot\|_1$, $\|\cdot\|_2$. We call them **equivalent** and write $\|\cdot\|_1 \sim \|\cdot\|_2$ if there is a positive number $a$ such that

$$a^{-1}\|x\|_1 \leq \|x\|_2 \leq a\|x\|_1, \quad x \in X. \tag{A.24}$$

Clearly, this is an equivalence relation, and a sequence $\{x_n\}$ converges in one norm if and only if it converges in the other.

**Theorem A.51.** *If $X$ is finite dimensional, all norms are equivalent.*

We state some important consequences.

**Corollary A.52.** *A finite dimensional normed vector space is always complete.*

**Corollary A.53.** *If $M$ is a finite dimensional subspace of a normed vector space, then $M$ is closed.*

**Corollary A.54.** *If $X$ is a finite dimensional normed vector space, then every bounded closed set $T$ in $X$ is compact.*

Corollary A.54 has a converse.

**Theorem A.55.** *If $X$ is a normed vector space and the surface of its unit sphere (i.e., the set $\|x\| = 1$) is compact, then $X$ is finite dimensional.*

**Lemma A.56.** *If $K$ is a bounded, closed, convex subset of $\mathbb{R}^n$, then for each $x \in \mathbb{R}^n$ there is a unique $y \in K$ such that*

$$\|x - y\| = d(x, K) = \inf_{z \in K} \|x - z\|.$$

## A.12  Weak convergence

**Definition A.57.** *A sequence $\{x_k\}$ of elements of a Banach space $X$ is said to converge **weakly** to an element $x \in X$ if*

$$x'(x_k) \longrightarrow x'(x) \text{ as } k \longrightarrow \infty \tag{A.25}$$

*for each $x' \in X'$.*

We abbreviate weak convergence by

$$x_k \rightharpoonup x.$$

In a Hilbert space, weak convergence is equivalent to

$$(x_k, y) \longrightarrow (x, y) \text{ as } k \longrightarrow \infty, \quad y \in X.$$

We shall see how this convergence compares with convergence in norm (sometimes called **strong** convergence for contrast). Clearly, a sequence converging in norm also converges weakly. We have

**Lemma A.58.** *A weakly convergent sequence is necessarily bounded.*

### A.13     Reflexive spaces

Let $X$ be a Banach space and let $x_0$ be any element of $X$. Set

$$F(x') = x'(x_0), \quad x' \in X'.$$

Then,

$$|F(x')| \le \|x_0\| \cdot \|x'\|.$$

This means that $F(x')$ is a bounded linear functional on $X'$. Hence, there is an $x_0'' \in X'' = (X')'$ such that

$$x_0''(x') = x'(x_0), \quad x' \in X'. \tag{A.26}$$

The element $x_0''$ is unique. For if $x_1''$ also satisfies (A.26), then

$$x_0''(x') - x_1''(x') = 0, \quad x' \in X',$$

showing that $x_0'' = x_1''$. We set $x_0'' = Jx_0$. Clearly, $J$ is a linear mapping of $X$ into $X''$ defined on the whole of $X$. Moreover, it is one-to-one. For if $Jx_0 = 0$, we see by (A.26) that $x'(x_0) = 0$ for all $x' \in X'$. Hence, $x_0 = 0$. In our new notation, (A.26) becomes

$$Jx(x') = x'(x), \quad x \in X, \ x' \in X'. \tag{A.27}$$

We also note that $J$ is a bounded operator. In fact, we have

$$\|Jx\| = \sup \frac{|Jx(x')|}{\|x'\|} = \sup \frac{|x'(x)|}{\|x'\|} = \|x\|, \tag{A.28}$$

where the least upper bound is taken over all nonvanishing $x' \in X'$.

**Definition A.59.** *We call $X$ **reflexive** if $R(J) = X''$, that is, if for every $x'' \in X''$ there is an $x \in X$ such that $Jx = x''$.*

**Lemma A.60.** *Every Hilbert space is reflexive.*

*Proof.* By the Riesz representation theorem (Theorem A.12), every $x' \in X'$ is of the form $(x, z)$, where $z \in X$. Hence, $X'$ can be made into a Hilbert space. Thus, elements of $X''$ are of the same form. □

We also have

**Theorem A.61.** *If $X$ is reflexive, then every bounded sequence has a weakly convergent subsequence.*

We now realize that weak convergence cannot be equivalent to strong convergence in a reflexive, infinite dimensional space. The reason is that if $X$ is reflexive, every sequence satisfying $\|x_n\| = 1$ has a weakly convergent subsequence (Theorem A.61). If this subsequence converged strongly, then it would follow that $X$ is finite dimensional (Theorem A.55). On the other hand, we have

**Theorem A.62.** *If $X$ is finite dimensional, then a sequence converges weakly if and only if it converges in norm.*

## A.14     Operators with closed ranges

**Theorem A.63.** *Let $X, Y$ be Banach spaces, and let $A$ be a one-to-one closed linear operator from $X$ to $Y$. Then a necessary and sufficient condition that $R(A)$ be closed in $Y$ is that*

$$\|x\| \leq C \|Ax\|, \quad x \in X. \tag{A.29}$$

*hold.*

**Theorem A.64.** *If $X, Y$ are Banach spaces, and $A$ is a closed linear operator from $X$ to $Y$, then $R(A)$ is closed in $Y$ if, and only if, there is a constant $C$ such that*

$$d(x, N(A)) = \inf_{z \in N(A)} \|x - z\| \leq C \|Ax\|, \quad x \in D(A) \tag{A.30}$$

*holds.*

**Theorem A.65.** *Let $X, Y$ be Banach spaces, and assume that $A \in B(X, Y)$. If $R(A)$ is closed in $Y$, then*

$$R(A') = N(A)^\circ, \tag{A.31}$$

*and hence $R(A')$ is closed in $X'$. Here $N(A)^\circ$ is the set of functionals in $X'$ which vanish on $N(A)$.*

**Definition A.66.** *Let $X, Y$ be Banach spaces. An operator $A \in B(X, Y)$ is said to be a* **Fredholm operator** *from $X$ to $Y$ if*

*(1) $\alpha(A) = \dim N(A)$ is finite,*
*(2) $R(A)$ is closed in $Y$,*
*(3) $\beta(A) = \dim N(A')$ is finite.*

The set of Fredholm operators from $X$ to $Y$ is denoted by $\Phi(X, Y)$. We have

**Lemma A.67.** *If $X = Y$ and $K \in K(X)$, then $I$–$K$ is a Fredholm operator.*

The set of operators for which (1) and (2) hold is denoted by $\Phi_+(X, Y)$. The set for which (2) and (3) hold is denoted by $\Phi_-(X, Y)$. Operators in $\Phi_+(X, Y)$ and $\Phi_-(X, Y)$ are called **semi-Fredholm operators**. The **index** of a semi-Fredholm operator is defined as

$$i(A) = \alpha(A) - \beta(A). \tag{A.32}$$

**Lemma A.68.** *For $K \in K(X)$, $i(I - K) = 0$.*

If $A \in \Phi(X, Y)$, then $i(A)$ is finite. If $A \in \Phi_+(X, Y)$ but not in $\Phi(X, Y)$, then $i(A) = -\infty$. If $A \in \Phi_-(X, Y)$ but not in $\Phi(X, Y)$, then $i(A) = \infty$.

**Theorem A.69.** *If $A$ is in $\Phi_\pm(X, Y)$ and $K \in K(X, Y)$, then $A + K \in \Phi_\pm(X, Y)$ and*

$$i(A + K) = i(A). \tag{A.33}$$

**Lemma A.70.** *Let $X$ be a normed vector space, and suppose that $X = N \oplus X_0$, where $X_0$ is a closed subspace and $N$ is finite dimensional. If $X_1$ is a subspace of $X$ containing $X_0$, then $X_1$ is closed.*

**Lemma A.71.** *Let $Y$ be a normed vector space, and let $R$ be a closed subspace. Then $R^\circ$ is of finite dimension $n$ if and only if there is an $n$ dimensional subspace $M$ of $Y$ such that $X = R \oplus M$.*

**Corollary A.72.** *If $R = R(A)$, then $R(A)^\circ = N(A')$. Hence, $\beta(A) = n$.*

**Theorem A.73.** *Assume that $X, Y, Z$ are Banach spaces. If $A \in \Phi(X, Y)$ and $B \in \Phi(Y, Z)$, then $BA \in \Phi(X, Z)$ and*

$$i(BA) = i(B) + i(A). \tag{A.34}$$

# Appendix B
## Measure and integration

### B.1    Measure zero

**Definition B.1.** *A* **cuboid** *(or* **parallelepiped** *or* **box***) $Q \subset \mathbb{R}^n$ is a set of the form*

$$Q = \{x \in \mathbb{R}^n : a_j < x_j < b_j, \ j = 1, \ldots, n\}.$$

*Its volume is given by*

$$vol \ Q = |Q| = (b_1 - a_1) \cdots (b_n - a_n).$$

**Definition B.2.** *A set $E \subset \mathbb{R}^n$ has* **measure zero** *if for each $\varepsilon > 0$ it can be covered by a sequence of cuboids of total volume $< \varepsilon$.*

This means that for each $\varepsilon > 0$ there is a sequence $\{Q_k\}$ of cuboids such that

$$E \subset \bigcup_{k=1}^{\infty} Q_k, \quad \sum_{k=1}^{\infty} |Q_k| < \varepsilon.$$

A statement is said to hold **almost everywhere** (abbreviated a.e.) if the set of points for which it is not true has measure 0. A denumerable union of sets of measure 0 has measure 0.

**Proposition B.3.** *The boundary of a cuboid having finite volume has measure 0.*

### B.2    Step functions

**Definition B.4.** *We call a function $\varphi(x)$ on $\mathbb{R}^n$ a* **step function** *if it has a constant value $c_k$ on each of a finite number $m$ of nonintersecting*

331

*cuboids $Q_k$ of finite volume and vanishes outside the closure of*

$$R = \bigcup_{k=1}^{m} Q_k.$$

*It may take any values on the boundary of R. We define the integral of $\varphi(x)$ to be*

$$\int \varphi(x)\, dx = \sum_{k=1}^{m} c_k |Q_k|.$$

**Proposition B.5.** *The sum or difference of step functions is a step function.*

**Definition B.6.** *The* **positive part** *$f^+(x)$ and* **negative part** *$f^-(x)$ of a function $f(x)$ are defined as*

$$f^{\pm}(x) = \max\{\pm f(x), 0\}.$$

*They are nonnegative functions.*

**Proposition B.7.** *The positive and negative parts of a step function are step functions.*

We have

**Lemma B.8.** *For every sequence of step functions which decrease to 0 a.e., their integrals converge to 0.*

**Lemma B.9.** *If $\{\varphi_n\}$ is a nondecreasing sequence of step functions which satisfy*

$$\int \varphi_n(x)\, dx \le C,$$

*then $\varphi_n(x)$ converges to a finite limit $f(x)$ a.e.*

## B.3    Integrable functions

**Definition B.10.** *If the function $f(x)$ is the limit a.e. of a sequence of step functions $\varphi_k$ satisfying the hypotheses of Lemma B.9, then we define the integral of $f(x)$ as*

$$\int f(x)\, dx = \lim_{k \to \infty} \int \varphi_k(x)\, dx.$$

For this definition to make sense, the limit must be independent of the sequence $\varphi_k$ as long as it satisfies the condition of Lemma B.9 and converges a.e. to $f(x)$. This is guaranteed by Lemma B.8. In fact, we have

**Lemma B.11.** *If the function $f(x)$ is the limit a.e. of sequences of step functions $\varphi_k$, $\psi_k$ satisfying the hypotheses of Lemma B.9, then*

$$\lim_{k\to\infty} \int \psi_k(x)\, dx = \lim_{k\to\infty} \int \varphi_k(x)\, dx.$$

*Proof.* For fixed $m$, the sequence of step functions $\{[\psi_m(x) - \varphi_k(x)]^+\}$ satisfies the hypotheses of Lemma B.8. Hence,

$$\int [\psi_m(x) - \varphi_k(x)]^+\, dx \to 0 \quad \text{as } k \to \infty.$$

This implies

$$\lim_{k\to\infty} \int [\psi_m(x) - \varphi_k(x)]\, dx \le 0,$$

or

$$\int \psi_m(x)\, dx \le \lim_{k\to\infty} \int \varphi_k(x)\, dx.$$

Since this is true for each $m$, we have

$$\lim_{m\to\infty} \int \psi_m(x)\, dx \le \lim_{k\to\infty} \int \varphi_k(x)\, dx.$$

Interchanging the two sequences, we obtain the desired result. □

Let $\tilde{S}$ denote the set of those function which are the limits a.e. of sequences of step functions satisfying the hypotheses of Lemma B.9. If $f_1, f_2 \in \tilde{S}$, then $f_1 + f_2 \in \tilde{S}$ and

$$\int [f_1 + f_2]\, dx = \int f_1\, dx + \int f_2\, dx.$$

**Definition B.12.** *We call a function $f(x)$ **integrable** (or **summable**) in the sense of Lebesgue if it is the difference a.e. of two functions $f_1, f_2 \in \tilde{S}$. We define its integral to be*

$$\int f(x)\, dx = \int f_1(x)\, dx - \int f_2(x)\, dx.$$

This definition is independent of the functions $f_1, f_2$. For, if $f = f_1 - f_2 = g_1 - g_2$, then $f_1 + g_2 = g_1 + f_2$, and consequently

$$\int f_1 \, dx + \int g_2 \, dx = \int g_1 \, dx + \int f_2 \, dx.$$

Thus,

$$\int f_1 \, dx - \int f_2 \, dx = \int g_1 \, dx - \int g_2 \, dx.$$

We denote the set of integrable functions by $L^1 = L^1(\mathbb{R}^n)$.

**Proposition B.13.** *If $f \in L^1$, then $|f|$ and $f^\pm = \max\{\pm f(x), 0\} \in L^1$.*

**Proposition B.14.** *If $f \in L^1$, then there is a sequence $\{\varphi_k\}$ of step functions converging to $f$ a.e. and such that*

$$\int |f(x) - \varphi_k(x)| \, dx \to 0 \quad \text{as} \quad k \to \infty.$$

**Theorem B.15.** *(Beppo-Levi) If $\{f_k\}$ is a nondecreasing sequence of functions in $L^1$ such that*

$$\int f_k \, dx \le C,$$

*then they converge a.e. to a summable function $f$, and*

$$\int f_k \, dx \to \int f \, dx.$$

**Theorem B.16.** *(Fatou) If $f_k \in L^1$ satisfy*

(a) $f_k(x) \ge 0$ a.e.,
(b) $f_k(x) \to f(x)$ a.e.,
(c) $\int f_k(x) \, dx \le c < \infty$,

*then $f \in L^1$ and*

$$\int f \, dx \le \liminf_{k \to \infty} \int f_k \, dx.$$

**Theorem B.17.** *If $f_k \in L^1$ and $f_k(x) \ge -W(x) \in L^1$, then*

$$\int [\liminf_{k \to \infty} f_k] \, dx \le \liminf_{k \to \infty} \int f_k \, dx.$$

*If $f_k(x) \le W(x) \in L^1$, then*

$$\limsup_{k \to \infty} \int f_k \, dx \le \int [\limsup_{k \to \infty} f_k] \, dx.$$

**Theorem B.18.** *(Lebesgue) If $f_k \in L^1$ satisfy*

  **(a)** $f_k(x) \to f(x)$ *a.e. and*
  **(b)** *there is a $g(x) \in L^1$ such that*

$$|f_k(x)| \le g(x) \text{ a.e.}, \quad k = 1, 2, \ldots,$$

*then $f \in L^1$ and*

$$\int f_k \, dx \to \int f \, dx.$$

## B.4    Measurable functions

**Definition B.19.** *A function on $\mathbb{R}^n$ is called **measurable** if it is the limit a.e. of a sequence of step functions.*

**Theorem B.20.** *Summable functions are measurable. If $f(x)$ is measurable, $g(x) \in L^1$ and*

$$|f(x)| \le g(x), \quad \text{a.e.},$$

*then $f(x) \in L^1$. If $f(x)$ and $g(x)$ are measurable, then $|f(x)|$, $f(x) \pm g(x)$, $f(x) \cdot g(x)$, $\max\{f(x), g(x)\}$ are all measurable. If $f(x) \ne 0$ a.e., then $1/f(x)$ is also measurable. If a sequence of measurable functions converges to $f(x)$ a.e., then $f(x)$ is measurable.*

## B.5    The spaces $L^p$

**Definition B.21.** *For $1 \le p < \infty$, we define $L^p = L^p(\mathbb{R}^n)$ to be the set of those measurable functions $f(x)$ such that $|f(x)|^p \in L^1$.*

**Theorem B.22.** *For each such p, the set $L^p$ is a Banach space with norm*

$$|f|_p = \left( \int |f|^p \, dx \right)^{1/p}.$$

**Theorem B.23.** *(Hölder) If $p > 1$, $q > 1$, $1/p + 1/q = 1$, $f \in L^p$, $g \in L^q$, then $fg$ is in $L^1$ and*

$$|fg|_1 \le |f|_p |g|_q.$$

*For these values of p, q the space $L^p$ is reflexive with dual space $L^q$.*

**Theorem B.24.** *If $f_k \in L^1$ satisfy (a) $f_k(x) \to f(x)$ a.e., (b) there is a sequence $\{g_k\} \subset L^1$ such that*

$$|f_k(x)| \le g_k(x) \text{ a.e.}, \quad k = 1, 2, \dots,$$

*and $g_k \to g$ in $L^1$, then $f \in L^1$ and*

$$\int f_k \, dx \to \int f \, dx.$$

**Theorem B.25.** *If $f_k \to f$ in $L^1$, then there is a subsequence converging to $f(x)$ a.e.*

**Theorem B.26.** *(Fubini) If $f \in L^1$, then*

$$\int f(x) \, dx = \int \cdots \int f(x_1, \dots, x_n) \, dx_1 \cdots dx_n,$$

*where the order of integration is immaterial.*

## B.6     Measurable sets

**Definition B.27.** *If $V \subset \mathbb{R}^n$, the* **characteristic function** *$\mu_V(x)$ of $V$ is defined to be equal to 1 or 0 depending on whether $x \in V$ or not. Thus*

$$\mu_V(x) = \begin{cases} 1, & x \in V, \\ 0, & x \notin V. \end{cases}$$

**Definition B.28.** *A set $V \subset \mathbb{R}^n$ is* **measurable** *if its characteristics function is measurable. If $\mu_V(x)$ is summable, then the measure of $V$ is defined to be*

$$m(V) = \int \mu_V(x) \, dx.$$

*Otherwise, $m(V) = \infty$.*

**Proposition B.29.** *Open and closed subsets of $\mathbb{R}^n$ are measurable.*

**Proposition B.30.** *A set has measure 0 if and only if it is measurable and its measure equals 0. A denumerable union or intersection of measurable sets is measurable. If the measurable sets $V_1, V_2, \dots,$ are disjoint, then*

$$m\left( \bigcup_{j=1}^{\infty} V_j \right) = \sum_{j=1}^{\infty} m(V_j).$$

*If $V_j \subset V_{j+1}$, then*

$$m\left(\bigcup_{j=1}^{\infty} V_j\right) = \lim_{j \to \infty} m(V_j).$$

*If $f(x)$ is measurable and $c$ is arbitrary, then the sets where*

$$f(x) \le c, \, f(x) < c, \, f(x) \ge c, \, f(x) > c$$

*are all measurable. If any one of these sets is measurable for arbitrary $c$, then $f(x)$ is measurable.*

**Theorem B.31.** *(Egoroff) If $E \subset \mathbb{R}^n$ is bounded and measurable, then for every sequence of functions converging point-wise on $E$ and every $\varepsilon > 0$ one can remove from $E$ a set of measure $< \varepsilon$ such that the convergence is uniform on the rest of $E$.*

**Theorem B.32.** *(Lusin) If $f(x)$ is defined on a measurable set $E \subset \mathbb{R}^n$, then it is measurable if and only if for every $\varepsilon > 0$ one can remove from $E$ an open set of measure $< \varepsilon$ such that $f(x)$ will be continuous on the rest of $E$.*

**Definition B.33.** *For any measurable set $\Omega \subset \mathbb{R}^n$ and for $1 \le p < \infty$, we let $L^p(\Omega)$ denote the set of those measurable functions $f(x)$ such that $f(x)\mu_\Omega(x)$ is in $L^p$. If $f \in L^1(\Omega)$, we define*

$$\int_\Omega f(x) \, dx = \int_{\mathbb{R}^n} f(x)\mu_\Omega(x) \, dx.$$

**Proposition B.34.** *The set $L^p(\Omega)$ is a Banach space with norm*

$$|f|_{p,\Omega} = \left(\int_\Omega |f(x)|^p \, dx\right)^{1/p}.$$

**Definition B.35.** *A function $F(x)$ defined on an interval $[a,b]$ is* **absolutely continuous** *if for every $\varepsilon > 0$ there is a $\delta > 0$ such that*

$$\sum_{k=1}^{N} [F(b_k) - F(a_k)] < \varepsilon$$

*whenever*

$$\sum_{k=1}^{N} (b_k - a_k) < \delta,$$

*where the $(a_k, b_k)$ are nonoverlapping intervals contained in $[a,b]$.*

**Theorem B.36.** *If $f(x)$ is a summable function in $[a, b]$, then*

$$F(x) = \int_a^x f(t)\,dt$$

*is absolutely continuous and satisfies*

$$F'(x) = f(x) \text{ a.e.}, \quad x \in [a, b].$$

*A function $F(x)$ is absolutely continuous in $[a, b]$ if and only if it is of this form, where $f(x)$ is summable in $[a, b]$.*

## B.7    Carathéodory functions

Let $\Omega$ be a bounded, measurable set in $\mathbb{R}^n$.

**Definition B.37.** *A function $f(x, t)$ on $\Omega \times \mathbb{R}$ is called a **Carathéodory function** if it is measurable on $\Omega$ for each $t \in \mathbb{R}$ and continuous on $\mathbb{R}$ for a.e. $x \in \Omega$.*

We shall need

**Theorem B.38.** *If $v(x)$ is continuous on $\overline{\Omega}$ and $f(x, t)$ is a Carathéodory function on $\Omega \times \mathbb{R}$, then the function $f(x, v(x))$ is measurable.*

*Proof.* Since $v(x)$ is continuous on a closed, bounded set in $\mathbb{R}^n$, there is an integer $N$ such that

$$|v(x)| \le N, \quad x \in \overline{\Omega}.$$

Let $I_0 = [-N, N]$, let $S$ be the set of measure 0 where $f(x, t)$ fails to be a continuous function of $t$ for $t \in I_0$, and let $\varepsilon > 0$ be given. For each integer $k > 0$ define

$$E_{\varepsilon,k} = \{x \in \Omega \backslash S : |t_1 - t_2| \le \frac{1}{k}, \ t_1, t_2 \text{ rational}, \ t_1, t_2 \in I_0,$$

$$\implies |f(x, t_1) - f(x, t_2)| \le \frac{\varepsilon}{3}\}.$$

Then $x \in \Omega \backslash (E_{\varepsilon,k} \cup S)$ if and only if there are rational numbers $t_1, t_2 \in I_0$ such that $|t_1 - t_2| \le 1/k$ and

$$|f(x, t_1) - f(x, t_2)| > \frac{\varepsilon}{3}.$$

For each choice of rational $t_1, t_2 \in I_0$ satisfying $|t_1 - t_2| \le 1/k$, the set of such points $x \in \Omega$ is measurable (Proposition B.30). Consequently, the set $\Omega \backslash (E_{\varepsilon,k} \cup S)$ is the denumerable union of measurable sets and, hence, measurable. This implies that $E_{\varepsilon,k}$ is measurable. Let

$$E_\varepsilon = \bigcup_{k=1}^\infty E_{\varepsilon,k}.$$

Since $E_{\varepsilon,k} \subset E_{\varepsilon,(k+1)}$, we see that

$$m(E_\varepsilon) = \lim_{k \to \infty} m(E_{\varepsilon,k}).$$

If $x \notin S$, then $f(x,t)$ is a continuous function in $t$ for $t \in I_0$. Hence, it is uniformly continuous there. Consequently, $x \in E_{\varepsilon,k}$ for $k$ sufficiently large. This shows that

$$\Omega \backslash E_\varepsilon \subset S.$$

Thus $m(\Omega \backslash E_\varepsilon) = 0$. This implies that

$$m(E_{\varepsilon,k}) \to m(E_\varepsilon) = m(\Omega)$$

as $k \to \infty$. Therefore, for each $\eta > 0$ there is an integer $r$ such that

$$m(E_{\varepsilon,r}) > m(\Omega) - \varepsilon \frac{\eta}{2}$$

and

$$|f(x,t_1) - f(x,t_2)| \leq \frac{\varepsilon}{3}$$

wherever $t_1, t_2 \in I_0$, $x \in E_{\varepsilon,r}$, and the $t_j$ are rational and satisfy $|t_1 - t_2| \leq 1/r$. Let $p = 4Nr$, and divide $I_0$ into $p$ intervals of length $1/2r$ each. Let $s_0, \ldots, s_p$ be their end points. By Lusin's theorem (Theorem B.32), there is a closed set $F_j \subset \bar{\Omega}$ for each $j$ such that

$$m(F_j) > m(\Omega) - \frac{\varepsilon \eta}{2p}$$

and $f(x,s_j)$ is continuous on $F_j$. Since $F_j$ is bounded and closed, it is compact, and there is a $\delta_j > 0$ such that

$$|f(x,s_j) - f(x',s_j)| < \frac{\varepsilon}{3}, \quad x, x' \in F_j, \ |x - x'| < \delta_j.$$

Let

$$F = \bigcap_{j=1}^p F_j.$$

Then $F$ is closed and satisfies

$$m(F) > m(\Omega) - \varepsilon \frac{\eta}{2}.$$

If we let $\delta_0$ be the smallest of the $\delta_j$, then we have

$$|f(x,s_j) - f(x',s_j)| < \frac{\varepsilon}{3}, \quad x, x' \in F, \ |x - x'| < \delta_0, \ j = 1, \ldots, p.$$

Since $v(x)$ is continuous on $\bar{\Omega}$, it is uniformly continuous there, and there is a $\tilde{\delta} > 0$ such that

$$|v(x) - v(x')| < \frac{1}{2r}, \quad x, x' \in \bar{\Omega}, \ |x - x'| < \tilde{\delta}.$$

Let $V = E_{\varepsilon, r} \cap F$. Then

$$m(V) > m(\Omega) - \varepsilon\eta.$$

Let $\delta = \min[\delta_0, \tilde{\delta}]$. If $|x - x'| < \delta$, then there is an $s_j$ such that $|v(x) - s_j|$ and $|v(x') - s_j|$ are both $< 1/r$. Consequently, for $x, x' \in V$ and $|x - x'| < \delta$, we have

$$\begin{aligned}
|f(x, v(x)) - f(x', v(x'))| &< |f(x, v(x)) - f(x, s_j)| \\
&+ |f(x, s_j) - f(x', s_j)| \\
&+ |f(x', s_j) - f(x', v(x'))| \\
&< \frac{\varepsilon}{3} + \frac{\varepsilon}{3} + \frac{\varepsilon}{3} = \varepsilon.
\end{aligned}$$

This shows that $f(x, v(x))$ is continuous if we remove from $\Omega$ an open set of measure less than $\eta$. To see this, let $\varepsilon_k = 1/2^k$. Then there is a set $V_k$ with $m(V_k) > m(\Omega) - \eta/2^k$ and a $\delta_k > 0$ such that

$$|f(x, v(x)) - f(x', v(x'))| < \frac{1}{2^k}$$

whenever $|x - x'| < \delta_k$ and $x, x' \in V_k$. Let

$$V = \bigcap_{k=1}^{\infty} V_k.$$

Then

$$m(V) > m(\Omega) - \eta \sum_{k=1}^{\infty} 2^{-k} = m(\Omega) - \eta.$$

Let $\rho$ be any positive number. Then there is a $k$ such that $2^{-k} < \rho$. For any $x, x' \in V$ satisfying $|x - x'| < \delta_k$ we have $x, x' \in V_k$, and consequently

$$|f(x, v(x)) - f(x', v(x'))| < \frac{1}{2^k} < \rho.$$

This shows that $f(x, v(x))$ is continuous on $V$. Since $m(V) > m(\Omega) - \eta$ and $\eta$ was arbitrary, we see that $f(x, v(x))$ is measurable by Lusin's theorem (Theorem B.32). □

# Appendix C
## Metric spaces

### C.1    Properties

If $X$ is a set, then a map $\rho(x, y)$ of $X \times X \to \mathbb{R}$ is called a **metric** on $X$ if it satisfies

(1)  $\rho(x, y) = 0 \iff x = y, \quad x, y \in X,$

(2)  $\rho(x, y) = \rho(y, x), \quad x, y \in X,$

and

(3)  $\rho(x, z) \leq \rho(x, y) + \rho(y, z), \quad x, y, z \in X.$

A set may have more than one metric. A **metric space** $(X, \rho)$ is a set $X$ together with a specific metric $\rho(x, y)$ defined on it. The elements of $X$ are usually called "points." A sequence of points $\{x_k\} \subset X$ converges in $(X, \rho)$ to a point $x \in X$ if

$$\rho(x_k, x) \to 0 \text{ as } k \to \infty.$$

In this case we write

$$x_k \to x \text{ in } (X, \rho).$$

It is called a **Cauchy** sequence in $(X, \rho)$ if

$$\rho(x_j, x_k) \to 0 \text{ as } j, k \to \infty.$$

The metric space $(X, \rho)$ is called **complete** if every Cauchy sequence in $(X, \rho)$ converges to an element $x \in X$. All Banach and Hilbert spaces including $\mathbb{R}^n$ are examples of complete metric spaces if we use the metric

$$\rho(x, y) = \|x - y\|, \quad x, y \in X.$$

If $(X, \rho), (Y, \sigma)$ are metric spaces, a map $f(x) : X \to Y$ is called **continuous** at $x_0 \in X$ if for every $\varepsilon > 0$ there is a $\delta > 0$ such that

$$\rho(x, x_0) < \delta \Longrightarrow \sigma(f(x), f(x_0)) < \varepsilon, \quad x \in X.$$

It is continuous on $X$ if it is continuous at each point of $X$. It is called **uniformly continuous** on $X$ if for each $\varepsilon > 0$ there is a $\delta > 0$ such that

$$\rho(x_1, x_2) < \delta \Longrightarrow \sigma(f(x_1), f(x_2)) < \varepsilon, \quad x_1, x_2 \in X.$$

Note that $\delta$ does not depend on $x_1, x_2$. A sequence of functions $f_k(x) : X \to Y$ converges point-wise in $(X, \rho)$ to a function $f(x)$ if

$$f_k(x) \to f(x), \quad x \in X.$$

It converges **uniformly** in $(X, \rho)$ if for every $\varepsilon > 0$ there is a number $N$ such that

$$\sigma(f_k(x), f(x)) < \varepsilon, \quad k > N, \; x \in X.$$

Note that $N$ does not depend of $x$.

A subset $Z \subset X$ is called **compact** in $(X, \rho)$ if each sequence $\{z_k\} \subset Z$ has a subsequence converging to a point $z \in Z$ in $(X, \rho)$. A subset $Z \subset X$ is called **bounded** if there is an $x_0 \in X$ such that

$$\rho(x, z_0) \leq C, \quad z \in Z.$$

A point $x_0 \in Z \subset X$ is called **interior** point of $Z$ if there is a $\delta > 0$ such that

$$\rho(x, x_0) < \delta \Longrightarrow x \in Z.$$

A subset $Z \subset X$ is called **open** if all of its points are interior points of $X$.

**Proposition C.1.** *Every union of open sets is open. Every finite intersection of open sets is open.*

A set $Z \subset X$ is called **closed** if its complement $X \backslash Z$ in $X$ is open.

**Proposition C.2.** *All intersections of closed sets are closed. All finite unions of closed sets are closed.*

An **open covering** of a set $Z \subset X$ is a collection of open sets in $(X, \rho)$ such that $Z$ is contained in their union.

**Theorem C.3.** *If $Z \subset X$ is compact in $(X, \rho)$, then from every open covering of $Z$ one can select a finite number of them that will cover $Z$.*

**Theorem C.4.** *Every bounded, closed subset of $\mathbb{R}^n$ is compact.*

The **closure** of a subset $Z$ of a metric space is the smallest closed set containing it, that is, the intersection of all closed sets containing $Z$.

**Theorem C.5.** *A metric space is complete if and only if every sequence $\{S_k\}$ of closed spheres satisfying $S_{k+1} \subset S_k$ having radii $r_k \to 0$ as $k \to \infty$ has a non–empty intersection.*

The space $C[a, b]$ is the metric space of real functions continuous on the closed interval $[a, b]$ with the metric

$$\rho(x, y) = \sup_{a \leq t \leq b} |x(t) - y(t)|.$$

A set $Z \subset C[a, b]$ is called **equicontinuous** if for each $\varepsilon > 0$ there is a $\delta > 0$ such that

$$|x(t) - x(t')| < \varepsilon$$

for all $x \in Z$ when $|t - t'| < \delta$.

**Theorem C.6.** *(Arzelà–Ascoli) A subset $Z \subset C[a, b]$ is compact if and only if it is closed, bounded, and equicontinuous.*

**Theorem C.7.** *A continuous map of a compact metric space into a metric space is uniformly continuous.*

**Theorem C.8.** *If a sequence of continuous functions from a metric space to a metric space converges uniformly, then the limit function is continuous.*

**Theorem C.9.** *If a real function is continuous on a compact metric space, then it attains its supremum and infimum.*

## C.2    Para-compact spaces

**Definition C.10.** *A collection $\Theta$ of sets is called a **refinement** of a collection $\Lambda$ of sets, if each member of $\Theta$ is a subset of one of the members of $\Lambda$.*

**Definition C.11.** *A collection of sets is called a **cover** of a set $X$ if the set $X$ is contained in the union of the members of the collection.*

**Definition C.12.** *A cover is called an* **open cover** *if every member of the collection is an open set.*

**Definition C.13.** *A cover of $X$ is called* **locally finite** *if each point of $X$ has a neighborhood which intersects only finitely many members of the cover.*

**Definition C.14.** *A set $X$ is called* **para-compact** *if each open cover has an open locally finite refinement.*

**Theorem C.15.** *Any subset of a metric space is para-compact.*

# Appendix D

## Pseudo-gradients

### D.1    The benefits

If you recall in Chapter 2, we introduced the concept of a pseudo-gradient. This was done in order to solve differential equations in which the right-hand side was not Lipschitz continuous. These equations came about when we tried to show that if the gradient of a $C^1$ function did not vanish, we could decrease the function. But the equations we wanted to use involved the gradient of the function, which was only known to be continuous and not Lipschitz continuous. Our approach was to substitute another function for the gradient which (a) was Lipschitz continuous and (b) allowed one to decrease the function when the gradient does not vanish. This was done by approximating the gradient by a smoother function. In $\mathbb{R}^2$ we used the Heine–Borel theorem to cover bounded sets by a finite number of small balls, construct a Lipschitz continuous function in each ball and then piece them together by means of a partition of unity.

However, in an infinite dimensional Hilbert space this approach does not seem to work. We do not have any difficulty constructing a Lipschitz continuous approximation in a ball or a finite number of balls. But we need to cover the space (or portion of the space) with balls that are locally finite (i.e., any small neighborhood intersects only a finite number of them). This is where the rub is. Can this be done in an arbitrary Hilbert space?

Fortunately, the answer is yes. However, we will need some powerful results from point set topology.

**Definition D.1.** *A collection* $\Theta$ *of sets is called a* **refinement** *of a* *collection* $\Lambda$ *of sets, if each member of* $\Theta$ *is a subset of one of the members of* $\Lambda$. *A collection of sets is called a* **cover** *of a set X if the set X is contained in the union of the members of the collection. It is called an* **open cover** *if every member of the collection is an open set. A cover of X is called* **locally finite** *if each point of X has a neighborhood which intersects only finitely many members of the cover. A set X is called* **para-compact** *if each open cover has an open locally finite refinement.*

What we need now is

**Theorem D.2.** *Any subset of a metric space is para-compact.*

Unfortunately, we do not have the time and space to prove this theorem. All we can do is refer to [Kelley, 1955] for a proof. We shall use it to prove the existence of a mapping $Y(u)$ needed to strengthen Theorem 2.5 (cf. Theorem D.5).

### D.2    The construction

We shall prove

**Lemma D.3.** *Let* $X(u)$ *be a continuous map from a Hilbert space E to itself, and let*

$$\hat{E} = \{u \in E : X(u) \neq 0\}.$$

*If* $\theta < 1$, *then there is a mapping* $Y(u)$ *from* $\hat{E}$ *to E which is locally Lipschitz continuous and satisfies*

$$\|Y(u)\| \leq 1, \quad \theta\|X(u)\| \leq (X(u), Y(u)), \quad u \in \hat{E}. \tag{D.1}$$

*Proof.* For each $u \in \hat{E}$, there is an element $h(u) \in E$ such that

$$\|h(u)\| = 1, \quad \|X(u)\| = (X(u), h(u)), \quad u \in \hat{E} \tag{D.2}$$

(just take $h(u) = X(u)/\|X(u)\|$). By the continuity of $X(u)$, for each $u \in \hat{E}$ there is a neighborhood $N(u)$ of $u$ such that

$$0 < \theta\|X(v)\| \leq (X(v), h(u)), \quad v \in N(u). \tag{D.3}$$

The collection $\{N(u)\}$ forms as open covering of $\hat{E}$. Since $\hat{E}$ is para-compact (Theorem D.2), this open cover has a locally finite open refinement $\{N_\tau\}$. Let $\{\psi_\tau\}$ be a locally Lipschitz continuous partition of unity subordinate to this refinement. We construct this partition of unity as follows. Our first step is to establish the following generalization of Lemma 2.3.

**Lemma D.4.** *If A is any set in a metric space (cf. Appendix C) and*

$$g(p) = d(p, A) = \inf_{q \in A} \rho(p, q),$$

*then*

$$|g(p) - g(p')| \le \rho(p, p').$$

*Proof.* If $q \in A$, then

$$\rho(p, q) \le \rho(p, p') + \rho(p', q)$$

and

$$d(p, A) \le \rho(p, q) \le \rho(p, p') + \rho(p', q).$$

Thus,

$$d(p, A) \le \rho(p, p') + d(p', A).$$

Consequently,

$$d(p, A) - d(p', A) \le \rho(p, p').$$

Interchanging $p$ and $p'$ gives

$$d(p', A) - d(p, A) \le \rho(p', p),$$

which produces the desired inequality. □

Now to the construction of the partition of unity. Let

$$g_\tau(p) = d(p, E \backslash N_\tau) = \inf_{q \notin N_\tau} \|p - q\|.$$

Then,

$$g_\tau(p) \equiv 0, \quad p \notin N_\tau,$$

and $g_\tau(p)$ is Lipschitz continuous by Lemma D.4. Define

$$\psi_\tau(p) = \frac{g_\tau(p)}{\sum_\nu g_\nu(p)}, \quad p \in N_\tau.$$

The denominator is positive and finite for each $p \in N_\tau$. The reason for this is that each $p \in N_\tau$ is contained in at least one, but not more than a finite number of $N_\nu$, and $g_\nu(p) = 0$ when $p$ is not in $N_\nu$.

For any $p \in \hat{E}$, let $\tilde{N}(p)$ be a small neighborhood of $p$. Then

$$\tilde{N}(p) \cap N_\tau \ne \phi$$

for only a finite number of $\tau$. Thus, there is only a finite number of $g_\tau$

which do not vanish on $\tilde{N}(p)$. Since each $g_\tau$ is locally Lipschitz continuous, the same is true of the denominator of each $\psi_\tau$. Thus the same is true of each $\psi_\tau$ itself.

For each $\tau$, let $u_\tau$ be an element for which $N_\tau \subset N(u_\tau)$. Let

$$Y(v) = \sum \psi_\tau(v)h(u_\tau).$$

Since $u_\tau$ is fixed on the support of $\psi_\tau$, $Y(v)$ is locally Lipschitz continuous. By (D.3)

$$\theta\|X(v)\| \leq (X(v), h(u_\tau)), \quad v \in N_\tau. \tag{D.4}$$

Thus

$$\|Y(v)\| \leq \sum \psi_\tau(v)\|h(u_\tau)\| = \sum \psi_\tau(v) = 1$$

and

$$(X(v), Y(v)) = \sum \psi_\tau(v)(X(v), h(u_\tau))$$
$$\geq \theta \sum \psi_\tau(v)\|X(v)\| = \theta\|X(v)\|.$$

This gives the desired result.                                       □

As a result of Lemma D.3, we can remove the requirement of local Lipschitz continuity from our theorems. This is accomplished by substituting the equation

$$\frac{d\sigma(t)v}{dt} = -Y(u) \tag{D.5}$$

for the equation

$$\frac{d\sigma(t)v}{dt} = -\frac{G'(\sigma(t)v)}{\max[\|G'(\sigma(t))\|_H, \delta]}, \tag{D.6}$$

in (2.48), where $Y(u)$ is the pseudo-gradient corresponding to $X(u) = G'(u)$. By using (D.5) in place of (2.48), we achieve the same goal, but a bit slower. This is similar to skiing down a smoother path even though it is not as steep as a more direct path which is not so smooth. As an example, by using this method we can obtain the following stronger version of the sandwich theorem.

**Theorem D.5.** *Let $M, N$ be closed subspaces of a Hilbert space $E$ such that $M = N^\perp$. Assume that at least one of these subspaces is finite dimensional. Let $G$ be a continuously differentiable functional on $E$ that* *satisfies*

$$m_0 = \sup_{v \in N} \inf_{w \in M} G(v + w) \neq -\infty \qquad (D.7)$$

*and*

$$m_1 = \inf_{w \in M} \sup_{v \in N} G(v + w) \neq \infty. \qquad (D.8)$$

*Then there is a sequence $\{u_k\} \subset E$ such that*

$$G(u_k) \to c, \ m_0 \leq c \leq m_1, \ G'(u_k) \to 0. \qquad (D.9)$$

In place of Theorem 2.8 we have

**Theorem D.6.** *Under hypotheses (1.78) and (2.28), there is at least one solution of (1.1),(1.2).*

In place of Theorem 2.23 we have

**Theorem D.7.** *Assume that (1.78), (2.28), and (2.80) hold. Then there is a nontrivial solution of (1.1),(1.2).*

In place of Theorem 2.26 we have

**Theorem D.8.** *Let $G$ be a continuously differentiable functional on a Hilbert space $H$. Assume that $G(0) = 0$ and that there are positive numbers $\varepsilon, \rho$ such that*

$$G(u) \geq \varepsilon \ \text{when} \ \|u\| = \rho. \qquad (D.10)$$

*Assume also that there is a nonzero element $\varphi_0 \in H$ such that*

$$G(r\varphi_0) \leq C_0, \quad r > 0, \qquad (D.11)$$

*for some constant $C_0$. Then there is a sequence $\{u_k\} \subset H$ such that*

$$G(u_k) \to c, \quad \varepsilon \leq c \leq C_0, \quad G'(u_k) \to 0. \qquad (D.12)$$

We replace Theorem 2.27 with

**Theorem D.9.** *Assume that either*

$$t^2 - 2F(x,t) \leq W(x) \in L^1(\Omega), \quad t > 0 \qquad (D.13)$$

*or*

$$t^2 - 2F(x,t) \leq W(x) \in L^1(\Omega), \quad t < 0. \qquad (D.14)$$

*Then under the hypotheses of Theorems 2.21 and 1.37, problem (1.1), (1.2) has at least one nontrivial solution.*

Theorem 2.6 can be replaced by

**Theorem D.10.** *Assume that (1.78) holds with*

$$1 \le \beta(x) \le 2, \quad \beta(x) \not\equiv 1, \beta(x) \not\equiv 2. \tag{D.15}$$

*If $G(u)$ is given by (1.63) and $G'$ is continuous, then there is a $u_0 \in H$ such that*

$$G'(u_0) = 0. \tag{D.16}$$

*In particular, if $f(x,t)$ is continuous in both variables, then $u_0$ is a solution of (1.1),(1.2) in the usual sense.*

In place of Lemma 2.19 we can state

**Lemma D.11.** *In addition to the hypotheses of Theorem D.6, assume that there are positive constants $\varepsilon$, $\rho$ such that*

$$G(u) \ge \varepsilon \tag{D.17}$$

*when*

$$\|u\|_H = \rho. \tag{D.18}$$

*Then there is a solution $u$ of (1.1),(1.2) satisfying (D.17).*

We can replace Theorem 3.22 by

**Theorem D.12.** *Assume that (1.78) holds with*

$$2 \le \beta(x) \le 5, \quad \beta(x) \not\equiv 2, \beta(x) \not\equiv 5. \tag{D.19}$$

*If $G_0(u)$ is given by (3.15), then there is a $u_0 \in H_0^1$ such that*

$$G_0'(u_0) = 0. \tag{D.20}$$

*In particular, if $f(x,t)$ is continuous in both variables, then $u_0$ is a solution of (1.1),(1.2) in the usual sense.*

In place of Theorem 3.23 we have

**Theorem D.13.** *Let $n$ be an integer $\ge 0$. Assume that (1.78) holds with $\beta(x)$ satisfying*

$$1 + n^2 \le \beta(x) \le 1 + (n+1)^2, \quad 1 + n^2 \not\equiv \beta(x) \not\equiv 1 + (n+1)^2. \tag{D.21}$$

*Then (3.1), (3.2) has a solution.*

We can replace Theorem 3.29 with

**Theorem D.14.** *Under the hypotheses of Theorems 3.25 and 3.28, if either*

$$t^2 - 2F(x,t) \leq W(x) \in L^1(\Omega), \quad t > 0$$

*or*

$$t^2 - 2F(x,t) \leq W(x) \in L^1(\Omega), \quad t < 0,$$

*then problem (3.1),(3.2) has at least one nontrivial solution.*

In place of Theorem 3.27 we have

**Theorem D.15.** *Assume that (1.78), (2.28), and (3.60) hold. Then there is a nontrivial solution of (3.1),(3.2).*

Concerning Theorem 10.32 we have

**Theorem D.16.** *Under the hypotheses of Theorem 10.26, there is a PS sequence satisfying*

$$G(u_k) \to a, \quad G'(u_k) \to 0.$$

In place of Theorem 10.34 we can write

**Theorem D.17.** *Under the hypotheses of Theorem 10.26, G has a critical point which is a solution of (10.54).*

For Theorem 10.39 we have

**Theorem D.18.** *Under the hypotheses of Theorems 10.26 and 10.37, G has a nontrivial critical point which is a solution of (10.54).*

In place of Theorem 10.40, we have

**Theorem D.19.** *Assume that (10.66) and (10.80) hold. Let $\lambda_m$ and $\lambda_{m+1}$ be consecutive eigenvalues of $-\Delta$. Assume that (10.82) holds with $\alpha(x)$ satisfying*

$$\lambda_m \leq \alpha \leq \lambda_{m+1}, \quad \lambda_m \not\equiv \alpha \not\equiv \lambda_{m+1}. \tag{D.22}$$

*Then (10.54) has a solution. If, in addition, (10.103) holds, then we are assured that it has a nontrivial solution.*

# Bibliography

Adams, R. A. *Sobolev Spaces*. New York, Academic Press, 1975.

Akhiezer, N. I. *The Calculus of Variations*. New York, Blaisdell, 1962.

Ambrosetti, A. and Prodi, G. *A Primer of Nonlinear Analysis*. Cambridge Studies in Advanced Mathematics, No. 34, 1993.

Ambrosetti, A. and Rabinowitz, P. H. Dual variational methods in critical point theory and applications. *J. Func. Anal.* **14** (1973) 349–381.

Berger, M. S. *Nonlinearity and Functional Analysis*. New York, Academic Press, 1977.

Berger, M. S. and Schechter, M. On the solvability of semilinear gradient operator equations. *Adv. Math.* **25** (1977) 97–132.

Burghes, D. N. and Downs, A. M. *Classical Mechanics and Control*. Chichester, Ellis Horwood, 1975.

Chabrowski, J. *Variational Methods for Potential Operator Equations. With Applications to Nonlinear Elliptic Equations*. de Gruyter Studies in Mathematics, 24. Berlin, Walter de Gruyter & Co., 1997.

Copson, E. T. *Metric Spaces*. Cambridge, UK, Cambridge University Press, 1968.

Dancer, E. N. On the Dirichlet problem for weakly nonlinear elliptic partial differential equations. *Proc. Royal Soc. Edinburgh* **76A** (1977) 283–300.

Ekeland, I. On the variational principle. *J. Math. Anal. Appl.* **47** (1974) 324–353.

Ekeland, I. and Temam, R. *Convex Analysis and Variational Problems*. Amsterdam, North Holland, 1976.

Elsgolc, L. E. *Calculus of Variations*. Oxford, Pergamon Press, 1961.

Fučík, S. Boundary value problems with jumping nonlinearities. *Casopis Pest. Mat.* **101** (1976) 69–87.

Fučík, S. and Kufner, A. *Nonlinear Differential Equations*. Amsterdam, Elsevier, 1980.

Gelfand, I. M. and Fomin, S. V. *Calculus of Variations*. Englewood Cliffs, NJ, Prentice-Hall, NJ, 1963.

Hardy, G. H. and Rogosinski, W. W. *Fourier Series*. Cambridge, UK, Cambridge University Press, 1968.

Ioffe, A. D. and Tihomirov, V. M. *Theory of Extremal Problems*. Amsterdam, North-Holland, 1979.

Kelley, J. L. *General Topology.* New York, Van Nostrand Reinhold, 1955.

Kolmogorov, A. N. and Fomin, S. V. *Introductory Real Analysis.* New York, Dover Publications, 1975.

Lax, P. D. On Cauchy's problem for hyperbolic equations and the differentiability of solutions of elliptic equations. *Comm. Pure Appl. Math.* **8** (1955) 615–633.

Lloyd, N. G. *Degree Theory.* Cambridge, UK, Cambridge University Press, 1978.

Mawhin, J. Topological degree and boundary value problems for nonlinear differential equations. *Topological methods for ordinary differential equations (Montecatini Terme, 1991),* 74–142, Lecture Notes in Math., 1537, Springer, Berlin, 1993.

Mawhin, J. and Willem, M. *Critical Point Theory and Hamiltonian Systems.* New York–Berlin, Springer–Verlag, 1989.

Nirenberg, L. *Topics in nonlinear functional analysis. With a chapter by E. Zehnder. Notes by R. A. Artino.* Lecture Notes, 1973–1974. Courant Institute of Mathematical Sciences, New York University, New York, 1974.

Nirenberg, L. Variational and topological methods in nonlinear problems. *Bull. Amer. Math. Soc. (N.S.)* **4** (1981) No. 3, 267–302.

Palais, R. S. and Smale, S. A generalized Morse theory. *Bull. Amer. Math. Soc.* **70** (1964) 165–172.

Rabinowitz, P. H. *Minimax Methods in Critical Point Theory with Applications to Nonlinear Partial Differential Equations.* Conf. Board of Math Sci., No. 65, Amer. Math. Soc., 1986.

Riesz, F. and Sz.-Nagy, B. *Functional Analysis.* New York, Frederick Ungar, 1955.

Schechter, M. Solution of nonlinear problems at resonance. *Indiana Univ. Math. J.* **39** (1990) 1061–1080.

Schechter, M. The Fučík spectrum. *Indiana Univ. Math. J.* **43** (1994) No. 4, 1139–1157.

Schechter, M. Resonance problems which intersect many eigenvalues. *Math. Nachr.* **183** (1997) 201–210.

Schechter, M. *Linking Methods in Critical Point Theory.* Boston, MA, Birkhäuser Boston, Inc., 1999.

Schechter, M. *Principles of Functional Analysis.* Providence, RI, American Mathematical Society, 2002.

Schechter, M. and Tintarev, K. Pairs of critical points produced by linking subsets with applications to semilinear elliptic problems. *Bull. Soc. Math. Belg. Sér.* B 44 (1992) no. 3, 249–261.

Silva, E. A. de B. Linking theorems and applications to semilinear elliptic problems at resonance. *Nonlinear Analysis TMA,* 16(1991) 455–477.

Smart, D. R. *Fixed Point Theorems.* Cambridge, UK, Cambridge University Press, 1974.

Vainberg, M. M. *Variational Methods for the Study of Nonlinear Operators.* San Francisco, Holden-Day, 1964.

Willem, M. *Minimax Theorems. Progress in Nonlinear Differential Equations and their Applications, 24.* Boston, MA, Birkhäuser Boston, Inc., 1996.

# Index

## Notes and Remarks

1. On page 69, line 5 should read:

$$T \leq \min(T_0, (R_0 - \|x_0 - x_1\|)/M_0), \quad K_0 T < 1.$$

2. On page 77, line 1 should read:

**Theorem 2.21.** *Assume that*

$$|f(x,t)| \leq C(|t|^q + 1), \quad x \in I, \, t \in \mathbb{R}$$

*for some $q > 1$, and that there is a $\delta > 0$*

3. On page 115, the functional $G$ should be replaced by $G_0$.

4. On page 117, equation (3.63) and the next sentence should read:

$$G_0(u) \geq \frac{1}{5}\|w\|_H^2, \quad \|u\|_H \leq \rho,$$

for $\rho > 0$ sufficiently small. Now suppose alternative (a) of the theorem did not hold.

5. On page 117, line 4 from the bottom should read:

$$\|v_k\|_H^2 = \pi c_k^2.$$

6. On page 118, line 8 from the bottom should read:

We note that (b) implies that every function $v \in V$ satisfying $v = c_k \sin x$ for $|c_k|$ sufficiently small is a solution of $G_0'(v) = 0$.

7. On page 126, Lemma 4.5 should read as follows:

**Lemma 4.5.** *If $G(u)$ is convex and l.s.c. on $E$, and $u_k \rightharpoonup u$, then*

$$G(u) \leq \liminf G(u_k).$$

8. On page 129, in the proof of Theorem 4.8, it is not necessary to temporarily assume that $G(u,v)$ is strictly convex. We merely let $w_\theta$ be any point in $M$ where $G(v_\theta, w)$ achieves its minimum and take a subsequence.

9. On page 131, equation (4.16) should read:

$$G(v, \bar{w}_R) \leq G(\bar{v}_R, \bar{w}_R) \leq G(\bar{v}_R, w), \quad v \in N_R, \, w \in M_R.$$

10. On page 132, it is not required to assume that $M$ is closed in Theorem 4.9 and Corollary 4.10.

## 2

11. On page 139, line 5 should read:

see that $x_0$ is a solution of $A(x) = 0$ iff it satisfies (4.21).

12. On page 146, line 10 from the bottom should read:

$$\frac{dT_\varepsilon}{d\varepsilon} = \int_0^{b_0} \frac{\partial F}{\partial z}(y, f_\varepsilon'(y))\eta'(y)\,dy = -\int_0^{b_0} \frac{d}{dy}\frac{\partial F}{\partial z}(y, f_\varepsilon'(y))\eta(y)\,dy.$$

13. On page 156, line 14 should read:

$$|G(u+v) - G(u) - (v, G_1'(u))_H| = \left|\int_0^1 (v, G_1'(u+tv) - G_1'(u))_H\,dt\right|$$

$$\leq \|v\|_H \int_0^1 \|G_1'(u+tv) - G_1'(u)\|_H\,dt.$$

14. On page 157, lines 10 and 12 should read:

$$T = \frac{1}{2}m(\dot{r}^2 + r^2\dot{\phi}^2)$$

and

$$L = T - V = \frac{1}{2}m(\dot{r}^2 + r^2\dot{\phi}^2) + mk/r,$$

respectively.

15. On page 171, line 5 should read:

(a) $d(I, \Omega, p) = 1$ if $p \in \Omega$. If both $p$ and $-p$ are in $\Omega$, then $d(\pm I, \Omega, p) = (\pm 1)^n$.

16. On page 172, line 9 from the bottom should read:

(c) states that if $\varphi_0, \varphi_1$ are homotopic and $p \notin h_t(\partial\Omega)$ for $0 \leq t \leq 1$, then

17. On page 183, lines 10 and 16 should read:

$$\|\sigma'(t)v\|_H \leq 1$$

and

$$\|P\sigma(t)v\|_H \geq \|v\|_H - \|v - P\sigma(t)v\|_H \geq R - t,$$

respectively. Line 3 from the bottom should read:

$$G(\sigma(t)v_0) \leq m_0 - \delta.$$

18. On page 190, replace Lemma 6.32 – Theorem 6.34 with:

**Lemma 6.32.** *Assume that $\varphi \in C^2(\bar{\Omega}, \mathbb{R}^n)$ and that $p_0, p_1$ are such that (6.20) has no solutions which are boundary points when $p = p(s) = (1 - s)p_0 + sp_1$, $0 \leq s \leq 1$. Assume also, that (6.20) has no solutions which are critical points when $p = p_0, p_1$. Then,*

$$d(\varphi, \Omega, p_0) = d(\varphi, \Omega, p_1).$$

3

*Proof.* By Theorem 6.29, (6.22) holds for $p = p_0, p_1$ and $\varepsilon > 0$ sufficiently small. Thus
(1)
$$d(\varphi, \Omega, p_1) - d(\varphi, \Omega, p_0) = \int_\Omega [j_\varepsilon(\varphi(x) - p_1) - j_\varepsilon(\varphi(x) - p_0)] \det \varphi'(x) \, dx.$$

We are going to show that there is a function $u(x) \in C_0^1(\Omega, \mathbb{R}^n)$ such that
$$\nabla \cdot u(x) = [j_\varepsilon(\varphi(x) - p_1) - j_\varepsilon(\varphi(x) - p_0)] \det \varphi'(x).$$

Since
$$\int_\Omega \nabla \cdot u(x) \, dx = 0,$$

the result will follow. First, we find a function $v(y) \in C_0^1(\mathbb{R}^n, \mathbb{R}^n)$ such that
$$\nabla \cdot v(y) = [j_\varepsilon(y - p_1) - j_\varepsilon(y - p_0)].$$
For this purpose, we take
$$F(y) = \int_0^1 j_\varepsilon(y - p(s)) \, ds, \quad v(y) = F(y)(p_1 - p_0).$$

Then
$$\begin{aligned}
\nabla \cdot v(y) &= \sum \frac{\partial v_i(y)}{\partial y_i} \\
&= \sum \frac{\partial F}{\partial y_i}(p_1 - p_0)_i \\
&= \sum \int_0^1 \frac{\partial j_\varepsilon(y - p(s))}{\partial y_i} \cdot (p_1 - p_0)_i \, ds \\
&= -\sum \int_0^1 \frac{\partial j_\varepsilon(y - p(s))}{\partial p(s)_i} \cdot (p_1 - p_0)_i \, ds \\
&= \int_0^1 \frac{d}{ds} j_\varepsilon(y - p(s)) \, ds \\
&= [j_\varepsilon(y - p_1) - j_\varepsilon(y - p_0)].
\end{aligned}$$

Let $A_{ji}(x)$ be the cofactor of $\partial \varphi_j / \partial x_i$ in $\det \varphi'(x)$. We take
$$u_i(x) = \sum_j v_j(\varphi(x)) A_{ji}(x).$$

Then
$$\nabla \cdot u(x) = \sum_{i,j,k} \frac{\partial v_j(\varphi(x))}{\partial x_k} \frac{\partial \varphi_k(x)}{\partial x_i} A_{ji} + \sum_{i,j} v_j(\varphi(x)) \frac{\partial A_{ji}(x)}{\partial x_i}.$$

Now
$$\sum_i \frac{\partial \varphi_k(x)}{\partial x_i} A_{ji} = \delta_{kj} \det \varphi'(x)$$

by the definition of $A_{ji}$. Moreover

$$\sum_i \frac{\partial A_{ji}(x)}{\partial x_i} = 0, \quad j = 1, \cdots, n.$$

To prove this, fix $m$, and let $g_q$ be the $n$ column vectors of length $n-1$ with components $\partial \varphi_p / \partial x_q$ and $p \neq m$. Let

$$A_q = \det[g_1, \cdots, g_n]$$

with $g_q$ missing. Let $h_{pq} = \partial g_q / \partial x_p = \partial g_p / \partial x_q$, and

$$B_{pq} = \det[h_{pq}, g_1, \cdots, g_n]$$

with $g_p$, $g_q$ both missing. Then $B_{pq} = B_{qp}$ and

$$\frac{\partial A_q}{\partial x_q} = \sum_{p<q} (-1)^{p+1} B_{pq} + \sum_{p>q} (-1)^p B_{pq}.$$

Thus,

$$\sum_q (-1)^q \frac{\partial A_q}{\partial x_q} = \sum_q \sum_{p<q} (-1)^{p+q+1} B_{pq} + \sum_q \sum_{p>q} (-1)^{p+q} B_{pq} = 0.$$

Since this is true for each $m$, the result follows. □

We now define $d(\varphi, \Omega, p)$ for those points $p$ such that (6.20) is satisfied by critical points but not by boundary points. Let $p_0$ be such a point. Let $B_0$ be a spherical neighborhood of $p_0$ such that (6.20) has no solutions in $\partial \Omega$ for $p \in B_0$. By Sard's theorem (Theorem 6.25), $B_0$ contains points $p$ such that no critical points satisfy (6.20). Let $p_1, p_2$ be any such points. Then by Lemma 6.32,

$$d(\varphi, \Omega, p_1) = d(\varphi, \Omega, p_2).$$

Thus, the degree is the same for all such points. We take $d(\varphi, \Omega, p_0)$ to be this constant value.

**Theorem 6.33.** $d(\varphi, \Omega, p)$ is constant on any component of $\mathbb{R}^n \backslash \varphi(\partial \Omega)$.

*Proof.* Assume that $\varphi \in C^1(\bar{\Omega}, \mathbb{R}^n)$ and that $p_0, p_1$ are such that (6.20) has no solutions which are boundary points when $p = p_0, p_1$. Assume also that there is a path $\gamma(s) \in \mathbb{R}^n$, $0 \leq s \leq 1$, such that $\gamma(0) = p_0$, $\gamma(1) = p_1$ and no boundary points satisfy (6.20) with $p = \gamma(s)$, $0 \leq s \leq 1$. Then,

$$d(\varphi, \Omega, p_0) = d(\varphi, \Omega, p_1).$$

By Theorem 6.29 for each $s \in [0,1]$ there is a $\delta_s > 0$ such that no boundary points satisfy

$$\varphi(x) = p$$

whenever $p$ satisfies $\|p - \gamma(s)\| < \delta_s$ and

$$d(\varphi, \Omega, p) = d(\varphi, \Omega, \gamma(s)).$$

By compactness, there is one $\delta > 0$ which will serve for all $s \in I$. Moreover, we can cover $I = [0, 1]$ with a finite number of intervals of length $> \eta > 0$ in which this holds in each interval. Let $0 = s_0 < s_1 < \cdots < s_m = 1$ be a partition of $I$ such that

$$\|\gamma(s_{k+1}) - \gamma(s_k)\| < \delta, \quad k = 0, 1, \cdots, m - 1.$$

Hence,

$$d(\varphi, \Omega, p_0) = d(\varphi, \Omega, s_0) = d(\varphi, \Omega, s_1) = \cdots = d(\varphi, \Omega, s_m) = d(\varphi, \Omega, p_1).$$

This completes the proof. $\qquad\square$

**Theorem 6.34.** *If no boundary points satisfy (6.20), then there is a $\delta > 0$ such that $d(\psi, \Omega, p) = d(\varphi, \Omega, p)$ when $\psi \in C^1(\bar{\Omega}, \mathbb{R}^n)$ and $\|\psi - \varphi\|_1 < \delta$.*

*Proof.* There is a neighborhood of $p$ such that no boundary points satisfy $\varphi(x) = p'$ for $p'$ in this neighborhood. Let $q$ be a point in the neighborhood for which $\varphi(x) = q$ is not satisfied by any critical point. Take $\delta > 0$ so small that $\|\psi - \varphi\|_1 < \delta$ implies $d(\psi, \Omega, q) = d(\varphi, \Omega, q)$ (Theorem (6.29)). If $x \in \partial\Omega$, then

$$|\psi(x) - p| \geq |\varphi(x) - p| - \delta.$$

Hence, we can take $\delta$ so small that $p, q$ are in the same component of $\mathbb{R}^n \setminus \psi(\partial\Omega)$. Consequently,

$$d(\psi, \Omega, p) = d(\psi, \Omega, q) = d(\varphi, \Omega, q) = d(\varphi, \Omega, p)$$

by Theorems 6.29 and 6.33. $\qquad\square$

19. On page 306, line 12 from the bottom should read:

    This implies that $\hat{u}_\tau = u_\tau$ a.e. for each $\tau$.

20. On page 307, the last line should read as follows: We apply Theorem 10.51 to the derivatives of $u$ up to order $\ell$.

Printed by Printforce, United Kingdom